Marine Greens

This book provides an in-depth overview of marine greens and their environmental and biotechnological applications. It addresses concepts such as the niche adaptation strategies of marine greens in their natural habitats, as well as their global distribution, and factors affecting their distribution and proliferation.

Marine greens are posited as an alternative to fossil fuels, mitigating global climate change and thereby fostering future environmental sustainability. In addition, the book explains the remediation of xenobiotics, wastewater, microplastics, marine debris, and marine green contamination. Expert authors from around the world explore the industrial and agricultural applications of marine greens in the production of enzymes and marine bioenergy, and what is needed to improve its production potential.

This is important reading for government and non-governmental organizations as well as industries and research institutions looking for ways to combat current industrial and environmental challenges.

Marine Greens

Environmental, Agricultural, Industrial and Biomedical Applications

Edited by
Sesan Abiodun Aransiola, Oluwafemi Adebayo Oyewole,
Naga Raju Maddela, Udeme Joshua Josiah Ijah, and
Bangeppagari Manjunatha

CRC Press
Taylor & Francis Group
Boca Raton London New York

CRC Press is an imprint of the
Taylor & Francis Group, an **informa** business

First edition published 2024
by CRC Press
2385 NW Executive Center Drive, Suite 320, Boca Raton FL 33431

and by CRC Press
4 Park Square, Milton Park, Abingdon, Oxon, OX14 4RN
CRC Press is an imprint of Taylor & Francis Group, LLC

ISBN: 9781032438856 (hbk)
ISBN: 9781032439815 (pbk)
ISBN: 9781003369738 (ebk)

DOI: 10.1201/9781003369738

Typeset in Times
by Deanta Global Publishing Services, Chennai, India

Contents

Foreword

Marine Greens: Environmental, Agricultural, Industrial, and Biomedical Applications provides complete information on the applications of marine greens and is a collection of individual, well-thought, and well-laid-out chapters containing various topics relevant to the benefits and applications of marine greens. This book is so relevant, as efforts are currently being made to advance the various fields of science and engineering and overcome problems relating to the environment, health, food security, and energy crisis. The findings in this book, therefore, will help to proffer solutions to environmental menaces including climate change, global warming, and flooding, thereby fostering environmental sustainability. Marine greens, especially algae, have been estimated to produce as much as 50% of all O_2 via photosynthesis; they also have the potential to produce bio-products. Ideally, biofuels and high-value products from marine algae could help to meet numerous international challenges, such as CO_2 emissions, fossil fuel consumption, climate change, and food safety. Marine greens regulate C and N metabolism. Carbon fixation takes place via oxygenic photosynthesis, and CO_2 is the primary carbon source.

Marine greens contain extremely rich and unique nutrients. They can accumulate large amounts of active substances, including proteins, lipids, and carbohydrates, making biomass rich in nutrients that improve values. Marine greens are a treasure house that provides fish, animals, and humans with a large supply of the nutrients they need. Each kind of seaweed has its own nutritional status, and marine algae also can meet the nutritional needs of any species. Marine greens have been used as feed for aquatic products and animal husbandry for decades. Ireland, the United Kingdom, South Africa, India, New Zealand, and other countries have established professional factories for marine algae feed. It is found that marine algae generally cause no toxic side effects to animals. They are very suitable for use as animal feed additives, which can improve the nutritional structure of feed and increase feed utilization. For these reasons and more, marine greens have numerous industrial and agricultural applications. Hence, this book has carefully considered the applications of marine greens in industrial, environmental, agricultural, and biomedical sectors. Marine greens have vast and untapped medicinal properties, such that algae-derived bioactive compounds with immunomodulatory activities, sulfated polysaccharides, and polyunsaturated fatty acids have a beneficial effect on the immune system, enzyme inhibition properties, antibacterial, antifungal, anti-inflammatory, antiviral, antioxidant, etc. The research and biomedical applications of marine greens have also been advanced in the treatment of various diseases, such as during the COVID-19 pandemic. In addition, they have probiotic/prebiotic properties in the intestine and modulate the gut microbiome; therefore, they can increase the effectiveness of vaccine production because of their wide array of medicinal values.

To justify the four sections in this book, the editors have done a great job by inviting 80 experienced experts from 10 different countries to contribute to this volume. Therefore, I am confident that the wealth of knowledge of all the contributors of chapters in this book is of valuable importance and will have a wide readership. Besides, it will be a good source of information for students, scientists, researchers, industrialists, agriculturists, medical personnel, engineers, and technicians for many years to come.

Professor Faruk Adamu Kuta
Vice Chancellor, Federal University of Technology
Minna, Nigeria

Preface

This book aims to provide an in-depth overview of marine greens with their environmental, agricultural, industrial, and biomedical applications. The book addresses general concepts as they relate to marine greens, including their niche, their adaptation strategies in their natural habitats, as well as their global distribution. It also reports the factors affecting the distribution and proliferation of marine greens. The book particularly describes the innovations of marine greens as a source of renewable energy due to the current global energy crisis. In addition, the book also explains the remediation of xenobiotics, wastewater, microplastics, marine debris, and marine green contamination.

Of note are the biotechnological applications of marine greens, especially due to the presence of bioactive compounds in marine greens that have antioxidant, anti-inflammatory, and anti-aging properties. Marine greens have been applied in nutraceuticals, cosmeceuticals, and pharmaceutical formulations, and as excipients in drug development. The bioactive compounds are also applicable to industrial and agricultural sectors, especially in the production of enzymes. With the current problems of drug-resistant pathogenic bacteria, new, safe, and cost-effective antibiotics that are effective against emerging and resistant bacterial strains are urgently needed. Several diseases, such as diabetes, have been controlled using marine greens. Marine greens are sources of food or food additives, and they have been linked to nutritional health benefits by serving as a source of essential vitamins and minerals, such as copper, iron, B vitamins, and electrolytes. Marine greens also contain iodine, tyrosine, and prebiotics for gut enrichment. They have the ability to lower cholesterol levels and protect cells from free radicals as well as detoxifying heavy metals from the body.

The book is subdivided into four sections: *Section 1—Environmental Applications of the Marine Greens, Section 2—Agricultural Applications, Section 3—Industrial Applications, and Section 4—Biomedical Applications of Marine Greens.* Section 1 consists of five chapters. Chapter 1 is designed to discuss the types, habitat, growth, and global distribution of marine greens. The abilities of marine greens in remediation of contaminants, microplastics, marine debris, and marine green contamination, marine greens' roles in climatic change and global warming mitigation, and marine green nanotechnology for remediation of wastewater are all discussed in Chapters 2, 3, 4, and 5, respectively. Section 2, which accommodates Chapters 6, 7, 8, 9, 10, and 11, reveals the applications of marine greens in aquaculture, marine greens as a source of biofertilizer, marine green macroalgae biomass production and application, marine green microalgae biomass production and applications, edible marine greens: uses and limitations, and marine greens as nutritious healthy foods, while Section 3 has three chapters: biotechnological potentials of marine bacteria in the production of enzymes, marine greens in biorefinery, and the potential of marine greens for biofuel production in Chapters 12, 13, and 14. The last section explores the biomedical applications of marine greens, further broken-down into six chapters of great importance. Chapters, 15, 16, 17, 18, 19, and 20 provide full discussions on the potential therapeutic role of marine greens for liver diseases; bioactive metabolites from marine greens and their health benefits; modulatory and neuroprotective effects of marine green spirulina on the brain and neurodegeneration; marine green biotechnology advances as bioactive compounds for medical use; potentially toxic metals in Africa, fresh and marine environment; marine green contamination, ecological and human health risks; and bioactive molecules of medicinal value in marine green algae.

The chapters were contributed by 80 academicians/scientists/researchers from 10 different countries (China, Ecuador, India, Nigeria, Canada, Egypt, Germany, the USA, the UK, Hong Kong) across the globe.

Sesan Abiodun Aransiola, PhD
Department of Microbiology,University of Abuja, Abuja, Nigeria

Oluwafemi Adebayo Oyewole, PhD
Department of Microbiology
Federal University of Technology, Minna, Nigeria

Naga Raju Maddela, PhD
Department of Biological Sciences
Technical University of Manabí, Portoviejo, Ecuador

Udeme Joshua Josiah Ijah, PhD
Department of Microbiology
Federal University of Technology, Minna, Nigeria

Bangeppagari Manjunatha, PhD
Sri Devaraj Urs Academy of Higher Education and Research (Deemed to Be University)
Kolar, Karnataka, India

Acknowledgments

We greatly acknowledge the support of the chapters' contributors for their valuable contributions and timely responses during this project. Without their enthusiasm and support, this volume would not have been ready in the scheduled time; therefore, we really appreciate their cooperation and collaboration. We also thank the anonymous reviewers for their constructive criticism, which helped us to improve the quality of this book by inviting experts to contribute additional chapters. We greatly acknowledge the Taylor & Francis Group Editorial and Production teams for their valuable support; without their guidelines, this project would not have been finished in such a very short time. It is our honor to work with them, honestly. Finally, we sincerely appreciate the management and colleagues at University of Abuja, Abuja Nigeria, Federal University of Technology, Minna, Nigeria, Universidad Técnica de Manabí, Portoviejo, Ecuador, and Sri Devaraj Urs Academy of Higher Education and Research, India for their unrestricted backing and for the establishment of treasured propositions at the time of book proposal and final book preparation.

Sesan Abiodun Aransiola, PhD
Department of Microbiology,University of Abuja,Abuja, Nigeria

Oluwafemi Adebayo Oyewole, PhD
Department of Microbiology
Federal University of Technology, Minna, Nigeria

Naga Raju Maddela, PhD
Department of Biological Sciences
Technical University of Manabí, Portoviejo, Ecuador

Udeme Joshua Josiah Ijah, PhD
Department of Microbiology
Federal University of Technology, Minna, Nigeria

Bangeppagari Manjunatha, PhD
Sri Devaraj Urs Academy of Higher Education and Research (Deemed to Be University)
Kolar, Karnataka, India

About the Editors

Sesan Abiodun Aransiola is a PhD holder in Environmental Microbiology in the Department of Microbiology, Federal University of Technology, Minna, Nigeria. He earned his first degree (B.Tech) and Master's Degree (M.Tech) from the same department in 2009 and 2014 respectively. He has demonstrated his research expertise in the production of vermicasts from vermicoposting (Vermitechnology) of organic wastes to assist plants in the remediation of soil polluted with heavy metals. Currently, he is a Lecturer at the Department of Microbiology, University of Abuja, Nigeria. His area of interest is Environmental Microbiology with research areas in phytoremediation, vermicomposting, biosorption and bioremediation of contaminated soil environment. He is an award-winning researcher and has authored more than 70 publications including book chapters, research and review articles of good international repute with high impact factor. He has co-edited scientific books of global interest which include microbial biotechnology for bioenergy (Elsevier). Ecological interplays in microbial enzymology, vermitechnology: economic, environmental and agricultural sustainability, prospects for soil regeneration and its impact on environmental protection (Springer). Microbial biofilms – applications and control (CRC Press/ Taylor & Francis Group) with many more in the process. Also, he has worked to the rank of an assistant chief scientific officer at Bioresources Development Centre, National Biotechnology Development Agency, Nigeria where he was involved in the uses of vinasse (a by-product of ethanol) as bio-fertilizer to reclaim polluted soil for agricultural purposes. Dr. Aransiola is a member of Nigerian Society for Microbiology and American Society for Microbiology, among others.

Oluwafemi Adebayo Oyewole, PhD, earned his BTech (1999–2004) and MTech (2009–2010) degrees in Microbiology from the Federal University of Technology Minna, Nigeria. He earned his PhD in Environmental Microbiology (2016) from the University of Portsmouth, UK. He is currently a Senior Lecturer in the Department of Microbiology, Federal University of Technology Minna, Nigeria. His current roles include teaching undergraduate and postgraduate students and research supervision in the areas of Environmental, Productive, and Applied Microbiology. He has been involved in several administrative responsibilities within the university. He is also engaging in community development activities and professional consultancy services outside of the university. He has over 100 research publications, won some research grants, has three patents from his research findings, and is a member of some professional associations.

Naga Raju Maddela, **PhD**, earned his MSc (1996–1998) and PhD (2012) in Microbiology from Sri Krishnadevaraya University, Anantapuramu, India. During his doctoral program in the area of Environmental Microbiology, he investigated the effects of industrial effluents/insecticides on soil microorganisms and their biological activities, and he has worked as Faculty in Microbiology since 1998, teaching undergraduate and postgraduate students. He worked as a Prometeo Investigator (fellowship received from the Secretariat of Higher Education, Science, Technology and Innovation) at Amazon State University, Ecuador during 2013–2015 and received a Postdoctoral Fellowship (2016–2018) from Sun Yat-sen University, China. He also received external funding from the China Postdoctoral Science Foundation in 2017, and internal funding from the Technical University of Manabí in 2020. He participated in national/international conferences and presented research data in China, Cuba, Ecuador, India, and Singapore. Currently, he is working as a Full Professor at the Faculty of Sciences and Health, Technical University of Manabí, Portoviejo, Ecuador. He has been actively publishing scientific articles, books (authored and edited), and chapters since 2007. As of now, he has published 60 articles, 10 books, and 40 book chapters.

Udeme Joshua Josiah Ijah, **PhD**, earned his BTech (1984) and MTech (1988) in Applied Microbiology from the Rivers State University of Technology, Port Harcourt, Nigeria, and his PhD (1993) in Microbiology with specialization in Environmental Microbiology from the University of Calabar, Cross River State, Nigeria. During his doctoral program, he investigated the Microbiology of Crude Oil Degradation in Soil, with specific focus on the effects of oil spills on soil physicochemical properties, microbial ecology, rates of crude oil biodegradation, and bioremediation of crude oil–polluted soil. He has worked as a faculty member and Adjunct Professor in several universities, where he has taught undergraduate and postgraduate students and graduated 54 MSc/MTech and 21 PhD students. He has received internal funding and internal sponsorship, worked in the area of Environmental Microbiology and Biotechnology, and participated in several national and international conferences. He is former sub-Dean, School of Science and Science Education, former Head of Department of Microbiology, and currently working as Professor in the Department of Microbiology, Federal University of Technology, Minna, Niger State, Nigeria. His current areas of research include: the role of microorganisms in mitigating climate change; bioremediation of microplastics in aquatic environments using microbial biofilm; and microbial products for environmental sustainability, energy, and water security. He has presented several conference papers and published 11 chapters, 5 proceedings, and 105 research papers in peer-reviewed journals.

Bangeppagari Manjunatha, PhD, is an Assistant Professor, Zebrafish Drug Screening Center, Department of Cell Biology and Molecular Genetics, Sri Devaraj Urs Academy of Higher Education and Research (Deemed to Be University), Kolar, Karnataka, India. He earned his MSc (2006) and PhD (2012) degrees from the Department of Zoology, Sri Krishnadevaraya University, Anantapur, Andhra Pradesh, India. During 2008–2012, he worked as a Junior and Senior Research Fellow in the prestigious UGC Sponsored Program for Meritorious Students in Science, Government of India. He later joined the Biology Institute of Shandong Academy of Sciences, China in 2012 and worked as a Postdoctoral Fellow until 2014. He has successfully handled one major research project (PROMETEO) funded by the Secretariat of Higher Education, Science, Technology and Innovation, Republic of Ecuador, South America; worked as a Principal Investigator at the University of the Armed Forces-ESPE; and served as an Assistant Professor in the same university until December 2016. He also worked as a Korea Research Fellowship-Brain Pool Postdoctoral Research Fellow at Pohang University of Science and Technology, South Korea from 2017 to 2021. He used zebrafish as an experimental disease model and made a significant contribution to the fields of environmental toxicology, ecotoxicology, and public health research in general from his postdoctoral experience: especially the effects of nanomaterials, including pristine graphene, graphene oxide, PM2.5 particles, agro pesticides, Chinese herbal compounds, and cosmetic agents such as hair dye and benzyl alcohol. He also conducted *in vivo* fish developmental and cardiovascular studies, and has experience in high-resolution confocal imaging, X-micro computed tomography, histopathology, transmission and scanning electron microscopy analysis, and biochemical, hematological, and enzymological techniques. During his research, he handled different experimental animals such as zebrafish, major carp, brine shrimp, rats, and guinea pigs. As an early career researcher, he has already published >80 papers in national and international journals and 3 eBooks (total number of citations: 836, H-index: 16; i10-index: 26), and participated/presented at 42 national and international seminars/conferences and workshops.

Contributors

Grace Onyukwo Abakpa
Department of Microbiology
Federal University of Health Sciences
Otukpo, Nigeria

O.P. Abioye
Department of Microbiology
Federal University of Technology Minna
Minna, Nigeria

Joshua Ibukun Adebomi
Chemical Engineering Department
University of Saskatchewan
Saskatoon, Saskatchewan, Canada

O.F. Adedugbe
Department of Chemical Sciences
Achievers University
Owo, Nigeria

Oyegbade Samuel Adeniyi
Department of Biological Sciences
College of Science and Technology
Covenant University
Ota, Nigeria

A.O. Adesina
Department of Chemistry
Federal University of Technology Akure
Akure, Nigeria

Odeseye Abdulhameed Adewale
Agricultural and Rural Management Training Institute
Kwara State, Nigeria

A.J. Adewumi
Department of Geological Sciences
Achievers University
Owo, Nigeria

Israel Emiezi Agarry
College of Food Science
Southwest University
Chongqing, China

Emmanuel Agiriga
Department of Food and Nutrition Sciences
Sheffield Hallam University
Sheffield, UK

Muazu Ahmed
Department of Biotechnology
Faculty of Health and Life Sciences
Coventry University
Coventry, UK

Ayodeji Johnson Ajibare
Department of Physiology
Faculty of Basic Medical Sciences
College of Medicine
Lead City University
Ibadan, Nigeria

S.A. Akande
Department of Mathematics
Federal University of Technology Minna
Minna, Nigeria

Fadekemi O. Akinhanmi
Department of Biological Sciences
Covenant University
Ota, Nigeria

Rasaq Olajide Akinsola
Department of Medicine
Division of Hematology & Oncology
Cedars Sinai Medical Center
Los Angeles, California, USA

Ulelu Jessica Akor
Department of Crop Science
Faculty of Agriculture
University of Abuja
Abuja, Nigeria

I.A. Amoo
Department of Chemistry
Federal University of Technology Akure
Akure, Nigeria

Olasore Abiodun Amos
Agricultural and Rural Management Training Institute
Kwara State, Nigeria

Sesan Abiodun Aransiola
Department of Microbiology, University of Abuja,Abuja, Nigeria

Madan Kumar Arumugam
Cancer Biology Laboratory
Centre for Molecular and Nanomedical Sciences
Sathyabama Institute of Science and Technology
Chennai, India

Abraham Olufemi Asuku
Department of Medical Biotechnology
Bioresources Development Centre
National Biotechnology Development Agency
Ogbomoso, Nigeria

Habiba Iliyasu Atta
Department of Microbiology
Faculty of Life Sciences
Ahmadu Bello University
Zaria, Nigeria

Oluwadurotimi S. Aworunse
Department of Biological Sciences
Covenant University
Ota, Nigeria

Opeyemi I. Ayanda
Department of Biological Sciences
Covenant University
Ota, Nigeria

Maryam Tayo Ayinla
Department of Physiology
Faculty of Basic Medical Sciences
College of Health Sciences
University of Ilorin
Ilorin, Nigeria

Zainab Azeez
Department of Microbiology
Federal University of Technology Minna
Minna, Nigeria

Babafemi Raphael Babaniyi
Bioresources Development Centre
National Biotechnology Development Agency
Ilorin, Nigeria

Gabriel Gbenga Babaniyi
Department of Agricultural Development and Management
Agricultural and Rural Management Training Institute
Ilorin, Nigeria

Musa Bishir
Department of Biology
Kogi State College of Education (Technical)
Kabba, Nigeria

Ademola Bisi-Omotosho
Department of Sustainability Management and Innovation
University of Westminster
London, UK

Ekene Chidozie
Department of Biology and Forensic Science
Admiralty University of Nigeria
Ibusa, Nigeria

Idris Abdullahi Dabban
Department of Biological Sciences
School of Applied and Natural Sciences
The Federal Polytechnic Bida
Bida, Nigeria

Oluwatosin Emmanuel Daramola
Department of Chemistry and Biochemistry
Texas Tech University
Lubbock, Texas, USA

I.S. Emumejakpor
Department of Environmental Management
University of Hertfordshire
Hatfield, UK

Okorie Maria Ezinne
Department of Public Health
Federal College of Dental Technology and Therapy
Trans-Ekulu, Nigeria

Umm-kulthum Abubakar Gimba
Department of Microbiology
Federal University of Technology Minna
Minna, Nigeria

Thiyagarajan Gopal
Centre for Laboratory Animal Research and Technology
Sathyabama Institute of Science and Technology
Chennai, India

Sudhakara Gujjala
Department of Biochemistry
Sri Krishnadevaraya University
Anantapur, India

Ibrahim Mohammed Hussaini
Department of Microbiology
Faculty of Life Sciences
Ahmadu Bello University
Zaria, Nigeria

Ibrahim Isiaka Hussein
Department of Conversion Technologies of Biobased Resources
University of Hohenheim
Stuttgart, Germany

Gordon C. Ibeanu
Department of Pharmaceutical Science
North Carolina Central University
Durham, North Carolina, USA

Kayode-Edwards Ifeoluwa Ihotu
Department of Biological Sciences
Covenant University
Ota, Nigeria

U.J.J. Ijah
Department of Microbiology
Federal University of Technology Minna
Minna, Nigeria

O.S. Ilesanmi
Department of Chemical Sciences
Achievers University
Owo, Nigeria

U.S. Ilyasu
Department of Microbiology
Federal University of Technology Minna
Minna, Nigeria

J.O. Jayeola
Fring Integrated Services Limited
Abuja, Nigeria

Srinivasan Kameswaran
Department of Botany
Vikrama Simhapuri University College
Kavali, India

R. Krishnamurthy
C.G. Bhakta Institute of Biotechnology
Uka Tarsadia University
Bardoli, India

Naga Raju Maddela
Department of Biological Sciences
Faculty of Sciences and Health
Technical University of Manabí
Portoviejo, Ecuador

Ch. Manjula
Department of Food and Nutritional Sciences
Acharya Nagarjuna University
Guntur, Andhra Pradesh, India

A. Maude
Department of Microbiology
Federal University of Technology Minna
Minna, Nigeria

Abdullahi Mohammed
Department of Biological Sciences
School of Applied and Natural Sciences
The Federal Polytechnic Bida
Bida, Nigeria

Ugonna Morikwe
Department of Pharmaceutical Microbiology & Biotechnology
Nnamdi Azikwe University
Awka, Nigeria
and
Department of Biology
North Carolina A&T State University
Greensboro, North Carolina, USA

Nathan Moses
Department of Biology and Forensic Science
Admiralty University of Nigeria
Ibusa, Nigeria

O.I. Musa
Department of Microbiology
Federal University of Technology Minna
Minna, Nigeria

S.I. Musa
Department of Biology and Forensic Science
Admiralty University of Nigeria
Ibusa, Nigeria
and
Department of Plant Biology and Biotechnology
University of Benin
Benin City, Nigeria

Adejoh Suleiman Ocholi
Department of Microbiology
Faculty of Life Sciences
Ahmadu Bello University
Zaria, Nigeria

Harrison Ogala
Department of Biology and Forensic Science
Admiralty University of Nigeria
Ibusa, Nigeria

O.D. Ogundele
Department of Chemical Sciences
Achievers University
Owo, Nigeria

Sheila Ojie
Department of Biology and Forensic Science
Admiralty University of Nigeria
Ibusa, Nigeria

Comfort Okoji
Department of Biology and Forensic Science
Admiralty University of Nigeria
Ibusa, Nigeria

Itoroabasi Joseph Okon
Department of Food and Industrial Biotechnology
National Biotechnology Development Agency
Abuja, Nigeria

V.A. Olagunju
Department of Chemistry and Biochemistry
Miami University
Oxford, Ohio, USA

Catherine Ohunene Olaitan
Department of Microbiology
Prince Abubakar Audu University
Anyigba, Nigeria

Bukola Rukayat Olowoyeye
Department of Microbiology
Federal University of Technology Akure
Akure, Nigeria

Isibor Patrick Omoregie
Department of Biological Sciences
Covenant University
Ota, Nigeria

Solomon Oni
Bioresources Development Centre (Isanlu)
National Biotechnology Development Agency
Abuja, Nigeria

Ifeoma Sonia Onukwu
Department of Biology and Forensic Science
Admiralty University of Nigeria
Ibusa, Nigeria

Abiodun Eniola Opeyemi
Department of Microbiology
University of Ibadan
Ibadan, Nigeria

Agbontaen David Osagie
Department of Public Health
University of South Wales
Cardiff, Wales, UK

Muritada Babatunde Oshinuga
Department of Environmental Sciences
Wayne State University
Detroit, Michigan, USA

Olubukunola Osisami
Department of Agricultural Biotechnology
National Biotechnology Development Agency
Abuja, Nigeria

D.A. Oyegoke
Department of Chemical Sciences
Achievers University
Owo, Nigeria

Oluwafemi Adebayo Oyewole
Department of Microbiology
Federal University of Technology Minna
Minna, Nigeria

Ramat Onyeneoyiza Raji
Department of Microbiology
Prince Abubakar Audu University
Anyigba, Nigeria

Bellamkonda Ramesh
Department of Biochemistry and Molecular Biology
University of Nebraska Medical Center
Omaha, Nebraska, USA

Tarek Reham
Department of Biotechnology
Cairo University
Cairo, Egypt

Adekunle Babajide Rowaiye
Department of Agricultural Biotechnology
National Biotechnology Development Agency
Abuja, Nigeria
and
Department of Pharmaceutical Science
North Carolina Central University
Durham, North Carolina, USA

J.O. Samuel
Department of Microbiology
Federal University of Technology Minna
Minna, Nigeria

A.T. Tiamiyyu
Department of Mathematics
The Chinese University of Hong Kong
Sha Tin, Hong Kong

David Adeiza Zakari
Department of Microbiology
Federal University of Technology Minna
Minna, Nigeria

Section I

Environmental Applications

1 Marine Greens
Types, Habitat, Growth, and Global Distribution

Idris Abdullahi Dabban, Ibrahim Mohammed Hussaini, Muazu Ahmed, Abdullahi Mohammed, and Tarek Reham

1.1 INTRODUCTION

Marine greens are algae and other photosynthetic organisms living in the bottom (benthic regions) or associated with the substrata of seas and oceans. They are among the most significant primary producers in tropical benthic ecosystems of the world's largest water bodies, including deep seas and oceans (Acosta-Calderón *et al.*, 2016). These groups of organisms have a simple reproductive system (Ilbanez *et al.*, 2012) with the ability to convert energy from sunlight into chemical energy, which is stored in the form of chemical compounds in a process known as photosynthesis (Shanab *et al.*, 2012). The biomass of marine greens is characterized by a diverse structure and function that respond, adapt, and protect them from various conditions of stress such as high salinity, osmotic stress, cold, heat, nitrogen deficiency, desiccation, photo-oxidation, and other harsh ecological conditions typical of deep water bodies (Shanab *et al.*, 2012). They are the oldest photosynthetic organisms, dating back 3.8 billion years (prokaryotic cyanophytes), and there are an estimated 280,000 known species (Stolz and Obermayer, 2005).

1.2 TYPES OF MARINE GREEN

1.2.1 Phaeophyta (Brown Algae)

Brown algae, which are often referred to as Phaeophytes, are a group of marine algae that typically have filamentous or thalloid morphologies and are brown in color because of the presence of fucoxanthin (Ellamie *et al.*, 2020). Fucoxanthin is the dominant xanthophyll pigment that obscures beta-carotene and other xanthophylls as well as chlorophylls A and C (there is no chlorophyll B), giving these algae their characteristic brown color. Brown algae are either brown or yellow-brown in appearance, and they can be found in the oceans and seas of both temperate and arctic climates. Examples include *Laminaria* and *Saccharina*, *Fucus*, *Sargassum muticum*, brown seaweeds, etc. (Ko *et al.*, 2014).

Their food reserves are often made up of complex polysaccharides, sugars, and higher alcohols. Laminaran is the main source of carbohydrates, and real starch is not present (in contrast to green algae) (Zhao *et al.*, 2007). Alginic acid, a heteropolysaccharide with a long chain, is combined with cellulose to create the cell walls (Ellamie *et al.*, 2020). Brown algae commonly feature root-like structures called "holdfasts" that are utilized to bind the algae to a surface while not actually being genuine roots. Seaweeds can grow in both freshwater and saltwater, while kelp (brown algae) can only grow in saltwater and is most frequently found around rocky shores (Zhao *et al.*, 2007).

DOI: 10.1201/9781003369738-2

1.2.2 Cyanobacteria (Blue-Green Algae)

Cyanobacteria are also called Cyanophyta, Cyanophyceae, blue-green bacteria, blue-green algae, or cyanoprokaryota. They are oxygen-producing photosynthetic prokaryotes that share habitats similar to multicellular (eukaryotic) algae (Kauff and Budel, 2011; Dvorak *et al.*, 2015). Examples include *Chroococcus subnudus*, *Pleurocapsa minor*, *Planktothrix agardhii*, *Arthrospira jenneri*, *Johanseninema constricum*, *Phormidium* sp., etc. Meanwhile, cyanobacteria have been found to exist in extreme habitats of the world (Seckbach, 2007). They have been estimated to have evolved about 3 billion years ago, which makes them the oldest photosynthetic oxygen-producing microorganisms. They contributed significantly to the increase in atmospheric oxygen (Bekker *et al.*, 2004;

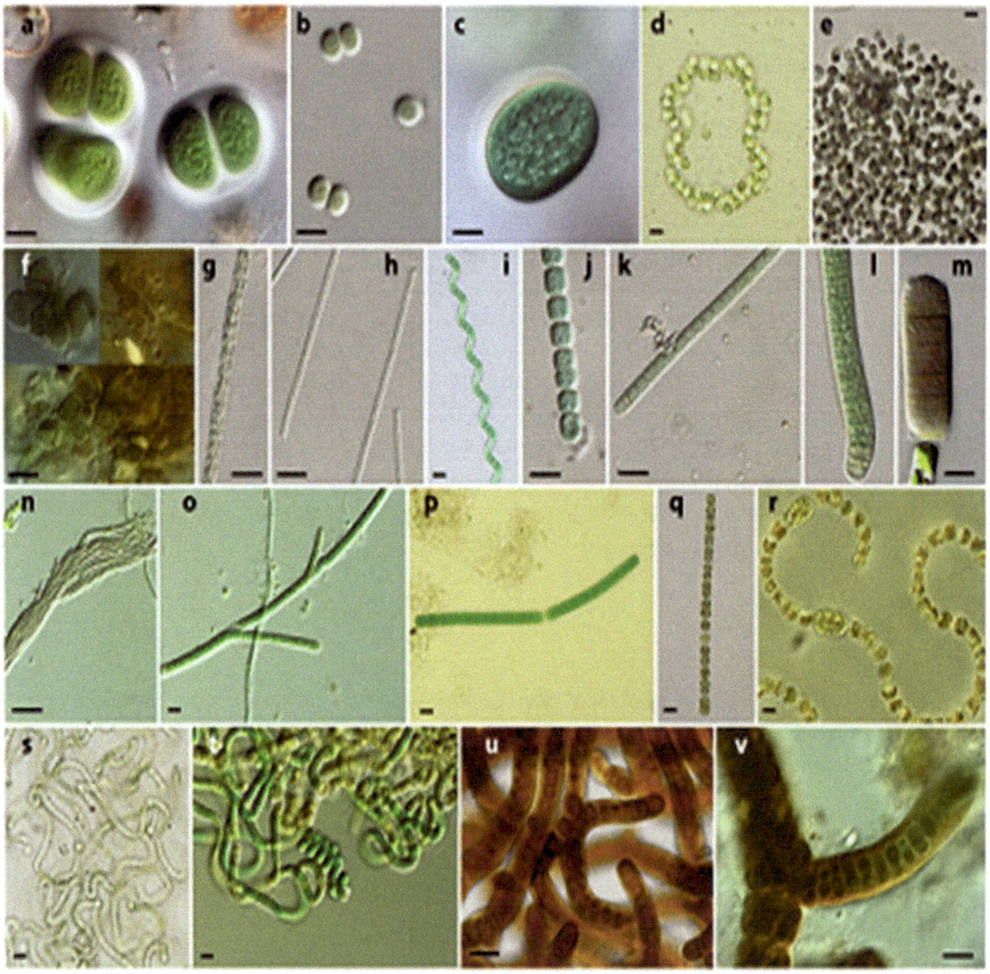

FIGURE 1.1 Morphological diversity of Cyanobacteria. Groups (orders) follow (Rippka *et al.*, 1979; Dvorak *et al.*, 2015). I. Chroococcales: a. *Chroococcus subnudus*, b. *C. limneticus*, c. *Cyanothece aeruginosa*, d. *Snowella litoralis*, e. *Microcystis aeruginosa*. II. Pleurocapsales: f. *Pleurocapsa minor.* III. Oscillatoriales: g. *Planktothrix agardhii*, h. *Limnothrix redekei*, i. *Arthrospira jenneri*, j. *Johanseninema constricum*, k. *Phormidium* sp., l., m. *Oscillatoria* sp., n. *Schizothrix* sp., o. *Tolypothrix* sp., p. *Katagnymene accurata*, IV. Nostocales: q. *Dolichospermum planctonicum*, r. *Dolichospermum* sp., s. *Nostoc* sp., t. *Nodularia moravica.* V. Stigonematales, u., v. *Stigonema* sp.

Kauff and Budel, 2011). The morphology and ultra-cellular structure of cyanobacteria vary remarkably from those of other unicellular and filamentous organisms.

Some cyanobacteria may possess microplasmodesmata, which are inter-cellular links considered to be a characteristic of the eukaryotic nature of the organisms (Nurnberg *et al.*, 2014; Dvorak *et al.*, 2015). In addition, the morphology and differentiation of the cells exhibited by some genera of cyanobacteria are such that they easily adapt to nitrogen fixation (heterocytes) as well as the resting phase of the cells (akinetes) (Whitton and Potts, 2000). Some known Cyanobacteria families include (Figure 1.1) Chroococcales, Pleurocapsales, Oscillatoriales, Nostocales, and Stigonematales (Dvorak *et al.*, 2015).

1.3 HABITAT OF MARINE GREENS

This is the habitat that supports marine life. Marine life depends on the amount of saltwater that is in the sea. The marine environment supports many kinds of these habitats and can be divided into coastal and open ocean habitats. Coastal habitats are found in the area that extends from as far as the tide comes in on the shoreline out to the edge of the continental shelf. Most marine life is found in coastal habitats, even though the shelf area occupies only 7% of the total ocean area. Open ocean habitats are found in the deep ocean beyond the edge of the continental shelf (Charette and Smith, 2010).

1.3.1 Pelagic Zone

On the basis of depth, intensity, and penetration of sunlight, types and characteristics of organisms present, as well as nutrient availability, the pelagic zone is further subdivided into Epilagic, Mesopelagic, Bathypelagic, and Abyssopelagic zones (Figure 1.2)

1.3.1.1 Epipelagic Zone

This is known as the photic zone, euphotic zone, or sunlight zone. It is the uppermost layer of a body of water that receives sunlight, allowing phytoplankton to perform photosynthesis. It undergoes a series of physical, chemical, and biological processes that supply nutrients to the upper water column. The photic zone is home to the majority of aquatic life due to the activity (primary production) of the phytoplankton.

1.4 THE GROWTH OF MARINE GREENS

1.4.1 Physical Growth Requirements of Marine Greens

1.4.1.1 Light

Light is a major requirement for marine algal growth, as it serves as an energy source, converting absorbed water and carbon dioxide into algal biomass during photosynthesis (Ozkurt, 2009; Nigam *et al.*, 2020). Light energy is used for fixing carbon dioxide to produce carbohydrates and protein (Blinová *et al.*, 2015). Intensity of light is also vital in nitrogen uptake and fixation, since the energy required for these processes is derived from photosynthesis (Hurd *et al.*, 2014; Jevne *et al.*, 2020).

The duration and intensity of light affect the photosynthetic process directly and determine the biochemical composition as well as the biomass production rate of algae (Krzemińska *et al.*, 2014). Optimum light intensity levels for most species of algae are about 200–400 μM photons/m^2/s (Schuurmans *et al.*, 2015). Low algal growth occurs when the light intensity is very low or very high (Sforza *et al.*, 2012; Ye *et al.*, 2012).

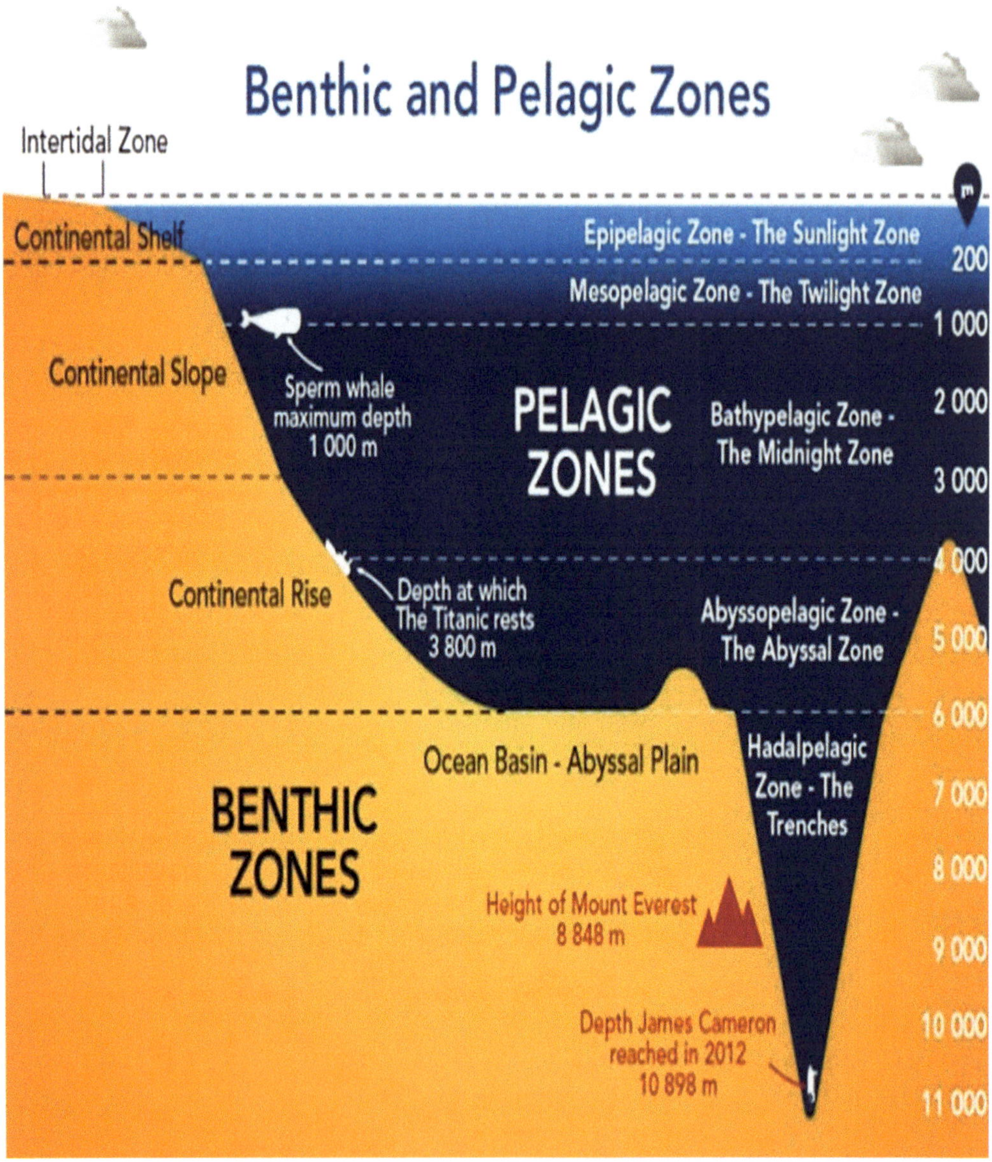

FIGURE 1.2 Habitat of marine greens

1.4.1.2 pH

The growth and metabolism of marine algae are also affected by the pH of their habitat, and optimal growth occurs in the pH range of 8.2–8.7 (Blinová *et al.*, 2015); however, they can grow in the pH range of 7–9 (Nigam *et al.*, 2020). Marine greens such as *Spirulina* survive and grow in alkaline lakes and grow favorably in a pH range of 8.5 to 11.0 (Vo *et al.*, 2015). Nutrient and carbon dioxide solubility in the environment and their subsequent availability for algal growth are determined by pH (Qiu *et al.*, 2017; Vale *et al.*, 2020). An increase in pH results in an increase in salinity, which is in turn harmful to algae (Juneja *et al.*, 2013).

1.4.1.3 Salinity

Many species of marine algae are able to tolerate changes in the salt concentration of their surrounding environment and grow optimally at a salinity relatively lower than that of their natural environment. The optimum salinity for the growth of marine algae is reported to range between 20 and 24 g/L (Perumal *et al.*, 2012). However, optimal production of *Spirulina* is achieved at a salt concentration of ˃30 g/L (Vo *et al.*, 2015)

1.4.1.4 Temperature

The rates of metabolism and growth of marine algae are affected by the temperature of their surrounding medium (Chowdury *et al.*, 2020). Temperature has a direct influence on the biochemical processes (such as photosynthesis) of the algae. Carbon assimilation is reduced when the temperature is low, while too high a temperature inactivates photosynthetic proteins and disturbs energy balance. All of these in turn affect and reduce the rate of photosynthesis (Khan *et al.*, 2018).

Generally, algae grow optimally at temperatures between 20°C and 35°C (Chowdury *et al.*, 2020). Some strains of *Chlorella*—a genus of marine algae—can tolerate and grow at a wider temperature range between 15°C and 40°C (Masojidek and Torzillo, 2008). The optimal growth temperature for *Spirulina* is reported to be in the range of 35°C to 39°C (Vo *et al.*, 2015). Also, some thermophiles (e.g., *Anacystis nidulans* and *Chaetoceros*) can survive at temperatures as high as 40°C, while algae found in hot springs survive temperatures close to 80°C (Covarrubias *et al.*, 2016).

1.4.1.5 Mixing and Aeration

Mixing and aeration are required for the growth of marine algae; they play a vital role in ensuring that nutrients, air, and CO_2 are distributed uniformly. They also prevent settling and aggregation of algal biomass by allowing light penetration and distribution (Show *et al.*, 2017). In addition to their role in nutrient dissolution and penetration of light, they also ensure proper exchange of gas (Zeng *et al.*, 2011).

1.4.2 Nutritional Growth Requirements of Marine Greens

The nutritional requirements of marine algae may vary depending on the species. However, their requirements for essential nutrients (macronutrients) and iron are the same for almost all species (Khan *et al.*, 2018).

1.4.2.1 Macronutrients

1.4.2.1.1 Carbon

Carbon is one of the macronutrients required for algal growth; hence, a carbon source is required for the growth of marine algae (Chowdury *et al.*, 2020). Carbon, as an important constituent of nucleic acids, carbohydrates, proteins, lipids, and vitamins, is required in high concentrations (Richmond, 2004). Photoautotrophic marine greens such as *Spirulina* use carbon dioxide (CO_2) as an inorganic carbon source and use light energy to photosynthesize (Blinová *et al.*, 2015). Assimilation of inorganic carbon by algae is through a diffusion process (pH 5.0–7.0) and an active transport (pH >7.0) process (Gonçalves *et al.*, 2017). Heterotrophs use organic carbon sources such as sugars, acids, and alcohols (Chowdury *et al.*, 2020). Mixotrophic algae can use both organic and inorganic carbon sources (Dahiya, 2014; Rashid *et al.*, 2014).

1.4.2.1.2 Nitrogen

Nitrogen is second to carbon as an important element or macronutrient for the growth of marine algae and is required in large quantities, being a vital constituent of proteins and nucleic acids (Chowdury *et al.*, 2020; Xu *et al.*, 2020). Furthermore, carbon and nitrogen play an important role in metabolism, as they are involved in the production of macromolecules (proteins and nucleic acids), vitamins, coenzymes, photosynthetic pigments (chlorophylls), phytohormones, and secondary metabolites (Richmond, 2004; Xu *et al*., 2020). The nitrogen source for algae in the marine ecosystem is either inorganic (nitrate [NO_3^-], nitrite [NO_2^-], or ammonium [NH^{4+}]) or organic (urea, amino acids, or as part of organic substances such as dissolved or particulate nitrogen) (Perez-Garcia *et al*., 2011; Hurd *et al*., 2014; Jevne *et al*., 2020; Aransiola *et al*., 2021). Irrespective of the nitrogen source, the growth rate of algae seems to be similar, and the assimilation of nitrogen is through active transport (Chowdury *et al*., 2020). *Spirulina* assimilates nitrate mainly as its source of nitrogen (Vo *et al*., 2015). Nitrate is used commonly as a nitrogen source in culture media because of its stability and less possibility of changing pH. Ammonia is used less frequently because of its toxicity to algae at concentrations ˃25 µM (Procházková *et al*., 2013).

Under limited nitrogen supply, reduced content of chlorophylls, reduced biomass production, and an increase in carotenoids occur, leading to cellular discoloration as well as build-up of organic compounds (polysaccharides, lipids, and some oils) (Chowdury *et al*., 2020). Under these conditions of nitrogen shortage, algae accumulate storage polysaccharides, and they revert to synthesis of proteins and pigments when the nitrogen level becomes sufficient (Xu *et al*., 2020).

1.4.2.1.3 Phosphorus

Phosphorus is an essential macronutrient required for the growth of marine algae and many metabolic functions of the cell, including energy transfer, synthesis of nucleic acids (Richmond, 2004), signal transduction, and photosynthesis as well as lipid and fatty acid biosynthesis (Ota *et al*., 2016; YangYang *et al*., 2018). It also plays a role in chlorophyll synthesis when present in high supply by stimulating the activity of enzymes involved in the synthesis (Xu *et al*., 2020; Auta *et al*., 2022). Phosphorus occurs naturally in the environment as pyrophosphate, polyphosphate, orthophosphate, and metaphosphate (Markou *et al*., 2014); however, orthophosphate is the preferred form in which it is absorbed (Richmond, 2004). Assimilation of phosphorus by algae depends on energy and takes place through active transport (Richmond, 2004; Gonçalves *et al*., 2017).

Phosphorus (in orthophosphate and polyphosphate forms) sometimes accumulates in excess of the amount required for growth by algae. When nutritional conditions become unfavorable, this excess phosphorus can be transformed into ATP, which is required to pump polyphosphate into the cell and convert phosphorus to polyphosphate (Chu *et al*., 2013; Maizatul *et al*., 2017; Solovchenko *et al*., 2020).

Other macronutrients required for algal growth are sodium (Na) and calcium (Ca). In addition, silicon is required by certain species of marine algae as a macronutrient (Khan *et al*., 2018).

1.4.2.2 Micronutrients

Micronutrients are essential in addition to macronutrients for adequate growth of marine algae. They are required in trace quantities; however, they strongly impact algal growth, since they influence the many activities that involve enzymes (Gardner-Dale *et al*., 2017; Khan *et al*., 2018). Micronutrients include sulfur (S), iron (Fe), chlorine (Cl), molybdenum (Mo), manganese (Mn), zinc (Zn), magnesium (Mg), copper (Cu), potassium (K), boron (B), and cobalt (Co) (Khan *et al*., 2018; Chowdury *et al*., 2020). Iron is involved in electron transport during photosynthesis (Chowdury *et al*., 2020).

1.4.3 Algal Culture Media for the Growth of Marine Greens

Growth media are used for growing marine greens in the laboratory. They are formulated to provide the required nutrients for growth and contain nitrogen, phosphorus, vitamins, and trace nutrient sources. Examples of commonly used media for growing marine algae are Walne medium and Guillard's F/2 medium (Table 1.2). These media contain similar constituents, namely, macronutrients, micronutrients (trace metals), and vitamins (Tables 1.1 and 1.2); however, the concentrations and ratios of these constituents vary (Nigam *et al*., 2020).

TABLE 1.1
Constituents of Walne Medium

Stock	Constituents	Quantities (g or mL)
Solutions A	Ferric chloride ($FeCl_3$)	0.8
(Nutrient solution)	Manganous chloride ($MnCl_2$, $4H_2O$)	0.4
	Boric acid (H_3BO_3)	33.6
	EDTA, di-sodium salt	45.0
	Sodium di-hydrogen orthophosphate (NaH_2PO_4, $2H_2O$)	20.0
	Sodium nitrate ($NaNO_3$)	100.0
Solution B	Zinc chloride ($ZnCl_2$)	2.1
(Trace metal solution)	Cobaltous chloride ($CoCl_2,6H_2O$)	2.0
	Ammonium molybdate ($(NH_4)_6Mo_7O_{24}$. $4H_2O$)	0.9
	Cupric sulfate ($CuSO_4$. $5H_2O$)	2.010.0
	Concentrated HCl	
Solution C	Vitamin B1	0.2
(Vitamin solution)	Vitamin B12	0.1

Adapted from Nigam, S., Sinha, S., Srivastava, A., and Srivastava, A., Cultivation and Production Techniques of Marine Algae Encyclopedia of Marine Biotechnology: Five Volume Set (ed. Se-Kwon Kim, 1st ed.). 2020. Copyright Wiley-VCH Verlag GmbH & Co. KGaA. Reproduced with permission.

TABLE 1.2
Constituents of Guillard's F/2 Medium

Stock solutions	Constituents	Final concentration (mg/L)
Nitrate and phosphate	$NaNO_3$	75
	$NaH_2PO_4.H_2O$	5
Silicate	$Na_2SiO_3.9H_2O$	30
Trace metal	$Na_2C_{10}H_{14}O_8N_2.H_2O$ (Na_2EDTA)	4.36
	$CoCl_2.6H_2O$	0.01
	$CuSO_4.5H_2O$	0.01
	$FeCl_3.6H_2O$	3.15
	$MnCl_2.4H_2O$	0.18
	$Na_2MoO_4.2H_2O$	0.006
	$ZnSO_4.7H_2O$	0.022
Vitamins	Vitamin B1 (Thiamine HCl)	0.1
	Vitamin B7 or H (Biotin)	0.0005
	Vitamin B12	0.0005

Adapted from Nigam, S., Sinha, S., Srivastava, A., and Srivastava, A., Cultivation and Production Techniques of Marine Algae Encyclopedia of Marine Biotechnology: Five Volume Set (ed. Se-Kwon Kim, 1st ed.). 2020. Copyright Wiley-VCH Verlag GmbH & Co. KGaA. Reproduced with permission.

1.4.4 Marine Greens Culture Technique

Marine greens are grown using either batch, continuous, or semi-continuous culture techniques (Nigam *et al.*, 2020). These growth culture techniques are used for the commercial production of marine algae in different parts of the world. They are shallow loop-like or circular structures/channels equipped with a paddlewheel or arm for mixing algae at slow speed. Open ponds, commonly

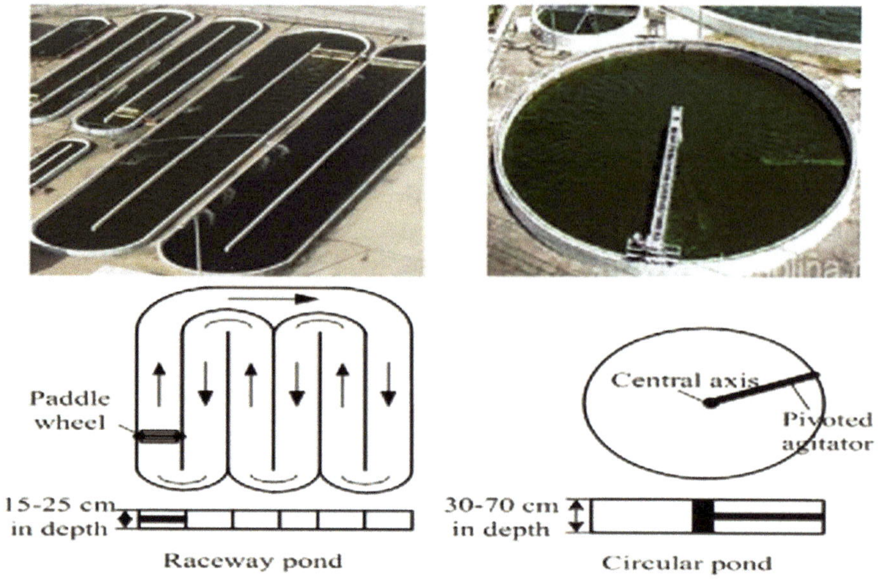

FIGURE 1.3 Pond types used in photoautotrophic production of algae. (From Kerestecioğlu, F.Ö. and Pekmezci, Y.T., *Int. J. Eng. Sci. Appl.*, 3, 52–70, 2019.)

used for algal cultivation, are classified into raceway ponds and circular ponds. Raceway ponds are used widely for the production of *Spirulina*, *Dunaliella*, and *Haematococcus* at commercial scale (Borowitzka, 2005; Nigam *et al.*, 2020), with reported daily productivity ranging between 14 and 50 gm^{-2} (Nigam *et al.*, 2020). Circular ponds are used for the commercial cultivation of *Chlorella* in Asia and have reported daily productivity ranging between 8.5 and 21 gm^{-2} (Nigam *et al.*, 2020) (Figure 1.3).

1.5 GLOBAL DISTRIBUTION OF MARINE GREENS

Phytoplankton live in the photic zone in the aquatic environment, which extends from the surface layer (neuston) to 250 m or deeper and is also dependent on local latitude. At any given time, the world's seas are estimated to contain up to 3.6×10^{25} phytoplankton, and they may produce more than 5×10^{13} kg of annual primary production, which serves as the foundation of the marine food web. The distribution and metabolic activity of several species of algae is influenced by diurnal, seasonal, vertical, regional, and other variations in nutrition supply, temperature, light, and other factors (Kumar and Singh, 2021). The euphotic zone is where 90% of marine life can be found. Examples of abiotic components are nutrients, organic debris, and dissolved oxygen. Specific nutrients, like nitrogen, phosphorus, silicon, and sulfur, have an impact on how quickly and densely an organism's population may grow. Competition and nutrient utilization are two major interactions that control the development of marine microbial ecosystems (Upadhyay *et al.*, 2019).

1.5.1 Global Distribution of Marine Microalgae

Microalgae, which include cyanobacteria, prochlorophytes, and eukaryotic algae, are within the diverse group of microscopic, phototrophic organisms (Kumar and Singh, 2021). *Thalassiosira pseudonana* (the diatom), *Heterosigma carterae* (a marine raphidophyte), and *Alexandrium minutum* (the dinoflagellate) are common marine microalgae that are found in the euphotic zone, which is the depth at which there is no net CO_2 assimilation; in this zone, the rate of CO_2 production is equal to the rate of CO_2 uptake (Upadhyay *et al.*, 2019).

Diatoms, with up to 10 million species, make up the largest category of microalgae found in marine habitats. Most of them are unicellular; however, filamentous species with a wide variety and abundance are also known to exist. They are categorized into three classes based on bilateral or radial symmetry: a) *Fragilariophyceae* (bilateral symmetry without raphes, including *Navicula, Synedra*, and *Fragilaria*); b) *Coscinodiscophyceae* (bilateral symmetry with raphes, including *Navicula, Nitzschia*, and *Phacodactylum*); c) *Bacillarophyceae* (radial symmetry, e.g., *Chaetoceros, Biddulphia*, and *Thalassiosira*) (Kumar and Singh, 2021). Malviya *et al.* (2016) in their study reported that except for the warmest and most hypersaline settings, diatoms are extensively distributed in practically all aquatic ecosystems. They can also exist as endosymbionts in dinoflagellates and foraminifers. The authors also reported the presence of diatoms in the North Atlantic Ocean, Mediterranean Sea, Red Sea, Indian Ocean, South Atlantic Ocean, Southern Ocean, and South Pacific Ocean.

1.5.2 Global Distribution of Marine Macroalgae

Marine macroalgae, often known as seaweeds, are a collection of macroscopic, multicellular photosynthetic organisms that are divided into three main classes—green (*Chlorophyta*), red (*Rhodophyta*), and brown (*Phaeophyta*)—depending on their cell structure and pigmentation (chemical). Seaweeds make up a significant portion of marine macrophytes, such as seagrass and mangroves, and they are the foundation of the aquatic food chain (Gupta *et al.*, 2017; Vitthal, 2021).

There are around 20,000 different types of seaweed distributed across the globe, and 221 of them are used commercially, including 110 species that are used to make phycocolloids and 145 species that are used as food. Indian marine macroalgae comprise 844 species of seaweed in 217 genera. The economic value of seaweeds in India reveals that only a small number of random studies have been conducted in Andhra Pradesh, Gujarat, Goa, Lakshadweep, Maharashtra, Odisha, and Tamil Nadu. Seaweeds grow luxuriantly around the Andaman and Nicobar Islands, Gujarat coast, Lakshadweep Island, and the southeast coast of Tamil Nadu from Mandapam to Kanyakumari, while there are plenty of seaweeds in the areas of Bombay, Raighad, Ratnagiri, Sindhudurgh, Goa, Karwar, Varkala, Kovalam, Vizhinjam, Visakhapatnam, and a few more locations like Chilka and Pulicat lakes (Vitthal, 2021).

1.6 CONCLUSION

Marine green organisms present great potential beneficial applications in medicine, industry, and the environment if properly exploited. Therefore, screening of these important organisms for their natural products, biomass quality, and applications should be of great interest to scientists around the world. Nonetheless, commercial applications of marine greens for food, feed, and fuels are limited by some setbacks, such as the development and management of safe and adequate production operations for the identification, bio-prospecting, selection, and improvement of strains of marine green organisms. Knowledge of biotechnology, genetic engineering, and omics technology could be useful to overcome the potential setbacks to the optimal utilization of marine greens for solving global problems.

REFERENCES

Acosta-Calderón, J.A., Mateo-Cid, L.E. and Mendoza-González, Á.C. (2016), An updated list of marine green algae (Chlorophyta, Ulvophyceae) from the biosphere reserve of Sian Ka'an, Quintana Roo, Mexico. *The Journal of Biodiversity Data*, 12(3), 1886. http://doi.org/10.15560/12.3.1886.

Aransiola, S.A., Ijah, U.J.J., Abioye, O.P. and Victor-Ekwebelem, M.O. (2021), Anammox in wastewater treatment. In N.R. Maddela, L.C. García Cruzatty and S. Chakraborty (Eds.), *Advances in the Domain of Environmental Biotechnology: Environmental and Microbial Biotechnology*. Springer, Singapore. https://doi.org/10.1007/978-981-15-8999-7_15.

Auta, H.S., Aboyeji, D.O., Bala, J.D., Abioye, O.P., Adabara, N.U., Aransiola, S.A., Hassana, A. and Azize, A. (2022), Marine microbial enzymes - An overview. In N.R. Maddela, S.A. Aransiola and R. Prasad (Eds.), *Ecological Interplays in Microbial Enzymology* (1st ed.). Springer Nature Singapore Pte Ltd. ISSN: 2662-1681. https://www.springer.com/series/16324.

Bekker, A., Holland, H.D., Wang, P.-L., Rumble, D., Stein, H.J., Hannah, J.L., Coetzee, L.L. and Beukes, N.J. (2004), Dating the rise of atmospheric oxygen. *Nature*, 427(6970), 117–120.

Blinová, L., Bartošová, A. and Gerulová, K. (2015), Cultivation of microalgae (*Chlorella vulgaris*) for biodiesel production. *Research Papers Faculty of Materials Science and Technology in Trnava Slovak University of Technology in Bratislava*, 23(36), 87–95.

Borowitzka, M.A. (2005), Culturing microalgae in outdoor ponds. In R.A. Andersen (Ed.), *Algal Culturing Techniques*. Elsevier, Academic Press, New York, pp. 205–218.

Charette, M. and Smith, W. (2010), The volume of earth's ocean. *Oceanography*, 23(2), 112–114. https://doi.org/10.5670%2Foceanog.2010.51.

Chowdury, K.H., Nahar, N. and Deb, U.K. (2020), The growth factors involved in microalgae cultivation for biofuel production: A review. *Computational Water, Energy, and Environmental Engineering*, 9(4), 185–215. https://doi.org/10.4236/cweee.2020.94012.

Chu, F.F., Chu, P.N., Cai, P.J., Li, W.W., Lam, P.K.S. and Zeng, R.J. (2013), Phosphorus plays an important role in enhancing biodiesel productivity of *Chlorella vulgaris* under nitrogen deficiency. *Bioresource Technology*, 134, 341–346.

Covarrubias, Y., Cantoral-Uriza, E.A., Casas-Flores, J.S. and García-Meza, J.V. (2016), Thermophile mats of microalgae growing on the woody structure of a cooling tower of a thermoelectric power plant in Central Mexico. *Revista Mexicana de Biodiversidad*, 87(2), 277–287.

Dahiya, A. (2014), *In Bioenergy: Biomass to Biofuel* (1st ed.), pp. 219–238.

Dvorak, P., Poulickova, A., Hasler, P., Belli, M., Casamatta, D.A. and Papini, A. (2015), Species concepts and speciation factors in cyanobacteria, with connection to the problems of diversity and classification. *Biodiversity and Conservation*, 24(4), 739–757. https://doi.org/10.1007/s10531-015-0888-6.

Ellamie, A.M., Fouda, W.A., Ibrahim, W.M. and Ramadan, G. (2020), Dietary supplementation of brown seaweed (*Sargassum latifolium*) Alleviates the environmental heat stress-induced toxicity in male Barki sheep (Ovisaries). *Journal of Thermal Biology*, 89, 102561.

Gardner-Dale, D.A., Bradley, I.M. and Guest, J.S. (2017), Influence of solids residence time and carbon storage on nitrogen and phosphorus recovery by microalgae across diel cycles. *Water Research*, 121, 231–239.

Gonçalves, A.L., Pires, J.C.M. and Simões, M. (2017), A review on the use of microalgal consortia for wastewater treatment. *Algal Research*, 24, 403–415. https://doi.org/10.1016/j.algal.2016.11.008.

Gupta, V., Jain, M. and Reddy, C.R.K. (2017), Macroalgal functional genomics: A missing area. In: Kumar, M., Ralph, P., (eds), *Systems Biology of Marine Ecosystems*. Springer,Cham, pp. 3–12.

Hurd, C.L., Harrison, P.J., Bischof, K. and Lobban, C.S. (2014), *Seaweed Ecology and Physiology* (2nd ed.), Cambridge University Press, Cambridge.

Ilbanez, E., Herrero, M., Mendiola, J.A. and Castro-Puyana, M. (2012), *In Marine Bioactive Compounds: Sources, Characterization and Applications* (ed. M. Hayes). Springer, pp. 55–98.

Jevne, L.S., Forbord, S. and Olsen, Y. (2020), The effect of nutrient availability and light conditions on the growth and intracellular nitrogen components of land-based cultivated *Saccharina latissima* (Phaeophyta). *Frontiers in Marine Science*, 7, 557460. https://doi.org/10.3389/fmars.2020.557460.

Juneja, A., Ceballos, R.M. and Murthy, G.S. (2013), Effects of environmental factors and nutrient availability on the biochemical composition of algae for biofuels production: A review. *Energies*, 6(9), 4607–4638. https://doi.org/10.3390/en6094607.

Kauff, F. and Budel, B. (2011), Phylogeny of cyanobacteria: An overview. *Progress in Botany*, 72, 209–224.

Kerestecioğlu, F.Ö. and Pekmezci, Y.T. (2019), Defining the problems in integration of microalgae photobioreactor systems to architecture. *International Journal of Engineering Science and Application*, 3(2), 52–70.

Khan, M.I., Shin, J.H. and Kim, J.D. (2018), The promising future of microalgae: Current status, challenges, and optimization of a sustainable and renewable industry for biofuels, feed, and other products. *Microbial Cell Factories*, 17(1), 36. https://doi.org/10.1186/s12934-018-0879-x.

Ko, S.J., Bu, Y., Bae, J., Bang, Y.-M., Kim, J., Lee, H., Lee, B.-J., Hyun, Y.H. and Park, J.-W. (2014), Protective effects of *Laminaria japonica* with probiotics on murine colitis. *Mediators of Inflammation*, 2014, 417814.

Krzemińska, I., Pawlik-Skowrońska, B., Trzcińska, M. and Tys, J. (2014), Influence of photoperiods on the growth rate and biomass productivity of green microalgae. *Bioprocess and Biosystems Engineering*, 37(4), 735–741. https://doi.org/10.1007/s00449-013-1044-x.

Kumar, A. and Singh, J.S. (2021), *Microalgae in Wastewater Remediation*. CRC Press, umber of random studies.

Maizatul, A.Y., Mohamed, R.M.S.R., Al-Gheethi, A.A. and Hashim, M.A. (2017), An overview of the utilisation of microalgae biomass derived from nutrient recycling of wetmarket wastewater and slaughterhouse wastewater. *International Aquatic Research*, 9(3), 177–193.

Malviya, S., Scalco, E., Audic, S., Vincent, F., Veluchamy, A., Poulain, J., Wincker, P., Iudicone, D., de Vargas, C., Bittner, L., Zingone, A. and Bowler, C. (2016), Insights into global diatom distribution and diversity in the world's ocean. *Proceedings of the National Academy of Sciences of the United States of America*, 113(11). https://doi.org/10.1073/pnas.1509523113.

Markou, G., Vandamme, D. and Muylaert, K. (2014), Microalgal and cyanobacterial cultivation: The supply of nutrients. *Water Research*, 65, 186–202. https://doi.org/10.1016/j.watres.2014.07.025.

Masojidek, J. and Torzillo, G. (2008), Mass production of freshwater microalgae. In S.E. Jorgenen and B.D. Fath (Eds.), *Ecological Engineering, Vol. 3, Encyclopaedia of Ecology*. Elsevier, Oxford, pp. 2226–2235. https://doi.org/10.1016/B978-008045405-4.00830-2.

Nigam, S., Sinha, S., Srivastava, A. and Srivastava, A. (2020), *Cultivation and Production Techniques of Marine Algae Encyclopedia of Marine Biotechnology: Five Volume Set* (ed. Se-Kwon Kim, 1st ed.). © 2020 John Wiley & Sons Ltd. Published 2020 by John Wiley and Sons Ltd.

Nurnberg, D.J., Mariscal, V., Parker, J., Mastroianni, G., Flores, E. and Mullineaux, C.W. (2014), Branching and intercellular communication in the Section V cyanobacterium Mastigocladus laminosus, a complex multicellular prokaryote. *Molecular Microbiology*, 91(5), 935–949.

Ota, S., Yoshihara, M., Yamazaki, T., Takeshita, T., Hirata, A., Konomi, M., Oshima, K., Hattori, M., Bisova, K., Zachleder, V. and Kawano, S. (2016), Deciphering the relationship among phosphate dynamics, electron-dense body and lipid accumulation in the green alga *Parachlorella kessleri*. *Scientific Reports*, 6, 25731.

Ozkurt, I. (2009), Qualifying of safflower and algae for energy. *Energy Education Science and Technology*, 23, 145–151.

Perez-Garcia, O., Escalante, F.M.E., de-Bashan, L.E. and Bashan, Y. (2011), Heterotrophic cultures of microalgae: Metabolism and potential products. *Water Research*, 45(1), 11–36. https://doi.org/10.1016/j.watres.2010.08.037.

Perumal, P., Prasath, B., Santhanam, P., Selvaraju, S. Ananth and Devi, A.S. (2012), Isolation and Culture of Microalgae, In: Perumal, S., A.R., T., Pachiappan, P. (eds), *Advances in Marine and Brackishwater Aquaculture*, New Delhi. pp 1–15.

Procházková, G., Brányiková, I., Zachleder, V. and Brányik, T. (2013), Effect of nutrient supply status on biomass composition of eukaryotic green microalgae. *Journal of Applied Phycology*, 26(3), 1359–1377.

Qiu, R., Gao, S., Lopez, P.A. and Ogden, K.L. (2017), Effects of pH on cell growth, lipid production and CO_2 addition of microalgae *Chlorella sorokiniana*. *Algal Research*, 28, 192–199. https://doi.org/10.1016/j.algal.2017.11.004.

Rashid, N., Saif Ur Rehman, M., Sadiq, M., Mahmood, T. and Han, J. (2014), Current status, issues and developments in microalgae derived biodiesel production. *Renewable and Sustainable Energy Reviews*, 40, 760–778.

Richmond, A. (2004), *Handbook of Microalgal Culture-Biotechnology and Applied Phycology*. Blackwell Science, Victoria.

Rippka, R., Deruelles, J., Waterbury, J.B., Herdman, M. and Stanier, R.Y. (1979), Generic assignments, strain histories and properties of pure cultures of cyanobacteria. *The Journal of General Microbiology*, 111(1), 1–61.

Schuurmans, R.M., van Alphen, P., Schuurmans, J.M., Matthijs, H.C.P. and Hellingwerf, K.J. (2015), Comparison of the photosynthetic yield of cyanobacteria and green algae: Different methods give different answers. *PLOS ONE*, 10(9), e0139061. https://doi.org/10.1371/journal.pone.0139061.

Seckbach, J. (2007), Algae and Cyanobacteria in extreme environments. In J. Seckbach (Ed.), *Cellular Origin, Life in Extreme Habitats and Astrobiology*, Vol. 11. Springer Science & Business Media, New York, pp. 1–811.

Sforza, E., Simionato, D., Giacometti, G.M., Bertucco, A. and Morosinotto, T. (2012), Adjusted light and dark cycles can optimize photosynthetic efficiency in algae growing in photobioreactors. *PLOS ONE*, 7(6), e38975. https://doi.org/10.1371/journal.pone.0038975.

Shanab, S.M.M., Mostafa, S.S.M., Shalaby, E.A. and Mahmoud, G.I. (2012), Aqueous extracts of microalgae exhibit antioxidant and anticancer activities. *Asian Pacific Journal of Tropical Biomedicine*, 2(8), 608–615.

Show, P.L., Tang, M.S.Y., Nagarajan, D., Ling, T.C., Ooi, C.-W. and Chang, J.-S. (2017), A holistic approach to managing microalgae for biofuel applications. *International Journal of Molecular Sciences*, 18(1), 215. https://doi.org/10.3390/ijms18010215.

Solovchenko, A., Gorelova, O., Karpova, O., Selyakh, I., Semenova, L., Chivkunova, O., Baulina, O., Vinogradova, E., Pugacheva, T., Scherbakov, P., Vasilieva, S., Lukyanov, A. and Lobakova, E. (2020), Phosphorus feast and famine in Cyanobacteria: Is luxury uptake of the nutrient just a consequence of acclimation to its shortage? *Cells*, 9(9), 1933.

Stolz, P. and Obermayer, B. (2005), Manufacturing microalgae for skin care. *Cosmetics & Toiletries*, 120, 99–106.

Upadhyay, R.K., Kumari, S., Kumar, P. and Rai, V. (2019), Spatial distribution of microalgae in marine systems: A reaction–diffusion model. *Ecological Complexity*, 39, 100771. https://doi.org/10.1016/j.ecocom.2019.100771.

Vale, M.A., Ferreira, A., Pires, J.C.M. and Goncalves, A.L. (2020), CO_2 capture using microalgae, Chapter 17. In *Biological Techniques for CO_2 Capture: Advances in Carbon Capture*. Elsevier Inc., pp. 381–405. https://doi.org/10.1016/B978-0-12-819657-1.00017-7.

Vitthal, W.A. (2021), Diversity and distribution of common seaweeds from the West Coast of Maharashtra in India. In: Thangadurai, D., Sangeetha, J., Elumalai, S., Thimmappa, S.C. (eds), *Phycobiotechnology: Biodiversity and Biotechnology of Algae and Algal Products for Food, Feed, and Fuel*. CRC Press Taylor and Francis Group, Boca Raton, Florida, United States, pp. 1–22.

Vo, T.-S., Ngo, D.-H. and Kim, S.-K. (2015), Nutritional and Pharmaceutical properties of Microalgal *Spirulina*, Chapter 19. In *Handbook of Marine Microalgae*. Elsevier Inc., pp. 299–308. https://doi.org/10.1016/B978-0-12-800776-1.00019-4.

Whitton, B.A. and Potts, M. (2000), *The Ecology of Cyanobacteria: Their Diversity in Time and Space*. Springer, Berlin.

Xu, N., Xu, K., Wang, W., Xu, Y., Ji, D., Chen, C. and Xie, C. (2020), Nutrient enrichment improves growth and food quality of two strains of the economic seaweed *Pyropia haitanensis*. *Frontiers in Marine Science*, 7, 544–582. https://doi.org/10.3389/fmars.2020.544582.

Yang, F.F., Xiang, W., Li, T. and Long, L. (2018), Transcriptome analysis for phosphorus starvation-induced lipid accumulation in *Scenedesmus* sp. *Scientific Reports*, 8(1), 1–11.

Ye, C.P., Zhang, M.C., Yang, Y.F. and Thirumaran, G. (2012), Photosynthetic performance in aquatic and terrestrial colonies of *Nostocfagelliforme* (Cyanophyceae) under aquatic and aerial conditions. *Journal of Arid Environments*, 85, 56–61.

Zeng, X., Danquah, M.K., Chen, X.D. and Lu, Y. (2011), Microalgae bioengineering: From CO_2 fixation to biofuel production. *Renewable and Sustainable Energy Reviews*, 15(6), 3252–3260.

Zhao, X., Xue, C.-H. and Li, B.F. (2007), Study of antioxidant activities of sulfated polysaccharides from *Laminaria japonica*. *Journal of Applied Phycology*, 20(4), 431–436.

2 Marine Greens
Abilities in Remediation of Contaminants

O.I. Musa, J.O. Samuel, U.S. Ilyasu, S.A. Akande, A.T. Tiamiyyu, U.J.J. Ijah, O.P. Abioye, and A. Maude

2.1 INTRODUCTION

"Marine greens" refers to the vast and diverse ecosystem of marine algae or seaweeds that exist in oceans, seas, and other bodies of saltwater. Algae are a type of plant that can grow both in the water and on land (Wang et al., 2014). However, marine green algae are only found in saltwater. They play a significant role in marine ecology, as they serve as the primary food source for many aquatic organisms and provide important nutrients to the oceanic food chain (Cardol et al., 2008).

Marine green algae play a vital role in maintaining the health and balance of marine ecosystems. They serve as a primary food source for many aquatic animals, including fish, crustaceans, and other invertebrates, and are also consumed by larger marine mammals such as manatees and sea otters (Edison et al., 2016). Additionally, they produce significant amounts of oxygen through photosynthesis, which helps to maintain the oxygen balance in marine environments. The algae also absorb carbon dioxide from the atmosphere and store it, thereby reducing the amount of carbon in the atmosphere and mitigating the effects of climate change (Edison et al., 2016).

Marine green algae are an essential component of marine ecosystems, providing crucial nutrients, oxygen, and ecological balance. The biodiversity and variety of these algae are crucial to the well-being of numerous marine species and help mitigate the effects of climate change by reducing carbon dioxide in the atmosphere. The preservation and conservation of these species are crucial to maintaining the health of our oceans and by extension, the planet (Ghoneim et al., 2014). Therefore, this review proposes to look at the ability of marine greens to remediate contaminants.

2.2 TYPES OF MARINE GREENS

Seaweed is a common name for many types of marine algae. Although they look like aquatic plants and can reach lengths of more than 150 feet, seaweeds are not plants (Aravind et al., 2020). Instead, marine algae are a diverse group of organisms belonging to the Protista kingdom, specifically the brown algae (*Phaeophyta*), the green algae (*Chlorophyta*), and the red algae (*Rhodophyta*). Algae aren't really plants, but they do share some key characteristics with their plant counterparts. For photosynthesis, marine algae rely on chlorophyll in much the same way as land plants do. Seaweeds, like land plants, have cell walls (Wang et al., 2014). However, plants have crucial characteristics that seaweeds lack, such as a root system, an internal circulatory system, and the ability to generate seeds and flowers (RothSchulze et al., 2018).

DOI: 10.1201/9781003369738-3

2.3 THE BROWN ALGAE (*PHAEOPHYTA*)

Phaeophyta (Figure 2.1) are a group of brown algae that include fucoxanthin, beta-carotene, and chlorophyll a and c. These algae, which are typically found in the ocean, are the most complex of all (Ali et al., 2017). The size of Phaeophyta can vary widely, from the micrometer scale to several meters. At most, they have a length of about 30 meters (Hakim and Patel, 2020)

Brown algae, or Phaeophyta, are a type of autotrophic, multicellular organism that belongs to the Chromophyta division The xanthophyll pigment fucoxanthin is present in them, along with the chlorophylls a and c. So, those that belong to the Phaeophyta group tend to be a unique shade of greenish brown (Remya et al., 2022). For Phaeophyta to thrive in the ocean's depths, the brown pigment is crucial. Most Phaeophyta live in saltwater; however, there are a select few that thrive in freshwater (Al-Homaidan et al., 2021). The majority of Phaeophyta are found in temperate zones of the northern hemisphere, while a few species can be found in tropical seas. As of now, scientists have identified between 1500 and 2000 different types of brown algae (Al-Homaidan et al., 2021).

2.3.1 Phaeophyta Classification

The classification of Phaeophyta is shown in Table 2.1.

2.4 THE GREEN ALGAE (*CHLOROPHYTA*)

Green algae, or Chlorophyta (Figure 2.2), are also commonly referred to as seaweed. Several species can be found on land, but they are much more commonly associated with water. Single-cell and multicellular forms, colonies, and coenocytic (one large cell) organisms are all possible. Chlorophyta convert sunlight into starch, which is then used as an energy reserve in the cells (Polle et al., 2014).

Chlorophylls (a and b) in the same ratio as in "higher" plants provide the green color, together with beta-carotene (a yellow pigment) and other xanthophylls (yellowish or brownish pigments).

FIGURE 2.1 Brown algae (*Phaeophyta*) (Mlblevins, 2009)

TABLE 2.1
Classification of Phaeophyta

Domain	**Eukaryota**
Kingdom	Protista
Phylum	Heterokontophyta (has chlorophyll a and c)
Class	Phaeophyceae (has pigment fucoxanthin)
Order	Laminarales, Fucales, Ectocarpales, Dictyotales
Family	Fucaceae
Genus	*Fucus*
Species	*serratus, distichus, vesiculosus, spiralis*

Source: Mlblevins, https://biologywise.com/phaeophyta-brown-algae, 2009.

FIGURE 2.2 Green algae (*Chlorophyta*) (Becker, 2013)

Carbohydrates, lipids, and oils found in higher plants make up the bulk of the reserves. The earliest known forms of plant life on Earth were probably green algae (Freitas and Loverde-Oliveira, 2013).

2.4.1 *Chlorophyta* Classification

The taxonomic class Chlorophyta has its roots in the kingdom Plantae. It could be a reference to a group under the kingdom Plantae that contains all known types of green algae. Chlorophyta and Charophyta are two separate phyla that were initially considered to be one giant group of green algae species (charophytes) (Cremen et al., 2019).

Green algae species that do well in salt water are called chlorophytes (members of the *Chlorophyta*), whereas their freshwater counterparts are called charophytes (i.e., belonging to *Charophyta*). Yet, chlorophytes can thrive in a wide variety of environments. Both freshwater and saltwater environments host these organisms. Some chlorophytes are adapted to survive in extreme conditions, such as species found in deserts, hypersaline areas, and polar regions (Cremen et al., 2019).

According to the original taxonomic system, the genus *Chlorophyta* contains over 7000 recognized species of green algae. By the second definition of taxonomy, only about 4300 species belong to the *Chlorophyta*, while the rest are now called charophytes (Cormaci et al., 2014). *Chlorophyta* is a division of kingdom Plantae that includes all green algae categorized into classes such as *Bryopsidophyceae*, *Chlorophyceae*, *Cladophorophyceae*, *Charophyceae*, *Dasycladophyceae*, *Klebsormidiophyceae*, *Prasinophyceae*, *Pleurastrophyceae*, *Trentepohliophyceae*, *Ulvophyceae*, *and Zygne* (Cormaci et al., 2014). According to more recent taxonomies, such as that of Leliaert et al. (2012), charophyte algae are distinct from chlorophytes and instead are classified as *Streptophyta* within the *Viridiplantae* (together with the embryophytes). It is important to stress, however, that new ways of classifying organisms will inevitably emerge as more research is conducted on a wider range of species (Cremen et al., 2019).

2.5 THE RED ALGAE (*RHODOPHYTA*)

The red color of these algae comes from the pigments phycoerythrin and phycocyanin, which obscure the presence of other pigments like chlorophyll a (no chlorophyll b), beta-carotene, and a variety of xanthophylls (Brawley et al., 2017). Flax seed starch and floridoside are the primary reserves; true starch, like that found in higher plants and green algae, is absent. Long-chained polysaccharides including cellulose, agar, and carrageenan are employed to build the walls. There are a number of unicellular forms from different origins, and more complex thalli are built up of filaments (Qiu et al., 2015) (Figure 2.3).

Red algae called coralline algae dot the ocean floor and secrete calcium carbonate onto their cell surfaces (*Corallina officinalis*). Bone regeneration therapy has made use of these corallines. The term *Corallina officinalis* refers to the old practice of using coralline algae as a vermifuge (Yoou et al., 2017).

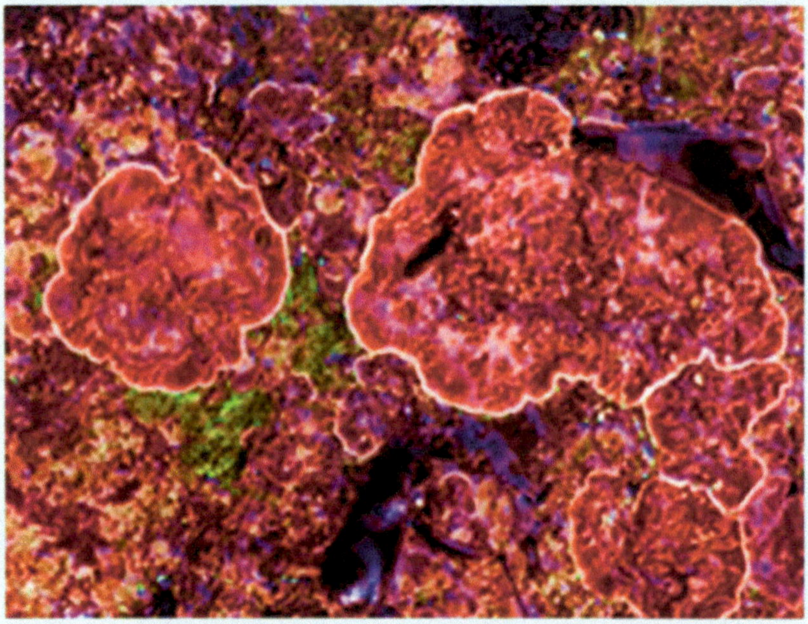

FIGURE 2.3 Red algae (*Rhodophyta*) (Freese and Lane, 2017)

Dulse (*Palmaria palmata*) and carrageen moss (*Chondrus crispus* and *Mastocarpus stellatus*) are two of the best-known edible red algae (Johnston et al., 2014).

It's true that there are many kinds of algae, and one of them is red. *Kappaphycus* and *Betaphycus* are the primary sources for the carrageenan utilized in most processed food today. This includes many of our favorite dairy products like yogurt, chocolate milk, and reconstituted puddings. Agar is made from red algae like *Gracilaria*, *Gelidium*, and *Pterocladia* and is used as a microbial growth medium, in the food industry, and in biotechnology (Rahelivao et al., 2015).

2.6 MARINE GREENS AND CONTAMINATION

Green marine algae have been shown to be very good at cleaning up pollution in marine environments. The algae are able to absorb, change, and even break down contaminants because of their different metabolic and physiological processes (Chekroun and Baghour, 2013; Babaniyi et al., 2023). Marine green algae, which are also called seaweeds, have been shown to be effective at cleaning up heavy metals, organic pollutants, and excessive nutrients in marine environments. Some of the things that marine green algae can do to help get rid of these pollutants are described in the following subsections.

2.6.1 Bioaccumulation

Marine green algae can accumulate and store contaminants within their tissues. This process is referred to as bioaccumulation and is a key mechanism through which algae can remove contaminants from the water. Once the algae have accumulated the contaminants, they can be harvested or allowed to die and sink to the bottom of the ocean, removing the contaminants from the water (Henriques et al., 2015). Bioaccumulation is the process by which substances, such as nutrients, metals, or organic compounds, accumulate in the tissues of marine green organisms (Mao et al., 2023). This accumulation occurs because the organisms are exposed to these substances in their environment and cannot metabolize or eliminate them efficiently. The mechanism of bioaccumulation is a result of the concentration of these substances in the tissues of the organisms over time, often due to their inability to metabolize or eliminate them efficiently (Mao et al., 2023).

One mechanism of bioaccumulation in marine green organisms is through the process of uptake and retention. Marine green organisms, such as algae or seaweed, take up substances from the surrounding water or sediments through their cell membranes. Some substances can also be absorbed through the surface of the organism or through the ingestion of food particles. Once the substances are taken up by the organism, they can accumulate in the tissues over time, leading to increased concentrations (Gojkovic et al., 2015). The degree of uptake and retention depends on several factors, including the chemical properties of the substance, the characteristics of the organism, and the environmental conditions. For example, substances that are hydrophobic, meaning they have a low solubility in water, tend to accumulate more readily in the tissues of marine green organisms (Zhang et al., 2016; Aransiola et al., 2021). This is because the organisms have a high lipid content in their cell membranes, which can attract and retain hydrophobic substances (Zhang et al., 2016). Another mechanism of bioaccumulation in marine green organisms is through the food chain; marine green organisms, such as plankton, are often the base of the marine food web (El-Shoubaky and Mohammad, 2016). When they consume substances that are present in the water or sediments, these substances can accumulate in their tissues. When higher-level organisms, such as fish, consume the plankton, they can accumulate even higher concentrations of these substances in their tissues (El-Shoubaky and Mohammad, 2016). This process is known as biomagnification and can lead to high concentrations of toxic substances in top predators, such as sharks or marine mammals (Sun et al., 2020). The level of bioaccumulation depends on the chemical properties of the substance, the characteristics of the organism, and the environmental conditions. For example, the temperature, salinity, and pH of the surrounding water can affect the uptake and retention of substances in marine green organisms. Additionally, factors such as the age, size, and metabolic rate of the organism can also affect the degree of bioaccumulation (Cao et al., 2015).

The mechanisms of bioaccumulation in marine green organisms are complex and can have significant impacts on the health of marine ecosystems and the organisms that rely on them. Understanding these mechanisms is important for developing effective strategies to mitigate the risks associated with exposure to toxic substances in the marine environment (Mondragón et al., 2023).

2.6.2 Nutrient Removal

Marine green algae can absorb excess nutrients such as nitrogen and phosphorus, which are often present in high concentrations in marine environments due to human activities such as agricultural runoff and wastewater discharge. By removing these excess nutrients, the algae can help reduce the risk of harmful algal blooms and improve water quality (Ramanan et al., 2015).

Green marine plants, including seaweeds, seagrasses, and other aquatic plants, play a critical role in removing nutrients from marine ecosystems (Ramanan et al., 2015). There are several mechanisms through which green marine plants remove nutrients, including the following.

2.6.2.1 Uptake by Roots or Thallus

Seagrasses and some species of seaweeds have specialized structures called roots that allow them to take up nutrients from the sediment and the water column. The roots of seagrasses and some seaweeds can also release oxygen, which can create an aerobic zone around the plant that enhances nutrient uptake. In addition to roots, some species of seaweeds have a thallus that can directly absorb nutrients from the water. The uptake of nutrients by green marine plants can reduce the concentration of these nutrients in the surrounding water, which can help prevent eutrophication and harmful algal blooms (Mantri et al., 2020).

2.6.2.2 Adsorption

Green marine plants, particularly seaweeds, have a large surface area relative to their volume, which makes them efficient at adsorbing nutrients from the water column. Seaweeds have specialized structures on their surface, such as hairs or ridges, which increase their surface area and enhance nutrient uptake. Once nutrients are adsorbed onto the surface of the seaweed, they can diffuse across the cell membrane and into the plant (El-Said et al., 2018).

2.6.2.3 Sedimentation

As green marine plants take up nutrients, they incorporate them into their tissues. Over time, this can lead to the accumulation of nutrients in the plant biomass, which can then sink to the bottom of the ocean when the plant dies or sheds its leaves. This process is known as sedimentation, and it can help remove nutrients from the water column and store them in the sediment. Seagrasses, for example, can accumulate large amounts of carbon and nutrients in their root systems, which can help stabilize sediments and prevent erosion (Green et al., 2016; Zhang et al., 2019).

2.6.2.4 Assimilation

Green marine plants can assimilate nutrients into their tissues and use them for growth and metabolism. This process is particularly important for nitrogen, which is a key limiting nutrient in many marine ecosystems. As green marine plants take up nitrogen, they can incorporate it into their biomass and transfer it to higher trophic levels in the food web. Seagrasses, for example, can assimilate nitrogen into their leaves and stems, which can then be consumed by herbivores and ultimately support higher-level predators (Sutherland and Craggs, 2017).

Green marine plants play an important role in removing nutrients from marine ecosystems through a combination of these mechanisms. By reducing the availability of nutrients in the water, they help to maintain the balance of the ecosystem and prevent harmful algal blooms and other negative impacts of nutrient pollution (Zhan et al., 2017).

2.6.2.5 Heavy Metal Removal

Marine green algae can also absorb and remove heavy metals from the water. Heavy metals such as mercury, lead, and cadmium are toxic to marine life and can accumulate in the tissues of aquatic organisms. Green algae can absorb these metals and store them in their tissues, reducing the concentration of heavy metals in the water and reducing the risk of toxicity to marine organisms (Danouche et al., 2021; Abioye et al., 2020).

Marine organisms such as algae, bacteria, and some types of shellfish have been found to have the ability to remove heavy metals from seawater. The mechanisms through which they do this can vary depending on the organism and the specific metal being removed, but two common mechanisms are adsorption and biomineralization.

2.6.2.5.1 Adsorption Or: Adsorption of Heavy Metal Ions

Adsorption is the process by which a substance (in this case, a heavy metal ion) is attracted and adheres to the surface of another substance (such as the cell surface of an organism). Many marine organisms have cell surfaces that contain functional groups (such as carboxyl, hydroxyl, or amino groups) that can bind to heavy metal ions. This process is often selective, meaning that different organisms may be able to remove different types of heavy metals. Once the metal ions are bound to the organism's surface, they can be removed from the seawater through sedimentation (settling to the bottom) or filtration (Cheng et al., 2019).

2.6.2.5.2 Biomineralization

Biomineralization is the process by which organisms form minerals within their tissues. Some marine organisms have the ability to precipitate heavy metals into biominerals such as carbonates, sulfides, or phosphates. This process can be a detoxification mechanism to prevent the metal ions from accumulating in the organism's tissues, or it can be a way to sequester the metals in a form that is less toxic to the environment. For example, some types of bacteria can produce sulfide minerals that can bind and remove heavy metal ions from seawater (Cheng et al., 2023).

2.7 ABILITIES OF MARINE GREENS IN REMEDIATION OF CONTAMINANTS

Algae play an important role in returning an environment that has been altered by various contaminants to its original state (Singhal et al., 2021). Phycoremediation is an algal-based emerging technology applied to remove various pollutants from water (Abioye et al., 2020). It is a low-cost, eco-friendly, and easily manageable remediation strategy. In this technique, algae are used as agents of remediation, and so far, a number of algae have been identified with high potential to detoxify various kinds of pollutants, such as nutrients, heavy metals, radionuclides, herbicides, and pesticides (Shackira et al., 2022). Both micro- and macroalgae are widely exploited to detoxify various contaminants from water bodies. Algae are a source of green energy, serve as a sink for CO_2, and are also a rich resource of economically important components, making phycoremediation a promising technology. Even though the detailed mechanism behind the phycoremediation technique is yet to be unraveled, processes like biosorption, accumulation, degradation, volatilization, and complexation have been proved to be operating in most of the species (Shackira et al., 2022).

This process can be carried out easily without disturbing human life or the environment during conduction and transportation. In algae-based bioremediation, algae fix carbon dioxide, release oxygen by photosynthesis, and increase biological oxygen demand in contaminated water (Singhal et al., 2021). They are highly adaptive, i.e., they can grow autotrophically, heterotrophically, or mixotrophically in the environment depending on the availability of substrate and light. Algae absorb nutrients like carbon, phosphate, and heavy metals from wastewater and produce new biomass, which is useful in the generation of bioenergy (Priya et al., 2022).

This can increase their chances of survival in harsh climates (Singhal et al., 2021) Microalgae can absorb many pollutants during photosynthesis in water (Nithin et al., 2020) Microalgae increase oxygen concentration in water, which increases the growth of several degraders, and oxygen production also decreases the need for external aeration, which is required for aerobic biodegradation. In association with heterotrophic microorganisms, microalgae can perform the degradation of several complex pesticides (the complexity of a pesticide depends on its chemical structure) (Chen and Wang, 2020).

Microalgae, mainly green algae belonging to the genera *Selenastrum*, *Scenedemus*, or *Chlorella*, have been demonstrated to be effective in the degradation of polycyclic aromatic hydrocarbons, such as naphthalene, phenanthrene, and pyrene (Ghosal et al., 2016), and in the immobilization of metals. The mechanisms enabling microalgae to remove toxic compounds, thus reducing their bioavailability and toxicity, mainly rely on the production of exopolysaccharides, which can mediate the uptake of contaminants on the cell surface and/or their complexation into less bioavailable forms (Casillo et al., 2021). Contaminants attached to the membrane or cell wall exopolysaccharides (depending on the microalgal taxa) can remain adherent or can be internalized and chelated by molecules belonging to the phytochelatin classes (Kaur and Reddy, 2019).

The possibility of exploiting the activity of microalgae for the degradation of aromatic compounds such as naphthalene was reported almost half a century ago. More recently, Lei et al. (2002) reported pyrene degradation ranging from 34% to 100% during seven days of treatment using green microalgae (*Chlamydomonas*, *Chlorella*, *Scenedesmus*, *Selenastrum*) or cyanophytes (*Synechocystis*). Similarly, other studies revealed that *Skeletonema costatum* and *Nitschia* sp. were effective in the removal of phenanthrene and fluoranthrene (Hong et al., 2008).

In addition, the green microalga *Chlorella vulgaris* displayed high potential in the remediation of waters contaminated by crude oil, with a bioremediation efficiency between 88% and 94% (Nweze and Aniebonam, 2009). Das and Deca (2019) identified *Chlorella vulgaris* BS1 as capable of degrading 98% of petroleum hydrocarbons at initial concentrations of 115 mg L^{-1} from water in 14 days (Al-Hussieny et al., 2020). It was also reported by Al-Hussieny et al. (2020) that five cyanophytes, namely, *Westiellopsis prolifica*, *Anabaena variabilia*, *Oscillatoria pranceps*, *Phormidium mucicola*, and *Lyngbya digueti*, were capable of reducing the concentrations of different hydrocarbon compounds from oil refinery waste waters by between 24% and 92%.

Marine microalgae–bacteria interactions remain to be further investigated to clarify the processes involved in hydrocarbon degradation as well as their actual potential to enhance bioremediation yields in bio-based approaches for the reclamation of contaminated marine sediments (Thompson et al., 2017).

2.8 CONCLUSION

In conclusion, the ability of marine green algae to remediate contaminants in marine environments is a promising area of research. While the effectiveness of algae-based remediation methods can vary depending on the type of contaminant, the species of algae, and the environmental conditions, there is significant potential for algae-based remediation methods to be used as a low-cost and environmentally friendly method for improving water quality and protecting marine ecosystems.

REFERENCES

Abioye, O. P., Ezegwu, B. U., Aransiola, S. A., & Ojeba, M. I. (2020). Phycoremediation of water contaminated with arsenic (As), cadmium (Cd) and lead (Pb) from a mining site in Minna, Nigeria. *European Journal of Biological Research*, *10*(1), 35–44. http://www.journals.tmkarpinski.com/index.php/ejbr/article/view/251.

Al-Homaidan, A. A., Al-Ghanayem, A. A., Al-Qahtani, H. S., Al-Abbad, A. F., Alabdullatif, J. A., Alwakeel, S. S., & Ameen, F. (2021). Effect of sampling time on the heavy metal concentrations of brown algae: A bioindicator study on the Arabian Gulf coast. *Chemosphere*, *263*, 127998.

Al-Hussieny, A. A., Imran, S. G., & Jabur, Z. A. (2020). The use of local blue-green algae in the bioremediation of hydrocarbon pollutants in wastewater from oil refineries. *Plant Archives*, *20*, 797–802.

Ali, A. Y., Idris, A. M., Ebrahim, A. M., & Eltayeb, M. A. (2017). Brown algae (Phaeophyta) for monitoring heavy metals at the Sudanese Red Sea coast. *Applied Water Science*, *7*(7), 3817–3824.

Aransiola, S. A., Ijah, U. J. J., Abioye, O. P., & Victor-Ekwebelem, M. O. (2021). Anammox in wastewater treatment. In N. R. Maddela, L. C. García Cruzatty, & S. Chakraborty (Eds.), *Advances in the Domain of Environmental Biotechnology: Environmental and Microbial Biotechnology*. Springer. https://doi.org/10.1007/978-981-15-8999-7_15.

Aravind, S., Kumar, P. S., Kumar, N. S., & Siddarth, N. (2020). Conversion of green algal biomass into bioenergy by pyrolysis. A review. *Environmental Chemistry Letters*, *18*(3), 829–849.

Babaniyi, B. R., Ogundele, O. D., Bisi-omotosho, A., Babaniyi, E. E., & Aransiola, S. A. (2023). Remediation approaches in environmental sustainability. In N. R. Maddela, L. K. W. Eller, & R. Prasad (Eds.), *Microbiology for Cleaner Production and Environmental Sustainability*. CRC Press. ISBN: 9781032496061. https://www.routledge.com/Microbiology-for-Cleaner-Production-and-Environmental-Sustainability/Maddela-Eller-Prasad/p/book/9781032496061.

Cao, D. J., Shi, X. D., Li, H., Xie, P. P., Zhang, H. M., Deng, J. W., & Liang, Y. G. (2015). Effects of lead on tolerance, bioaccumulation, and antioxidative defense system of green algae, Cladophora. *Ecotoxicology and Environmental Safety*, *112*, 231–237.

Cardol, P., Bailleul, B., Rappaport, F., Derelle, E., Béal, D., Breyton, C., Bailey, S., Wollman, F.A., Grossman, A., Moreau, H., & Finazzi, G. (2008). An original adaptation of photosynthesis in the marine green alga Ostreococcus. *Proceedings of the National Academy of Sciences*, *105*(22), 7881–7886.

Casillo, A., Fabozzi, A., Russo Krauss, I., Parrilli, E., Biggs, C. I., Gibson, M. I., & Corsaro, M. M. (2021). Physicochemical approach to understanding the structure, conformation, and activity of mannan polysaccharides. *Biomacromolecules*, *22*(4), 1445–1457.

Chekroun, K. B., & Baghour, M. (2013). The role of algae in phytoremediation of heavy metals: A review. *Journal of Materials and Environmental Science*, *4*(6), 873–880.

Chen, H., & Wang, Q. (2020). Microalgae-based nitrogen bioremediation. *Algal Research*, *46*, 101775.

Cheng, S. Y., Show, P. L., Lau, B. F., Chang, J. S., & Ling, T. C. (2019). New prospects for modified algae in heavy metal adsorption. *Trends in Biotechnology*, *37*(11), 1255–1268.

Cheng, Y., Zhang, T., Chen, S., Li, F., Qing, R., Lan, T., & Liu, N. (2023). Unusual uranium biomineralization induced by green algae: Behavior investigation and mechanism probe. *Journal of Environmental Sciences*, *124*, 915–922.

Cormaci, M., Furnari, G., & Alongi, G. (2014). Flora marina bentonica del Mediterraneo: Chlorophyta. *Bullettin of the Gioenia Academy of Natural Sciences of Catania*, *47*(377), FP11–FP436.

Cremen, M. C. M., Leliaert, F., West, J., Lam, D. W., Shimada, S., Lopez-Bautista, J. M., & Verbruggen, H. (2019). Reassessment of the classification of Bryopsidales (Chlorophyta) based on chloroplast phylogenomic analyses. *Molecular Phylogenetics and Evolution*, *130*, 397–405.

Danouche, M., El Ghachtouli, N., & El Arroussi, H. (2021). Phycoremediation mechanisms of heavy metals using living green microalgae: Physicochemical and molecular approaches for enhancing selectivity and removal capacity. *Heliyon*, *7*(7), e07609.

Das, B., & Deka, S. (2019). A cost-effective and environmentally sustainable process for phycoremediation of oil field formation water for its safe disposal and reuse. *Scientific Reports*, *9*(1), 1–15.

Edison, T. N. J. I., Atchudan, R., Kamal, C., & Lee, Y. R. (2016). Caulerpa racemosa: A marine green alga for eco-friendly synthesis of silver nanoparticles and its catalytic degradation of methylene blue. *Bioprocess and Biosystems Engineering*, *39*(9), 1401–1408.

El-Said, G. F., El-Sadaawy, M. M., & Aly-Eldeen, M. A. (2018). Adsorption isotherms and kinetic studies for the defluoridation from aqueous solution using eco-friendly raw marine green algae, Ulva Lactuca. *Environmental Monitoring and Assessment*, *190*, 1–15.

El-Shoubaky, G. A., & Mohammad, S. H. (2016). Bioaccumulation of gasoline in brackish green algae and popular clams. *The Egyptian Journal of Aquatic Research*, *42*(1), 91–98.

Freitas, L., & Loverde-Oliveira, S. (2013). Checklist of green algae (Chlorophyta) for the state of Mato Grosso, Central Brazil. *Check List*, *9*(6), 1471–1483.

Ghoneim, M. M., El-Desoky, H. S., El-Moselhy, K. M., Amer, A., Abou El-Naga, E. H., Mohamedein, L. I., & Al-Prol, A. E. (2014). Removal of cadmium from aqueous solution using marine green algae, Ulva lactuca. *The Egyptian Journal of Aquatic Research*, *40*(3), 235–242.

Ghosal, D., Ghosh, S., Dutta, T. K., & Ahn, Y. (2016). Current state of knowledge in microbial degradation of polycyclic aromatic hydrocarbons (PAHs): A review. *Frontiers in Microbiology*, 7, 1369.

Gojkovic, Ž., Garbayo, I., Ariza, J. L. G., Márová, I., & Vílchez, C. (2015). Selenium bioaccumulation and toxicity in cultures of green microalgae. *Algal Research, 7*, 106–116.

Green, D. S., Boots, B., Sigwart, J., Jiang, S., & Rocha, C. (2016). Effects of conventional and biodegradable microplastics on a marine ecosystem engineer (Arenicola marina) and sediment nutrient cycling. *Environmental Pollution, 208*(B), 426–434.

Hakim, M. M., & Patel, I. C. (2020). A review on phytoconstituents of marine brown algae. *Future Journal of Pharmaceutical Sciences, 6*, 1–11.

Henriques, B., Rocha, L. S., Lopes, C. B., Figueira, P., Monteiro, R. J., Duarte, A. D. C., & Pereira, E. (2015). Study on bioaccumulation and biosorption of mercury by living marine macroalgae: Prospecting for a new remediation biotechnology applied to saline waters. *Chemical Engineering Journal, 281*, 759–770.

Hong, Y. W., Yuan, D. X., Lin, Q. M., & Yang, T. L. (2008). Accumulation and biodegradation of phenanthrene and fluoranthene by the algae enriched from a mangrove aquatic ecosystem. *Marine Pollution Bulletin, 56*(8), 1400–1405.

Johnston, E. T., Lim, P. E., Buhari, N., Keil, E. J., Djawad, M. I., & Vis, M. L. (2014). Diversity of freshwater red algae (Rhodophyta) in Malaysia and Indonesia from morphological and molecular data. *Phycologia, 53*(4), 329–341.

Kaur, T., & Reddy, M. S. (2019). Recent developments in ectomycorrhizal research. *Advancing Frontiers in Mycology & Mycotechnology: Basic and Applied Aspects of Fungi*, https://doi.org/10.1007/978-981-13-9349-5_12.Corpus ID: 208588831.

Lei, A. P., Wong, Y. S., & Tam, N. F. Y. (2002). Removal of pyrene by different microalgal species. *Water Science and Technology: A Journal of the International Association on Water Pollution Research, 46*(11–12), 195–201.

Leliaert, F., Smith, D. R., Moreau, H., Herron, M. D., Verbruggen, H., Delwiche, C. F., & De Clerck, O. (2012). Phylogeny and molecular evolution of the green algae. *Critical Reviews in Plant Sciences, 31*(1), 1–46.

Mantri, V. A., Kazi, M. A., Balar, N. B., Gupta, V., & Gajaria, T. (2020). Concise review of green algal genus Ulva Linnaeus. *Journal of Applied Phycology, 32*(5), 2725–2741.

Mao, W., Li, M., Xue, X., Cao, W., Wang, X., Xu, F., & Jiang, W. (2023). Bioaccumulation and toxicity of perfluorooctanoic acid and perfluorooctane sulfonate in marine algae Chlorella sp. *Science of the Total Environment*, 870, 161882. https://doi.org/10.1016/j.scitotenv.2023. 161882

Mlblevins. (2009). Phaeophyta (brown algae). Retrieved February 20, 2023, from https://biologywise.com/phaeophyta-brown-algae.

Mondragón, M. E., Luzardo, O. P., Henríquez-Hernández, L. A., Rodríguez-Hernández, Á., Zumbado, M., Martínez, J. R. R., ... Islas, C. G. R. (2023). Trophic behavior of inorganic elements in nesting sea turtles (Chelonia mydas, Eretmochelys imbricata, and Caretta caretta) in Quintana Roo: Biomagnification and biodilution effect in blood and scute tissues. *Marine Pollution Bulletin, 187*, 114582.

Nithin, B. R., Bhuyar, P., Trejo, M., Rahim, M. H. A., Maniam, G. P., & Govindan, N. (2020). Culturing of green photosynthetic microalgae (Chlorella sp.) using palm oil mill effluent (POME) for future biodiesel production. *Maejo International Journal of Energy and Environmental Communication, 2*(1), 1–8.

Nweze, N. O., & Aniebonam, C. (2009). Bioremediation of petroleum products impacted freshwater using locally available algae. *Biological-Research, 7*(1), 484–490.

Polle, J. E., Neofotis, P., Huang, A., Chang, W., Sury, K., & Wiech, E. M. (2014). Carbon partitioning in green algae (Chlorophyta) and the enolase enzyme. *Metabolites, 4*(3), 612–628.

Priya, A. K., Jalil, A. A., Vadivel, S., Dutta, K., Rajendran, S., Fujii, M., & Soto-Moscoso, M. (2022). Heavy metal remediation from wastewater using microalgae: Recent advances and future trends. *Chemosphere, 305*, 135375.

Qiu, H., Price, D. C., Yang, E. C., Yoon, H. S., & Bhattacharya, D. (2015). Evidence of ancient genome reduction in red algae (Rhodophyta). *Journal of Phycology, 51*(4), 624–636.

Rahelivao, M. P., Gruner, M., Andriamanantoanina, H., Andriamihaja, B., Bauer, I., & Knölker, H. J. (2015). Red algae (Rhodophyta) from the coast of Madagascar: Preliminary bioactivity studies and isolation of natural products. *Marine Drugs, 13*(7), 4197–4216.

Ramanan, R., Kang, Z., Kim, B. H., Cho, D. H., Jin, L., Oh, H. M., & Kim, H. S. (2015). Phycosphere bacterial diversity in green algae reveals an apparent similarity across habitats. *Algal Research, 8*, 140–144.

Remya, R. R., Samrot, A. V., Kumar, S. S., Mohanavel, V., Karthick, A., Chinnaiyan, V. K., & Muhibbullah, M. (2022). Bioactive potential of brown algae. *Adsorption Science and Technology, 2022.* Article ID 9104835 | https://doi.org/10.1155/2022/9104835

Roth-Schulze, A. J., Pintado, J., Zozaya-Valdés, E., Cremades, J., Ruiz, P., Kjelleberg, S., & Thomas, T. (2018). Functional biogeography and host specificity of bacterial communities associated with the Marine Green Alga Ulva spp. *Molecular Ecology, 27*(8), 1952–1965.

Shackira, A. M., Sarath, N. G., & Puthur, J. T. (2022). Phycoremediation: A means for restoration of water contamination. *Environmental Sustainability, 5*(1), 25–38.

Singhal, M., Jadhav, S., Sonone, S. S., Sankhla, M. S., & Kumar, R. (2021). Microalgae based sustainable bioremediation of water contaminated by pesticides. *Biointerface Research in Applied Chemistry, 12*(1), 149–169.

Sun, T., Wu, H., Wang, X., Ji, C., Shan, X., & Li, F. (2020). Evaluation on the biomagnification or biodilution of trace metals in global marine food webs by meta-analysis. *Environmental Pollution, 264*, 113856.

Sutherland, D. L., & Craggs, R. J. (2017). Utilising periphytic algae as nutrient removal systems for the treatment of diffuse nutrient pollution in waterways. *Algal Research, 25*, 496–506.

Thompson, H., Angelova, A., Bowler, B., Jones, M., & Gutierrez, T. (2017). Enhanced crude oil biodegradative potential of natural phytoplankton-associated hydrocarbonoclastic bacteria. *Environmental Microbiology, 19*(7), 2843–2861.

Wang, L., Wang, X., Wu, H., & Liu, R. (2014). Overview on biological activities and molecular characteristics of sulfated polysaccharides from marine green algae in recent years. *Marine Drugs, 12*(9), 4984–5020.

Yoon, H. S., Nelson, W., Lindstrom, S. C., Boo, S. M., Pueschel, C., Qiu, H., & Bhattacharya, D. (2017). Rhodophyta. In *Handbook of the Protists: Second Edition* (pp. 89–133). Springer International Publishing. https://doi.org/10.1007/978-3-319-28149-0_33ShareExport

Zhan, J., Rong, J., & Wang, Q. (2017). Mixotrophic cultivation, a preferable microalgae cultivation mode for biomass/bioenergy production, and bioremediation, advances and prospect. *International Journal of Hydrogen Energy, 42*(12), 8505–8517.

Zhang, C., Lu, J., & Wu, J. (2019). Adsorptive removal of polycyclic aromatic hydrocarbons by detritus of green tide algae deposited in coastal sediment. *Science of the Total Environment, 670*, 320–327.

3 Microplastics, Marine Debris, and Marine Green Contamination

Babafemi Raphael Babaniyi, Joshua Ibukun Adebomi, Bukola Rukayat Olowoyeye, and Oluwatosin Emmanuel Daramola

3.1 INTRODUCTION TO MICROPLASTICS

Plastics are synthetic materials composed of long chains of carbon molecules bonded with hydrogen, nitrogen, oxygen, and sulfur (Zainudin et al., 2020; Azeem et al., 2021). Plastics can be categorized based on their size, i.e., microplastics (>25 mm), mesoplastics (5–25 mm), microplastics (MPs) (0.1–5 mm), and nanoplastics (NPs) (<100 nm) (Alimi et al., 2018). Production of plastic worldwide is approximately 6300 million tons, of which 79% is deposited in landfills and other ecological networks (Geyer et al., 2017). Global production of plastic was 368 million tons in 2019; China recorded the production of 114.08 million tons of plastic, and production was 58.88 million tons in Europe (Lau et al., 2020; Plastics Europe, 2020. Furthermore, 275 million metric tons of land-based plastic waste was estimated from 192 coastal countries, which resulted in 4.8 to 12.7 million metric tons being found in the ocean environment in 2010 (Jambeck et al., 2015; Amelia et al., 2021). Additionally, the process of fragmentation and degradation of plastic litter in the marine ecological network facilitates the formation of secondary noxious microplastics throughout the marine network. About 60 to 99 million metric tons of microplastic waste was recorded in 2015 (Lebreton and Andrady, 2019; Auta et al., 2022). An estimated 3 to 6 million metric tons of microplastic was disposed of as litter in the environment in 2015 (Ryberg et al., 2019). The shape of each type of microplastic depends on the polymeric unit it is made from, because producers tend to manufacture plastic with distinct properties, such as resistance, flexibility, toughness, or durability (Espí et al., 2006). However, plastics are made up of monomers that are potentially hazardous to the marine ecosystem. A perfect example is polyurethane; this polymer is used to manufacture foams, but it is composed of harmful monomers that affect humans and soil biota, and can potentially hamper the functionality of the marine ecosystem due to the uncontrollable influx and production of several million tons of plastics annually, which undoubtedly adds to the concentration in the soil and the marine environment (Yudi et al., 2020).

In addition, plastics floating on the sea surface can block the sun's rays and prevent the absorption of sunlight by algae, which can lead to chlorosis. Furthermore, microplastics have the capacity to cross the cell membrane of *Chlamydomonas* and distort the expression of genes responsible for growth (Sun et al., 2016). In nanometer form, microplastics are capable of passing through the blood–brain barrier in the medaka to enter the brain, liver, and testicles, which will result in the death of the fish (Chelsea et al., 2014). According to Rossi et al.'s evaluation of the impact of nano-sized polystyrene on the activities of biofilms by using a molecular simulation investigation procedure, the results revealed that it is possible to insert nanoparticles into membranes, change the structure of membranes, reduce the activity of membrane proteins, and thus, change the function of cells (Rossi et al., 2014). This chapter addresses microplastic debris in the marine ecosystem and

 DOI: 10.1201/9781003369738-4

TABLE 3.1
Various Microplastic Pollutants Bioaccumulated in Marine Organisms

Microplastic	Pollutant	Organism	Class	Reference
PE and PP	PBDEs and PCBs	Zooplankton	Bivalvia	Yeo et al. (2020)
PS	PAHs	Marine copepod (*Calanus finmarchicus*)	Hexanauplia	Sørensen et al. (2020)
DPE and PP	PCB	Rabbitfish (*Siganus rivulatus*)	Actinopterygii	van der Hal et al. (2020)
Plastic	Lower chlorinated PCBs/pesticides	North Pacific lanternfish (*Tarletonbeania taylori*)	Actinopterygii	Gassel and Rochman (2019)
PE	Ag	Rainbow trout (*Oncorhynchus mykiss*)	Actinopterygii	Khan et al. (2017)

PAH Polycyclic aromatic hydrocarbon
PCB Polychlorinated biphenyls
PBDE Polybrominated biphenyl ethers
PP Polypropylene
PE Polyethylene

marine greens, with an emphasis on issues arising due to the ever-increasing plastic contaminants in the marine environment (Table 3.1).

3.2 MARINE DEBRIS

Marine plastic waste and microplastic pollution rank among the top ten environmental issues facing the planet today and have a significant impact on both the sustainability of marine economies and the health of marine environments. Strong action must be taken to essentially stop the trend of intensifying marine pollution as well as to advance the establishment of a scientific and methodical marine control system for plastic waste and microplastic pollution (Yang et al., 2021). Babaniyi et al. (2022) stated that marine debris threatens the health of lakes and streams, where fish and other aquatic species depend on large amounts of oxygen to live. Olagoke et al. (2022) also pointed out that textile businesses are thought to be the main contributors to pollution in soil and water bodies. It is estimated that several million metric tons of improperly disposed plastic waste enter the ocean each year from coastal areas (Egger et al., 2020). However, microplastics have been operationally defined as synthetic or semisynthetic materials built of polymers and additives that are <5 mm in diameter, with nanoplastics defined as having an upper size limit of 1000 nm (Turner, 2021). However, microplastics can be further divided into primary particles such as pre-production pellets, clothing fibers, and exfoliating beads used in cosmetic products that are less than 5 mm in size before entering the environment, and secondary particles that are created when larger debris is broken down in place into fragments, films, foams, and fibers that are less than 5 mm in size. Microplastics may also be classified according to additional physical or chemical attributes, such as polymer type, color, roughness, transparency, and shape (Turner, 2021; Du et al., 2022).

3.3 MICROPLASTIC DEBRIS

Marine microscopic plastic (microplastic) debris is a contemporary societal problem that highlights the difficulty of balancing the practicality of plastic in everyday living with the risk of harming the environment through irresponsible disposal. The term "microplastics" refers to plastic debris that is less than 5 mm long, or roughly the size of a sesame seed. Plastic debris can be of any form or size. Consequently, plastic pollution in the oceans is a known environmental problem on a global scale. Public concern about this pollution is growing as a result of its scope and our better knowledge of its potential effects. Every year, millions of tons of plastic reach the ocean, and in the years to come, the amount is anticipated to rise (Lebreton and Andrady, 2019). Plastic materials in the marine environment can decompose or shatter into smaller fragments over time, mostly as a result of weathering and mechanical forces like wave action and sand abrasion (Barrett et al., 2020). Microplastics (MPs) are described as plastics that are between 5 mm and 1 μm in size. Due to their size, MPs can have a negative impact on organisms and are readily ingested by a wide variety of marine species, from high trophic species to low trophic species. Freshwater and marine ecosystems are both impacted by microplastic debris, and numerous studies have shown that contamination in aquatic environments is occurring globally (Alvarez-Hernández et al., 2019; Xiong et al., 2019; Jeyasanta et al., 2020; Eibes and Gabel, 2022). Rivers are believed to be a significant pathway for microplastics to travel in order to reach the sea, according to field research and modeling projections. However, very few studies have examined the quantity and occurrence of microplastics in waterways (Eibes and Gabel, 2022).

3.4 FORMATION MECHANISM OF MICROPLASTIC DEBRIS IN MARINE ENVIRONMENT

Microplastics can appear in many different forms, such as larger plastic pieces that have broken up, resin pellets used in the production of plastic, or microbeads, which are tiny, manufactured plastic beads used in cosmetic and health goods. However, Amelia et al. (2020) and Verla et al. (2019a) claim that three plausible mechanisms exist by which chemicals adhere to microplastic particles: (1) adsorption onto microplastics as hydrophobic adsorbents, (2) biofilm growth assisted, and (3) plastic additives and related chemicals contained within resins.

3.4.1 Biofilm Growth Assisted

Biofilm is a sorbent for metals, according to a review study on the immobilization of metal by biofilms (van Hullebusch et al., 2003). Quorum sensing (QS) is a method used by microorganisms to create biofilms. When microbial cells adhere to one another on a living or non-living surface, they create a matrix of extracellular polymeric substance (EPS), and QS is known to regulate gene expression in this process (Johansen et al., 2019). Biofilms are everywhere in nature and are frequently found on aqueous or buried hard surfaces, such as microplastics accumulating in the sea surface microlayer (SML) of seawater (Sun et al., 2018).

3.5 MARINE GREENS

Marine macroalgae are basic, internally plant-like organisms that typically inhabit coastal regions. Various communities of scarlet, brown, and green macroalgae make up the majority of them. Marine greens, also known as sea greens, edible algae, sea vegetables, or seaweed, are a type of edible marine algae, which means they're technically not really plants. They don't rely on roots for uptake; their leaves absorb water and nutrients through their surfaces. In a variety of agricultural fields, they are regarded as superb natural biosources (Kholssi et al., 2022). They are very adept at enhancing the chemical and physical characteristics of soil. The production of a wide variety of biologically active biocidal compounds against pathogens that infect plants is another characteristic of marine macroalgae. Unfortunately, the majority of the literature on marine macroalgae and their derivatives that is currently accessible focuses primarily on their pharmaceutical applications, while their potential application in the development of sustainable agriculture is still frequently seen as a secondary objective (Singh et al., 2023).

3.6 CONTAMINATION OF MARINE GREENS

Pollution of marine greens is caused by:

i. Chemical spills
ii. Oil spills
iii. Operational discharges
iv. Parasiticides
v. Decommissioning of disused installations
vi. Seabed mining
vii. Dredging of sediment and dumping at sea

3.6.1 Chemical Spills

The most essential mode of transport for the majority of chemicals is shipping. These chemicals are known as Hazardous and Noxious Substances (HNSs), broadly meaning any substance (except oil) that when it enters the marine environment poses a health risk to marine and human life and has the potential to harm living organisms, destroy property, and stop other permissible uses of the sea (IMO, 2000).

The top 100 chemicals being transported in European waters have been identified in Europe by the HASREP project (Response to Harmful Substances spilled at sea); it states that the 15 bulk HNSs that are mostly shipped are phosphoric acid, acetic acid, sulfuric acid, ammonia, ethanol, phenol, molasses, methanol, benzene and its mixtures, sodium hydroxide solution, xylenes, styrene, methyl tert-butyl ether (MTBE), animal fat, and palm and other vegetable oils (HASREP, 2005).

3.6.2 Operational Discharges

Pollution due to operational discharges is difficult to assess compared with that caused by enormous, catastrophic spills. Pollution involves contaminants such as oil, chemicals from fire-extinguishers and refrigerating equipment, lubricants, detergents, and cleaners. To evaluate the environmental impacts of this pollution, the types of pollutants that are released are used. Nonetheless, the damage done by this type of contaminant seems to be neglected by both the shipping industry and researchers. Thus, vital information concerning it is limited (Honkanen et al., 2013; McLaughlin et al., 2014).

3.6.3 Parasiticides

To delouse farmed fish, many types of antiparasitic agents are commonly used; these include benzamide and avermectins such as flubenzurons and emamectin benzoate, pyrethroids such as cypermethrin and deltamethrin, and organophosphates such as azamethiphos. They are used orally by administering into feed or dissolving in water (Olsvik et al., 2014; Samuelsen et al., 2014). After their usage, these therapeutic agents are released into the oceanic body using different media, such as in particulate form via uneaten feed and fecal material, or in soluble form through urine, and dispersed throughout the water column. Consequently, from these treatments, both the seabed and water column receive metabolites and active ingredients (Telfer et al., 2006).

3.6.4 Decommissioning of Disused Installations

The problem of decommissioning (essentially a reclamation process) arises when oil or gas fields reach the end of their operating life. One of the factors to consider when decommissioning is the environmental hazards of the process. This process will result in the release of contaminants from the piles of drill cuttings accumulated on the seabed and from re-suspended sediments. Furthermore, it will lead to ecosystem disturbance (Schroeder and Love, 2004; OGP, 2012). If the installations (structures) were to be kept in the ocean, this could lead to marine pollution, such as corrosion of structures left behind and leaching of contaminants such as toxic substances (heavy metals, residue oil, polychlorinated biphenyls [PCBs], and so on) (Lakhal et al., 2009; Adedayo, 2011). A massive steel installation is made of 90% steel, 2% aluminum, and 0.3% copper. When such structures deteriorate, the surrounding environment will ultimately be exposed to chemicals such as lead, iron, mercury, and cadmium (Adedayo, 2011).

3.7 REDUCING POLLUTION OF MARINE GREENS

The health of the ocean, industries, and communities is important, and as such, ways to reduce marine pollution need to be addressed. The ocean is so vast, that cleaning it can be an intimidating task. Regardless, one of the successful ways to protect it and reduce the waste being dumped in the ocean is effective management of pollution at its source (DeGeorges et al., 2010; Rochman, 2016; Simmonds et al., 2014).

Thus, to reduce the amount of pollution that enters the coastal environment, systems need to be placed where they will have the greatest impact on reducing pollution, such as:

i. At the source of the pollutant
ii. Along the way, when the pollutant is released
iii. At the sink, when the pollutant has entered the sea

Above all, if rapid and effective reduction of ocean pollution is to occur, then all these actions need to be implemented.

3.7.1 At the Sources

The most effective way to reduce and prevent marine pollution is by remediating it at its multitude of sources, particularly for municipally based pollutants and land-based and sea-based industries. Illustrations include reduced frequency and timing of seismic blasting leading to a decrease in underwater noise pollution at the source, reductions in production on a per-item basis due to changes in packaging materials, and less agricultural runoff in coastal waters as a result of reduction in fertilizer usage (Bennett et al., 2001). The advantage of acting at the source is that pollutants cannot enter the marine environment when they are not developed or used initially. Action can occur at the source by incorporating various steps such as the replacement of technologies and products for more environmentally friendly alternatives, prevention of contaminants, incentives, outreach campaigns, and the introduction of bans.

3.7.2 Along the Way

In this step, once pollution has been released from the source and is in transit to the marine environment, the reduction of marine pollution is achieved by implementing various approaches along the way. Point-source pollution can be remediated by acting along the way, which provides the opportunity to target some pollutants effectively. Infrastructure such as wastewater treatment plants (WWTPs) and gross pollutant traps (GPTs) can be used to reduce municipally based pollutants along the way, but some pollution from light or sound might be more difficult to minimize using these techniques. WWTPs have the capacity to capture litter, pharmaceuticals, and excess nutrients that are carried through sewerage and wastewater systems. However, one of the disadvantages of using this method is that there is more likelihood of leakage to the environment and more production because of pollution management en route. Furthermore, the setup used for remediation often requires ongoing maintenance, can become physically blocked or result in increased risk to human health and the environment (e.g., flooding during heavy rainfall) if not managed properly, and most importantly, it is expensive. Land- and sea-based pollution require different approaches, which may lead to challenges; however, it offers opportunities for resolution (Alexander et al., 2020).

3.7.3 At the Sinks

Removal of pollution in the ocean is needed at the sinks. This method is the most expensive, challenging, and the least likely to produce positive outcomes. More than 70% of the earth's surface is covered by ocean, with its depth reaching as far as 10 km. Consequently, this makes it easy for pollutants to spread and logistically and economically difficult to clean. Regardless of this disadvantage, collecting pollutants and cleaning the marine environment is the most feasible approach. One example of successful usage of this method is the remediation of excessive nutrients, which includes integrated multi-trophic aquaculture (Buck et al., 2018). Sea-based and municipally-based industry contaminants are commonly removed during clean-up events. For example, community members often volunteer to remove and clean pollutants from coastal environments and wildlife during large oil spills. Thus, such incidents increase awareness among people about the issues of marine pollution, and if data are recorded, this can provide a standard with which changes can be compared. Efforts need to be prioritized in areas with high accumulation of pollution to address pollution at sinks (e.g., oil spills). To attain long-term results for repeated removal or cleaning of contaminants, the pollution upstream needs to be properly managed along the route or at the source.

3.8 CONCLUSION

The accumulation of microplastics in marine ecology is a major concern currently, and investigations addressing their impacts on single species are on the rise, along with studies on their prolonged effects on the marine community as a whole. The effects of microplastics are more quickly noticeable in freshwater, because sediments act as accumulators of microplastics, although the sources of microplastics should be carefully considered due to aggregation and biofouling processes and subsequent settling, which establishes hotspot areas that unleash risk to benthic organisms. The potential of benthic macroinvertebrates to ingest microplastics depends on the organism's makeup, such as feeding habits or developmental stages; it also depends on the properties of the plastic particles and environmental conditions. Studies revealed a reduction in the growth of the amphipod *Gammarus pulex* after 28 days of exposure to polystyrene microplastics. Consequently, the continuous alteration of genes and differences in sensitivity to microplastic particles could result in structural changes in the community setup and associated disproportionate responses.

REFERENCES

Adedayo, A.M. (2011). Environmental risk and decommissioning of offshore oil platforms in Nigeria. *NIALS Journal of Environmental Law*, 1, 1–30.

Alimi, O.S., Farner Budarz, J., Hernandez, L.M., Tufenkji, N. (2018). Microplastic and Nanoplastic in aquatic environments: Aggregation, deposition, and enhanced contaminant transport. *Science of the Total Environment*, 52, 1704–1724.

Álvarez-Hernández, C., Cairós, C., López-Darias, J., Mazzetti, E., Hernández-Sánchez, C., González-Sálamo, J., Hernández-Borges, J. (2019). Microplastic debris in beaches of Tenerife (Canary Islands, Spain). *Marine Pollution Bulletin*, 146, 26–32.

Amelia, T.S.M., Wan Mohd Khalik, W.M.A., Ong, M.C., Shao, Y.T., Pan, H.-J., Bhubalan, K. (2021). Marine microplastics as vectors of major ocean pollutants and its hazards to the marine ecosystem and humans. *Review: Progress in Earth and Planetary Science*, 8(12), 2–26. https://doi.org/10.1186/s40645-020-00405-4.

Auta, S.H., Abioye, O.P., Aransiola, S.A., Bala, J.D., Chukwuemeka, V.I., Hassan, A., Aziz, A., Fauziah, S.H. (2022). Enhanced microbial degradation of PET and PS microplastics under natural conditions in mangrove environment. *Journal of Environmental Management*, 304, 114273. https://doi.org/10.1016/j.jenvman.2021.114273.

Azeem, I., Adeel, M., Ahmad, M.A., Shakoor, N., Jiangcuo, G.D., Azeem, K., Ishfaq, M., Shakoor, A., Ayaz, M., Xu, M., Rui, Y. (2021). Uptake and accumulation of nano/microplastics in plants: A critical review. *Nanomaterials*, 11(11), 2935. https://doi.org/10.3390/nano11112935.

Babaniyi, G.G., Oyemade, B., Orija, D. (2022). Effects of aquatic (freshwater and marine) pollution on microbial enzyme activities. In: N.R. Maddela, S.A. Aransiola and R. Prasad (eds.), *Ecological Interplays in Microbial Enzymology* (pp. 379–403). Singapore: Springer Nature Singapore.

Baloch, G.N., Tariq, S., Ehteshamul-Haque, S., Athar, M., Sultana, V., Ara, J. (2013). Management of root diseases of eggplant and watermelon with the application of asafoetida and seaweeds. *Journal of Applied Botany and Food Quality*, 86(1) 138–142.

Barrett, J., Chase, Z., Zhang, J., Holl, M.M.B., Willis, K., Williams, A., Hardesty, B.D., Wilcox, C. (2020). Microplastic pollution in deep-sea sediments from the Great Australian Bight. *Frontiers in Marine Science*, 7, 808–832.

Bennett, E.M., Carpenter, S.R., Caraco, N.F. (2001). Human impact on erodable phosphorus and eutrophication: A global perspective: Increasing accumulation of phosphorus in soil threatens rivers, lakes, and coastal oceans with eutrophication. *BioScience*, 51(3), 227–234. https://doi.org/10.1641/0006-3568(2001)051[0227:HIOEPA]2.0.CO;2.

Buck, B., Troell, M., Krause, G., Angel, D., Grote, B., Chopin, T. (2018). State of the art and challenges for offshore integrated multitrophic aquaculture (IMTA). *Frontiers in Marine Science*, 5, 165. https://doi.org/10.3389/fmars.2018.00165.

Chengjun, S., Fenghua, J., Jingxi, L., Li, Z. (2016). Research progress on the source, distribution and ecological environment impact of microplastics in the ocean. *Advances in Marine Science*, 34(4), 449–461.

DeGeorges, A., Goreau, T.J., Reilly, B. (2010). Land-sourced pollution with an emphasis on domestic sewage: Lessons from the Caribbean and implications for coastal development on Indian Ocean and pacific coral reefs. *Sustainability*, 2(9), 2919–2949. https://doi.org/10.3390/su2092919.

Du, Y., Liu, X., Dong, X., Yin, Z. (2022). A review on marine plastisphere: Biodiversity, formation, and role in degradation. *Computational and Structural Biotechnology Journal*, 20, 975–988

Egger, M., Nijhof, R., Quiros, L., Leone, G., Royer, S.J., McWhirter, A.C., Kantakov, G.A., Radchenko, V.I., Pakhomov, E.A., Hunt, B.P.V., Lebreton, L. (2020). A spatially variable scarcity of floating microplastics in the eastern North Pacific Ocean. *Environmental Research Letters*, 15(11), 114056.

Eibes, P.M., Gabel, F. (2022). Floating microplastic debris in a rural river in Germany: Distribution, types and potential sources and sinks. *Science of the Total Environment*, 816, 151641.

Espí, E., Salmerón, A., Fontecha, A., García, Y., Real, A.I. (2006). PLastic films for agricultural applications. *Journal of Plastic Film and Sheeting*, 22(2), 85–102.

Geyer, R., Jambeck, J.R., Law, K.L. (2017). Production, use, and fate of all plastics ever made. *Science Advances*, 3(7), e1700782.

Giulia, R., Barnoud, J., Monticelli, L. (2014). Polystyrene nanoparticles perturb lipid membranes. *Journal of Physical Chemistry Letters*, 5(1), 241–246.

HASREP. (2005). Response to harmful substances spilled at sea. Task 2 Risk assessment methodology for the transport of hazardous and harmful substances in the European Union maritime watersbulk. *Cedre*. 32 p.

Hedge, L.H., Knott, N.A., Johnston, E.L. (2009). Dredging related metal bioaccumulation in oysters. *Marine Pollution Bulletin*, 58(6), 832–840.

Honkanen, M., Häkkinen, J., Posti, A. (2013). Assessment of the chemical concentrations and the environmental risk of tank cleaning effluents in the Baltic Sea. *WMU Journal of Maritime Affairs*. http://doi.org/10.1007/s13437-013-0042-9.

IMO (International Maritime Organization). (2000). Protocol on preparedness, response and co-operation to pollution incidents by hazardous and noxious substances (OPRC-HNS Protocol).

Jambeck, J.R., Geyer, R., Wilcox, C., Siegler, T.R., Perryman, M., Andrady, A., Narayan, R., Law, K.L. (2015). Plastic waste inputs from land into the ocean. *Science*, 347(6223), 768–771.

Jeyasanta, K.I., Sathish, N., Patterson, J., Edward, J.P. (2020). Macro-, meso-and microplastic debris in the beaches of Tuticorin district, Southeast coast of India. *Marine Pollution Bulletin*, 154, 111055.

Johansen, M.P., Cresswell, T., Davis, J., Howard, D.L., Howell, N.R., Prentice, E. (2019). Biofilm-enhanced adsorption of strong and weak cations onto different microplastic sample types: Use of spectroscopy, microscopy and radiotracer methods. *Water Research*, 158, 392–400.

Kholssi, R., Lougraimzi, H., Grina, F., Lorentz, J.F., Silva, I., Castaño-Sánchez, O., Marks, E.A. (2022). Green agriculture: A review of the application of micro-and macroalgae and their impact on crop production on soil quality. *Journal of Soil Science and Plant Nutrition*, 22(4), 4627–4641.
Lakhal, S.Y., Khan, M.I., Islam, M.R. (2009). An "olympic" framework for a green decommissioning of an offshore oil platform. *Ocean & Coastal Management*, 52(2), 113–123.
Lau, W.W.Y., Shiran, Y., Bailey, R.M., Cook, E., Stuchtey, M.R., Koskella, J., Velis, C.A., Godfrey, L., Boucher, J., Murphy, M.B. (2020). Evaluating scenarios toward zero plastic pollution. *Science*, 369, eaba9475.
Lebreton, L., Andrady, A. (2019). Future scenarios of global plastic waste generation and disposal. *Palgrave Communications*, 5(1), 1–11.
Lozano, Y.M., Lehnert, T., Linck, L.T., Lehmann, A., Rillig, M.C. (2020). Microplastic shape, concentration and polymer type affect soil properties and plant biomass, bioRxiv Preprint. https://doi.org/10.1101/2020.07.27.223768.
McLaughlin, C., Falatko, D., Danesi, R., Albert, R. (2014). Characterizing shipboard bilgewater effluent before and after treatment. *Environmental Science and Pollution Research International*, 21(8), 5637–5652.
OGP (International Association of Oil & Gas Producers). (2012). Decommissioning of offshore concrete gravity-based structures (CGBS) in the OSPAR maritime area/other global regions. Report No. 484, November 2012.
Olagoke, O.V., Babaniyi, G.G., Afolabi, F.J., Babaniyi, B.R. (2022). Soil microbial enzymes and mitigation of heavy metal uptake by plants. In: N.R. Maddela, S.A. Aransiola and R. Prasad (eds.), *Ecological Interplays in Microbial Enzymology* (pp. 215–246). Singapore: Springer Nature Singapore.
Olsvik, P.A., Ørnsrud, R., Lunestad, B.T., Steine, N., Fredriksen, B.N. (2014). Transcriptional responses in Atlantic salmon (Salmo salar) exposed to deltamethrin, alone or in combination with azamethiphos. *Comparative Biochemistry and Physiology. Part C*, 162, 23–33.
Plastics Europe. Plastics Europe Market Research Group (PEMRG) and Conversion Market & Strategy GmbH, Final Report. 2020. https://www.plasticseurope.org/en/resources/publications/4312-plastics-facts-2020 (accessed 21 December 2022).
Rochman, C.M. (2016). Strategies for reducing ocean plastic debris should be diverse and guided by science. *Environmental Research Letters*, 11(4), 041001. https://doi.org/10.1088/1748-9326/11/4/041001.
Rochman, C.M., Kurobe, T., Flores, I., Teh, S.J. (2014). Early warning signs of endocrine disruption in adult fish from the ingestion of polyethylene with and without sorbed chemical pollutants from the marine environment. *Science of the Total Environment*, 15; 493, 656–661 .
Ryberg, M.W., Hauschild, M.Z., Wang, F., Averous-Monnery, S., Laurent, A. (2019). Global environmental losses of plastics across their value chains. *Resources, Conservation and Recycling*, 151, 104459.
Samuelsen, O.B., Lunestad, B.T., Farestveit, E., Grefsrud, E.S., Hannisdal, R., Holmelid, B., Tjensvoll, T., Agnalt, A.L. (2014). Mortality and deformities in European lobster (Homarus gammarus) juveniles exposed to the anti-parasitic drug teflubenzuron. *Aquatic Toxicology*, 149, 8–15.
Schroeder, D.M., Love, M.S. (2004). Ecological and political issues surrounding decommissioning of offshore oil facilities in the Southern California Bight. *Ocean & Coastal Management*, 47(1–2), 21–48.
Simmonds, M.P., Dolman, S.J., Jasny, M., Parsons, E.C.M., Weilgart, L., Wright, A.J., Leaper, R. (2014). Marine noise pollution - Increasing recognition but need for more practical action. *Journal of Ocean Technology*, 9, 71–90.
Singh, S., Kumar, V., Datta, S., Singh, S., Dhanjal, D.S., Kaul, N., Ramamurthy, P.C. (2023). Marine algae as green agriculture. In: V. Kumar, K. Tsatsaragkou and N. Asim(eds.), *Green Chemistry in Agriculture and Food Production*, CRC Press, 1–12, ISBN: 9780429289538.
Sun, M., Ye, M., Jiao, W., Feng, Y., Yu, P., Liu, M., Jiao, J., He, X., Liu, K., Zhao, Y., Wu, J., Jiang, X., Hu, F. (2018). Changes in tetracycline partitioning and bacteria/phage-comediated ARGs in microplastic-contaminated greenhouse soil facilitated by sophorolipid. *Journal of Hazardous Materials*, 345, 131–139.
Telfer, T.C., Baird, D.J., McHenery, J.G., Stone, J., Sutherland, I., Wislocki, P. (2006). Environmental effects of the anti-sea lice (Copepoda: Caligidae) therapeutant emamectin benzoate under commercial use conditions in the marine environment. *Aquaculture*, 260(1–4), 163–180.
Turner, A. (2021). Paint particles in the marine environment: An overlooked component of microplastics. *Water Research X*, 12, 100110.
van Hullebusch, E.D., Zandvoort, M.H., Lens, P.N. (2003). Metal immobilisation by biofilms: Mechanisms and analytical tools. *Reviews in Environmental Science and Bio/Technology*, 2(1), 9–33.
Xiong, X., Wu, C., Elser, J.J., Mei, Z., Hao, Y. (2019). Occurrence and fate of microplastic debris in middle and lower reaches of the Yangtze River–from inland to the sea. *Science of the Total Environment*, 659, 66–73.

Yang, Y., Chen, L., Xue, L. (2021). Looking for a Chinese solution to global problems: The situation and countermeasures of marine plastic waste and microplastics pollution governance system in China. *Chinese Journal of Population, Resources and Environment*, 19(4), 352–357.

Zainudin, B.H., Wong, T.W., Hamdan, H. (2020). Pectin as oral colon-specific nano-and microparticulate drug carriers. In: M. Al Ali Almaadeed, D. Ponnamma and M. A. Carignano (eds.) *Polymer Science and Innovative Applications* pp. 257–286).

4 Marine Greens
Roles in Climate Change and Global Warming Mitigation

Umm-kulthum Abubakar Gimba, Sesan Abiodun Aransiola, Oluwafemi Adebayo Oyewole, and Naga Raju Maddela

4.1 INTRODUCTION

Marine greens (also known as sea greens) are types of plants or marine algae that grow in or near the ocean (Leslie, 2022). They have leaves that can absorb water and nutrients directly through their surfaces. Generally, the main components of marine greens are lipids, proteins, nucleic acids, carbohydrates, and carotenoids such as astaxanthin, fucoxanthin, and lutein (Chen and Xu, 2020). The specific composition of marine greens, however, is dependent on the strain of algae and can also be influenced by method of cultivation (Williams, 2022). Marine greens are being cultivated in natural and artificial ponds because of their immense benefits. Two major kinds of commercial cultivation systems for marine greens have been developed. They are outdoor open ponds and photobioreactor or closed systems. Outdoor open pond systems, for example flat panels, raceway ponds, and bubble columns are most frequently used in commercial cultivation of marine greens, although closed systems (or photobioreactors) have been shown to be the most efficient systems due to the fact that they minimize CO_2 losses and have minimal nutrient requirements (Cheng et al., 2019; Singh et al., 2019). Additionally, outdoor open pond systems are usually the most viable for cultivation of marine greens even in environments where the microbial consortia are complex and dynamic (Carroppo and Pagliara, 2022). The CO_2 fixation efficiency of the open and closed systems was investigated by Aslam and Mughal (2016). They observed that in the open pond system, CO_2 fixation efficiency was about 42%, compared with the closed photobioreactor system, where CO_2 efficiency was about 85%.

4.2 ROLE OF MACROALGAE IN CLIMATE CHANGE AND GLOBAL WARMING MITIGATION

Since the dawn of industrialization, an unprecedented amount of greenhouse gases, notably CO_2 (which accounts for 68% of greenhouse gas [GHG] emissions), has been released into the environment due to anthropogenic activities such as burning of fossil fuels, factory farming, industrial activities, mining, and waste generation (Even et al., 2022; Farghali et al., 2022; Leh-Togi Zobeashia et al., 2018). Emissions from these sources have increased the global atmospheric CO_2 concentration. It has been estimated that global fossil CO_2 emissions from 2011 to 2020 were about 9.49 Pg C yr^{-1} (equivalent to about 34.8 Pg CO_2 yr^{-1}) with a growth rate of 0.4% yr^{-1}, although the highest CO_2 emissions in 2019 were about 9.71 Pg C yr^{-1} (equivalent to 35.6 Pg CO_2 yr^{-1}) (Wu et al., 2023). These emissions have resulted in climate change and global warming, which has had a catastrophic impact on humans, plants, wildlife, soils, and even waters.

Since the 2015 Paris Agreement on climate change, which called for the removal of carbon from the atmosphere and the lowering of emissions, the need for an environmentally friendly, efficient, and economical solution to replace the burning of fossil fuels and thus mitigate the effects of global

DOI: 10.1201/9781003369738-5

warming has been emphasized (Troell et al., 2022). Different strategies have been propounded to reverse the current trends of CO_2 emissions, with little success. Hence, the search for solutions to mitigate climate change and global warming is continuing. Physical methods such as ocean storage, biochar burial, and geological sequestration; chemical methods such as chemical scrubbing and mineral carbonization; and biological land-based processes such as agriculture, reforestation, and photosynthetic microorganisms have been explored with little success due to their numerous disadvantages (Kassim and Meng, 2017; Ould and Caldwell, 2022). In light of this, ocean-based strategies for combating global warming and climate change have been highlighted as a way forward (as opposed to land-based strategies). This is in recognition of the significant role the ocean plays as a site of intense carbon sequestration and storage, thus regulating CO_2 emissions into the atmosphere (Wu et al., 2023). The marine environment alone has taken up more than 560 Pg CO_2 since the beginning of industrialization. It absorbs 25 million tons of anthropogenic CO_2 per day. Additionally, it absorbs 30% of the carbon dioxide emitted globally, thus being the second-largest reservoir of organic carbon (Gao and Hader, 2020; Wu et al., 2022). The search for biological solutions to curb the impact of climate change and global warming has led to the discovery of carbon sinks such as microorganisms, soil, plant biomass, and algal biomass. Marine macro- and microalgal biomass has been highlighted as an efficient carbon sink with numerous advantages (Mehrotra and Pathak, 2020).

Many macroalgae species have been studied and shown to be excellent at sequestering carbon, which helps reduce the impact of climate change and global warming. Mehrotra and Pathak (2020) defined carbon sequestration as a process whereby carbon dioxide or other forms of carbon are captured and stored over an extended period of time. This eliminates the negative impact of CO_2 (a potent GHG that contributes to climate change caused by global warming). Cultivation of marine greens such as kelp, wakame, and nori is highlighted as an important strategy to reduce CO_2 emissions in China. For example, about 700 million tons of carbon per year can be fixed by these marine greens in the global continental shelf (Zhang et al., 2017). Therefore, carbon fixation by marine greens can have a significant impact on reducing carbon emissions, with kelp contributing more than 70% of all carbon fixed by cultivated marine greens, followed by nori, wakame, and *Gracilaria*. These marine greens convert dissolved inorganic carbon and CO_2 into organic carbon (Zhang et al., 2017).

Furthermore, according to Banerjee et al. (2020), carbon sequestration by marine greens will significantly reduce carbon dioxide emissions because they can be used to capture carbon from stationary sources, which can then be used to produce useful chemicals and even generate energy. Mashoreng et al. (2019) demonstrated in their study that marine greens, particularly maricultured seaweed species (*Kappaphycus alvarezii*, green and brown strains of *Eucheuma spinosum*) and pond-cultured seaweeds (*Gracilaria verrucosa* and *Caulerpa racemosa*) can be used as carbon sequestrants. Carbon sequestration was estimated at 57.64 tons CO_2 ha y^{-1} for maricultured seaweeds and 12.38 tons CO_2 ha y^{-1} for pond-cultured seaweeds. The total annual carbon sequestration from seaweed cultivation was estimated at 2,656,625 tons CO_2 per year from maricultured seaweeds and 621,377 tons CO_2 per year from pond-cultured seaweeds.

Another important role marine greens play in the mitigation of climate change and global warming is dampening wave energy in ocean environments. According to Duarte et al. (2017), cultivation of marine greens, notably seaweeds, can dampen wave energy, which helps prevent coastal erosion. This critical role helps protect shorelines by providing coastal protection structures. Reports by Duarte et al. (2017) highlighted that Norwegian kelp forests (predominantly *Laminaria hyperborea*), for example, reduced wave heights by about 60%. Similarly, Wood et al. (2017) reported that a mass of macroalgae has an effect on the sea currents, as it reduces water velocity and internal waves, as witnessed on a shellfish farm in New Zealand, where water currents were reduced by up to 63%.

Besides dampening wave energy, marine greens such as macroalgae or seaweeds have been shown to play a crucial role in reducing ocean acidification (which contributes to climate change and global warming). The ocean surface layers have become 26% more acidic compared with pre-industrial times, leading to ocean acidification. This is because the ocean absorbs about one-third of CO_2 emissions from anthropogenic sources (Roberts et al., 2017). Dense macroalgae farms (such as kelp forests supporting a high biodiversity of calcifiers, such as mollusks, lobsters, and crabs) cause the ocean to have high pH during daylight (which is beneficial to calcifiers). Elevation of pH in ocean waters helps in supplying oxygen to the waters and locally reducing the effects of ocean acidification and de-oxygenation (Duarte et al., 2017).

Marine greens such as seaweeds can also play a unique role in climate change and global warming mitigation by reducing emissions from the agricultural sector. Methane (a more potent GHG than carbon dioxide) emitted from livestock industries can be reduced by the addition of seaweeds into feedstock for animals (Sultana et al., 2022). In a laboratory setting, the addition of 5% of organic matter content of *Asparagopsis* to feedstock was observed to reduce the amount of methane gas produced (through enteric fermentation by ruminants and during the decomposition of their manure) by as much as 95% (Klinger, 2021). This highlights that modifying the feedstock using seaweeds could be a potential approach to reduce global emissions of methane gas from the agricultural sector.

Another critical role that macroalgae play in reducing climate change and global warming due to agricultural activities is seen in reducing the intense emissions associated with production of industrial fertilizers. Duarte et al. (2017) highlighted that managing macroalgae production to mitigate climate change is imperative, as macroalgae aquaculture has also been observed to impact the agricultural sector (which accounts for about 30% of GHG emissions) by reducing emissions from land-based food production systems. The impact of synthetic fertilizer can also be reduced by substituting synthetic fertilizers with seaweed compost or nutrient-rich seaweed biochar. Additionally, algal biochar has been demonstrated to help carbon sequestration. It also improves soil amelioration by enhancing the water-holding capacity and nutrient status of many soils. Also, the addition of algal biochar to the soil provides a substrate for enhanced microbial activity (Shukla et al., 2017).

Marine greens, mainly macroalgae, play a significant role in circulating organic matter in shallow beaches, deep-sea sediments, and intertidal rocky regions. Being primary producers in the marine ecosystem, seaweeds can sequester carbon, producing organic carbon, which forms the base of the entire marine food web. They can also store huge amounts of carbon within their biomass to transfer to other habitats, thus reducing CO_2 emissions, which would have contributed to climate change and global warming (Behera et al., 2022). In Singapore, macroalgae accounted for about 650 Mg of the carbon biomass stock, which surpassed the aboveground carbon found in seagrass meadows. Dominant macroalgae species found along the Singaporean coast were *Ulva* and *Sargassum*.

4.3 PATHWAYS OF CO_2 SEQUESTRATION BY MARINE MACROALGAE

Marine greens (for example macroalgae) absorb CO_2 from water in the process of photosynthesis as they grow. Because macroalgae communities are autotrophic, they generate more organic carbon and are thus able to capture most of the CO_2 in marine vegetated habitats. Subsequently, when macroalgae die, CO_2 is stored in sediments in the ocean floor, where it is less likely to be released back into the atmosphere (Duarte et al., 2017). In marine aquaculture, the fixed carbon is usually in four forms: removable carbon (i.e. carbon fixed in cultured organisms), dissolved organic carbon (DOC), particulate organic carbon, and carbon buried in sediments (Zhang et al., 2017). Marine greens, notably seaweeds, being primary producers, can release substantial amounts of DOC in marine environments. Kelp mariculture has been found to significantly increase DOC concentration in a seaweed farming area in China (Li et al., 2022) Figure 4.1 shows the pathways of carbon sequestration and transport in marine greens such as kelp in a maricultured environment.

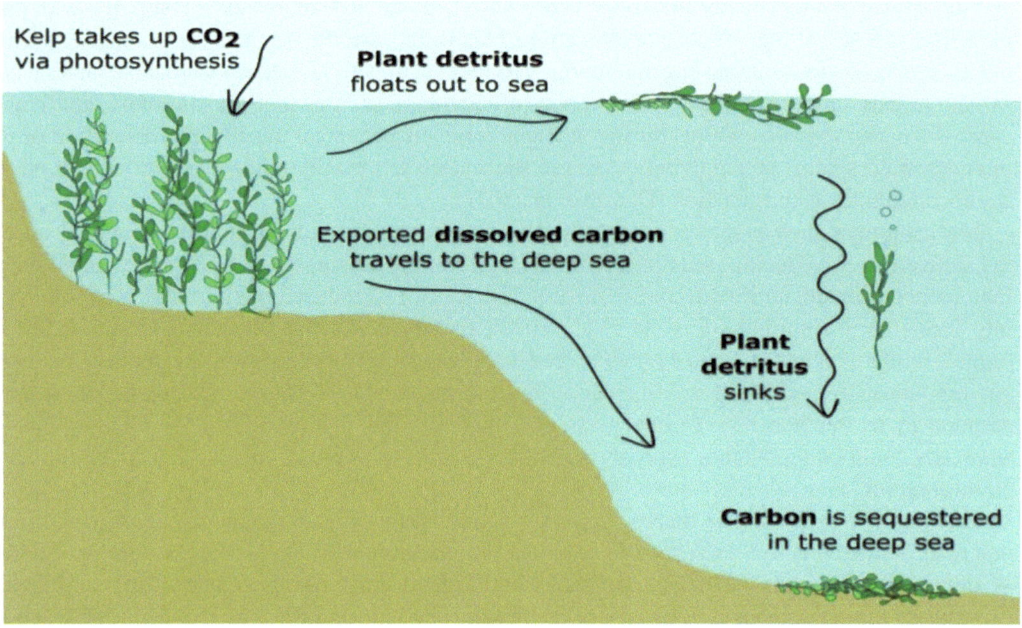

FIGURE 4.1 Carbon sequestration pathways and transport by macroalgae. (Adapted from Krause-Jensen et al. (2018).)

4.4 ROLE OF MARINE MICROALGAE IN CLIMATE CHANGE AND GLOBAL WARMING MITIGATION

Marine microalgae include simple prokaryotic blue-green algae (cyanobacteria) and eukaryotic forms such as diatoms and red and green algae (Carroppo and Pagliara, 2022). They are carbon-neutral organisms that capture light energy from the sun using their chlorophyll and/or other photosynthetic pigments (Mehrotra and Pathak, 2020). Compared with other biological methods, microalgal photosynthesis produces higher carbon dioxide fixation rates, it does not lead to environmental alterations, and there is no requirement for further disposal of residues from captured CO_2 (Yun et al., 2016). Microalgae can either be unicellular or have a simple multicellular structure. Their size ranges from 0.2 to 200 micrometers (Mustafa et al., 2021; Carroppo and Pagliara, 2022). Marine microalgae are able to convert up to 80% of CO_2 into oxygen (Arul Delpin and Jayaprakashvel, 2018). In addition to microalgal CO_2 fixation efficiency, microalgal biomass can be used as a raw material in producing different bioproducts, such as biofuels, cosmetics, pharmaceuticals, nutraceuticals, fine chemicals, food and animal feed supplements for domesticated farm animals such as cattle and chickens, or aquaculture feed for fishes such as salmon and shrimp (Yun et al., 2016; Yan et al., 2021). Based on studies by Saravanan et al. (2022), CO_2 bio-mitigation and biofuel production is an excellent strategy to reduce CO_2 emissions, thus tackling global warming.

In addition to their role as carbon sequestrants, marine microalgae and cyanobacterial species can play important roles in reducing global warming and climate change by serving as carbon fixers (Ghosh and Kiran, 2017). Among the carbon-fixing organisms, microalgae and cyanobacteria are able to grow faster and have 10 to 50 times higher carbon dioxide fixation efficiency than terrestrial plants (this is in addition to their biomass being useful in biofuel, pharmaceutical, cosmetics, and nutraceutical production) (Chen et al., 2020). Moreira and Pires (2016) reported that cultivating 1 kilogram of microalgae can help fix up to 2 kilograms of CO_2. *Chlorella vulgaris* and *Anabaena* sp. have both been reported to have a high CO_2 fixation rate of 6.24 $gL^{-1}d^{-1}$ and 1.45 $gL^{-1}d^{-1}$, respectively. Similarly, *Chlorella vulgaris* has been reported to have a CO_2 fixation rate as high as 18% v/v.

Microalgae, being a leading producer in the benthic food chain, play significant roles in marine ecosystems due to their ability to harness solar energy. Microalgae have been observed to efficiently conduct photosynthesis by acquiring inorganic carbon even from very low atmospheric CO_2 concentrations (Onyeaka et al., 2021). Similarly, microalgae are able to capture light and convert it into chemical energy. Under optimum conditions of growth, they channel most of the energy captured into cell division (6–12-h cycle). This makes them attractive in biological carbon fixation because they can reduce atmospheric CO_2, thus playing a crucial role in mitigating climate change and reducing the effects of global warming. Because microalgae are carbon fixers, they are able to utilize CO_2 from the atmosphere (in the form of carbonates such as $NaHCO_3$ and Na_2CO_3) for cellular growth (with cellular lipids up to 80% in some species) and biomass production (Goswami et al., 2021). The genera *Scenedesmus, Euglena, Anabaena, Chlorococcum, Spirulina, Chlorella, Botryococcus*, and *Haematococcus* have the greatest carbon fixation potential (Banerjee et al., 2020). They are considered for biological sequestration of carbon due to their CO_2 concentrating mechanism (CCM) (Banerjee et al., 2020). The bio-fixation of CO_2 and the production of microalgal biomass depends on the following factors: (i) the characteristics of the microalgae species, (ii) the physiochemical process effects, and (iii) the impact of the culture system on the microalgae species (Ighalo et al., 2022). Chen and Xu (2020) reported that in open ponds, the average biomass production rate of microalgae was 20 gm^{-2} per day, which is equivalent to a carbon sequestration intensity of 11,280 t CO_2 km^{-2} per year. Similarly, in 2020, Zhao and Su developed a model that demonstrated the potential of microalgae to reduce atmospheric CO_2. Based on their simulations, cultivating 5300 km^2 of microalgae in open or closed culture systems can reduce annual emissions by about 0.5393 Gt CO_2 and yield 324.33 million tons of microalgal biomass.

Microalgae and cyanobacteria are able to grow fast, have high CO_2 tolerance, require low light intensity, are able to tolerate extreme environmental stresses, and have high photosynthetic ability and a high biomass production rate (due to their exponential growth) (Banerjee et al., 2020).

Another significant role microalgae play as excellent carbon sequestrants in mitigating the effect of climate change and global warming is based on their ability to absorb large amounts of CO_2 and flue gases containing nitrous oxide, sulfur dioxide, and other hazardous compounds from stationary sources such as generating stations and processing plants where CO_2 is being emitted, which can then be converted into biofuels (de Morais et al., 2019; Aransiola et al., 2023). Biological sequestration of carbon dioxide through microalgal systems will help reduce atmospheric CO_2. The species usually employed for CO_2 sequestration are *Chlorella* sp., *Spirulina* sp., *Chlorococcum littorale, Dunaliella tertiolecta, Isochrysis galbana, Chlamydomonas reinhardtii, Acutodesmus obliquus, Botryococcus braunii, Nannochloropsis, Dunaliella salina, and Scenedesmus* sp. (Guldhe et al., 2015; Arul and Jayaprakashvel, 2018). Anguselvi et al. (2019) used microalgae to capture CO_2 from industries. Freshwater algae (*Oscillatoria*) were cultured in a closed pond system and monitored based on growth rate, photosynthesis rate, and tolerance to temperature and gases (NOx, SOx, and gaseous hydrocarbons). They observed that higher CO_2 capture occurred from 16–32 h at optimum pH of 7–9 and temperature of 25–30°C. They also observed that the fresh microalgae utilized 90–99% CO_2 from a natural gas processing plant (3–15% of CO_2 from flue gas as a carbon source). Similarly, Dinsehbabu et al. (2019) reported that microalgae have been demonstrated to trap CO_2 from flue gases released into the atmosphere. Additionally, findings by Ahiahonu et al. (2021) revealed that *Chlorella sorokiniana* and *Tetradesmus reginae* were observed to biosequester CO_2 at a rate of 23.05 ± 0.711 gCO_2 $L^{-1}day^{-1}$ for *C. sorokiniana* and 31.52 ± 0.600 gCO_2 $L^{-1}day^{-1}$ for *Tetradesmus reginae.*

Furthermore, a power plant in Spain has been designed to biofix CO_2 directly from a flue gas chimney through microalgae cultivation. This has led to the production of biomass (about 100 tons per year). Similarly, in the United States, ethanol production was achieved by cultivating microalgae to curb emissions from power plants (De Morais et al., 2019).

Chlorella sp., *Nannochloropsis* sp., and *Scenedesmus dimorphus* have also been reported to tolerate these gases. The ability of these microalgae to utilize these compounds for their growth makes them potential candidates to achieve zero pollution from emissions due to anthropogenic activities that are contributing to climate change (Yen et al., 2015; Singh et al., 2019)

Brasil, Silva, and Siqueira (2017) opined that with the increase in human population and the consequent growing demand for energy, microalgae biorefineries can play a crucial role as a more sustainable alternative to fossil fuels by reducing excessive dependency on fossil fuels. This is especially vital due to growing concerns about the CO_2 emissions due to carbon-based fossil fuels. In addition, microalgal biofuel production promotes CO_2 sequestration and future energy production (Jaiswal et al., 2020).

4.5 PATHWAYS OF CO_2 SEQUESTRATION BY MICROALGAE

There are three major pathways through which microalgae acquire carbon: via direct uptake of HCO_3, via active CO_2 transport. and using the carbonic anhydrase enzyme. Microalgae can also use simple diffusion to take up CO_2 (Onyeaka et al., 2021). CO_2 fixation in microalgae is usually through carbon concentration mechanisms or the Calvin Benson cycle (Goswami et al., 2022). Microalgae are generally fast-growing and can adapt to photoautotrophic, heterotrophic, and mixotrophic environments (Banerjee et al., 2020). Microalgal species with heterotrophic metabolism survive in dark environments where there is no light to obtain carbon and energy by absorbing organic complexes in a medium (Goswami et al., 2022). Using this metabolic pathway, the pentose phosphate pathway (PPP), which allows organic molecules to flow through cell walls and be transformed into lipids and other metabolites through respiration, is employed by microalgae. In some instances, however, strains of microalgae have been observed to utilize light as a source of energy (photoheterotrophy). This allows them to grow faster and have high biomass, protein, and lipid. Glucose is the organic carbon source most often employed in the development of heterotrophic microalgae (Goswami et al., 2022).

In auto-phototrophic metabolism, most microalgae utilize the autotrophic mode of carbon assimilation. Microalgae utilize inorganic carbon such as HCO_3^- and CO_2, which they obtain from the atmosphere, for conversion into organic compounds (Mehrotra and Pathak, 2020). Autotrophic assimilation of carbon dioxide takes place via a photosynthetic mechanism through two phases: the first phase is light dependent, while the second phase is light independent. The light-dependent phase exploits light energy to produce a large amount of ATP (an energy-storing molecule) and nicotinamide adenine dinucleotide (NADPH). In addition, the chlorophyll dimer is utilized by the two photosystems known as PS I and PS II to transmit sunlight toward the electron transport chain. This leads to the generation of around 1.3 molecules of ATP per pair of electron compounds (Mehrotra and Pathak, 2020). The second phase is the light-independent or dark reaction phase. This usually occurs in the absence of light, although it can occur in the presence of light as well. In this phase, carbon fixation reactions take place by means of energy-storage products (ATP and NADPH) produced during the light-dependent phase (Iverson, 2006). This fixation process takes place through a set of reactions called the Calvin Benson cycle. Carbon is absorbed in the Calvin Benson cycle in three stages: carboxylation, reduction, and regeneration (in this sequence). In the carboxylation phase, CO_2 is absorbed and fixed into an organic compound/sugar via the enzyme 5-bisphosphate carboxylase (RuBisCo). Ribulose-1 is then converted into ribulose-1 and 5-bisphosphate (RuBP) to generate two molecules of 3-phosphoglycerate (3-PGA). The 3-PGA is then phosphorylated to glyceraldehyde 3-phosphate (G-3-P) via 3-phosphoglycerate kinase and glyceraldehyde phosphate dehydrogenase.. Thereafter, RuBP is replenished (due to its weak linkage of ribulose-1,5-biphosphate [RuBP] with CO_2) by a series of chemical reactions, and the fixation cycle advances to the reduction and regeneration stages (Zhou et al., 2017).

4.6 CULTIVATION OF MARINE GREENS

Marine greens such as microalgae are able to dissolve inorganic carbon in the forms of CO_2, H_2CO_3, HCO_3, and CO. Marine greens capture CO_2 via photosynthesis and convert it into organic compounds using ATP and NADPH. Different types of photobioreactors have been designed to grow marine greens. The design of a typical photobioreactor considers important parameters such as carbon and nutrient level, pH, temperature, light intensity, and the dark and light cycle (Arul and Jayaprakashvel, 2018). This is because the structure of a photobioreactor significantly affects the CO_2 utilization efficiency and CO_2 fixation efficiency of marine greens (Li et al., 2013). Marine greens can be cultivated using two major types of cultivation system: the open-pond cultivation system and the closed-pond system. A brief discussion of each cultivation system follows.

4.6.1 Open Pond Systems

The open pond systems are of two major types: natural water bodies, such as ponds and lakes, and artificial ponds, such as fiber-reinforced plastic tanks and cemented raceways (Shukla et al., 2017). Artificial pond systems for commercial microalgae production are designed typically with an area of 0.2 to 0.5 hectares and a width of 0.25 meters. The open pond system is usually cheap and easy to operate and maintain. The high surface area per volume ratio of open ponds allows them to absorb huge amounts of CO_2. This makes them suitable for handling large quantities of flue gases from stationary sources such as power plants (de Morais et al., 2019). Some of the disadvantages of this cultivation system include: (i) the culture is easily contaminated; (ii) it requires a lot of space to set up; and (iii) the open structure makes it prone to water loss due to evaporation (Cheah et al., 2015; De Morais et al., 2019).

4.6.1.1 Types of Open Pond Systems

4.6.1.1.1 Raceway Ponds

These are shallow ponds with a width of 15 to 25 meters. The pond has a paddle wheel attached for agitation (to ensure nutrient homogenization and air circulation). It also has baffles to control and guide the flow of culture. Intermittent gas sparging is also supplied to raceway ponds to increase CO_2 fixation efficiency (Li et al., 2013). It is used for the commercial cultivation of *Dunaliella salina*, *Spirulina platensis*, *Hematococcis* sp., and *Chlorella* sp. (Cheah et al., 2015). According to Acedo et al. (2020), green algal species such as *Chlorella sorokiniana* cultivated in semi-automated raceway ponds were efficiently utilized to enhance carbon capture. The microalgal yield was about 0.4 gL^{-1} ash-free dry weight with a lipid content of 24%. They highlighted that microalgal farms can be excellent sites for carbon capture and the production of algal biofuel.

4.6.1.1.2 Multi-Layer Bioreactor

This cultivation system has multiple layers of tanks placed one on top of the other. The culture media flows from one tank to the other through piping systems. Some advantages of this system are that it requires less space, it is cheap to operate and maintain, and it is easy to scale up (Cheah et al., 2015).

4.6.1.2 Closed Pond Systems or Photobioreactors

These cultivation systems are designed to provide a controlled environment with suitable parameters for the cultivation of microalgae. This system reduces CO_2 losses by prolonging CO_2 mass transfer efficiency and retention time, thus increasing photosynthetic efficiency and biomass production rate (Arul and Jayaprakashvel, 2018). Reports by Ghosh and Kiran (2017) stated that closed pond systems for microalgal cultivation over a surface area of about 60 miles can fix carbon dioxide from the exhaust of a 1-gigawatt power plant. This can also lead to the generation of about 14,000 gallons of biofuel per acre per year.

4.7 CONCLUSION

To achieve the targets set by agreement at the Conference of the Parties (COP21) meetings in Paris, there is an urgent need to reduce carbon dioxide emissions to near zero by mid-century and to remove carbon dioxide from the atmosphere in the latter half of this century to reverse the current trends leading to global warming and climate change. Among the various techniques of CO_2 mitigation, marine greens such as macroalgae and microalgae are key to achieving these targets because they are effective and efficient bio-sequestrants of CO_2 and other noxious gases from the atmosphere. The role of marine greens in the mitigation of climate change and global warming should therefore never be underestimated.

REFERENCES

Ahiahonu, E.K., Anku, W.W., Roopnarain, A., Green, E., Govender, P.P. and Serepa-Dlamini, M.H., 2021. Bioprospecting wild South African microalgae as a potential third-generation biofuel feedstock, biological carbon-capture agent and for nutraceutical applications. *Biomass Conversion and Biorefinery, 3*, pp. 1–16.

Anguselvi, V., Masto, R.E., Mukherjee, A. and Singh, P.K., 2019. CO_2 capture for industries by algae. In: Y.K. Wong (ed), *Algae*. IntechOpen. ISBN: 978-1-83880-563-0.

Aransiola, S.A., Victor-Ekwebelem, M. O., Leh-Togi Zobeashia, S. S. and Maddela, N.R., 2023. Sources and techniques for biofuel generation. In M.P. Shah (ed.), *Green Approach to Alternative Fuel for a Sustainable Future*. Elsevier Inc. https://doi.org/10.1016/B978-0-12-824318-3.00026-6.

Arul Delpin, S. S. and Jayaprakashvel, M., 2018. Carbon sequestration potential of marine microalgae. *Oceanography & Fisheries Open Access Journal, 6*(5), pp. 141–142.

Aslam, A. and Mughal, T.A., 2016. A review on microalgae to achieve maximal carbon dioxide (CO2) mitigation from industrial flue gases. *International Journal of Research in Advent Technology, 4*(9), pp. 1–18.

Banerjee, I., Dutta, S., Pohrmen, C.B., Verma, R. and Singh, D., 2020. Microalgae-based carbon sequestration to mitigate climate change and application of nanomaterials in algal biorefinery. *Octa Journal of Biosciences, 8*, pp. 129–136.

Behera, D.P., Vadodariya, V., Veeragurunathan, V., Sigamani, S., Moovendhan, M., Srinivasan, R., Kolandhasamy, P. and Ingle, K.N., 2022. Seaweeds cultivation methods and their role in climate mitigation and environmental cleanup. *Total Environment Research Themes, 3*, p. 100016.

Caroppo, C. and Pagliara, P., 2022. Microalgae: A promising future. *Microorganisms, 10*(8), p. 1488.

Cheah, W.Y., Show, P.L., Chang, J.S., Ling, T.C. and Juan, J.C., 2015. Biosequestration of atmospheric CO_2 and flue gas-containing CO_2 by microalgae. *Bioresource Technology, 184*, pp. 190–201.

Chen, Y. and Xu, C., 2020. Exploring new blue carbon plants for sustainable ecosystems. *Trends in Plant Science, 25*(11), pp. 1067–1070.

Chen, Y., Xu, C. and Vaidyanathan, S., 2020. Influence of gas management on biochemical conversion of CO_2 by microalgae for biofuel production. *Applied Energy, 261*, p. 114420.

Cheng, J., Zhu, Y., Zhang, Z. and Yang, W., 2019. Modification and improvement of microalgae strains for strengthening CO_2 fixation from coal-fired flue gas in power plants. *Bioresource Technology, 291*, p. 121850.

de Morais, M.G., de Morais, E.G., Duarte, J.H., Deamici, K.M., Mitchell, B.G. and Costa, J.A.V., 2019. Biological CO_2 mitigation by microalgae: Technological trends, future prospects and challenges. *World Journal of Microbiology and Biotechnology, 35*(5), pp. 1–7.

Dineshbabu, G., Vijayan, D., Uma, V.S., Makut, B.B. and Das, D., 2019. Microalgal systems for integrated carbon sequestration from flue gas and wastewater treatment. In: S.K. Gupta and F. Bux (eds.), *Application of Microalgae in Wastewater Treatment: Volume 2: Biorefinery Approaches of Wastewater Treatment*, pp. 339–370. Springer, Cham.

Duarte, C.M., Wu, J., Xiao, X., Bruhn, A. and Krause-Jensen, D., 2017. Can seaweed farming play a role in climate change mitigation and adaptation? *Frontiers in Marine Science, 4*, p. 100.

Even, C., Hadroug, D., Boumlaik, Y. and Simon, G., 2022. Microalgae-based bioenergy with carbon capture and storage quantified as a negative emissions technology. *Energy Nexus, 7*, 100117.

Farghali, M., Mohamed, I.M., Osman, A.I. and Rooney, D.W., 2022. Seaweed for climate mitigation, wastewater treatment, bioenergy, bioplastic, biochar, food, pharmaceuticals, and cosmetics: A review. *Environmental Chemistry Letters, 21*, pp. 97–152 .

Gao, K. and Häder, D.P., 2020. Photosynthetic performances of marine microalgae under influences of rising CO_2 and solar UV radiation. *Microbial Photosynthesis, 1*, pp. 139–150.

Ghosh, A. and Kiran, B., 2017. Carbon concentration in algae: Reducing CO_2 from exhaust gas. *Trends in Biotechnology, 35*(9), pp. 806–808.

Goswami, R.K., Agrawal, K. and Verma, P., 2021. An overview of microalgal carotenoids: Advances in the production and its impact on sustainable development. In: N. Srivastava and M. Srivastava (eds.), *Bioenergy Research: Evaluating Strategies for Commercialization and Sustainability*, pp. 105–128.

Goswami, R.K., Mehariya, S., Karthikeyan, O.P., Gupta, V.K. and Verma, P., 2022. Multifaceted application of microalgal biomass integrated with carbon dioxide reduction and wastewater remediation: A flexible concept for sustainable environment. *Journal of Cleaner Production*, *339*, 130654.

Guldhe, A., Bhola, V., Rawat, I. and Bux, F., 2015. Carbon dioxide sequestration by microalgae: Biorefinery approach for clean energy and environment. *Algae and Environmental Sustainability*, *7*, pp. 147–154.

Jaiswal, A., Babu, V., Baishya, B. and Jaiswal, P., 2020. Biochemical pathways regulated by algae to mitigate global carbon emissions: A review. *Journal of Environmental Pathology Toxicology and Oncology*, *39*(4).

Kassim, M.A. and Meng, T.K., 2017. Carbon dioxide (CO_2) biofixation by microalgae and its potential for biorefinery and biofuel production. *Science of the Total Environment*, *584*, pp. 1121–1129.

Kostas, E.T., Adams, J.M., Ruiz, H.A., Durán-Jiménez, G. and Lye, G.J., 2021. Macroalgal biorefinery concepts for the circular bioeconomy: A review on biotechnological developments and future perspectives. *Renewable and Sustainable Energy Reviews*, *151*, 111553.

Krause-Jensen, D., Lavery, P., Serrano, O., Marbà, N., Masque, P. and Duarte, C.M., 2018. Sequestration of macroalgal carbon: The elephant in the Blue Carbon room. *Biology Letters*, *14*(6), 20180236.

Leh-Togi Zobeashia, S.S., Aransiola, S.A., Ijah, U.J.J. and Abioye, O.P., 2018. Anaerobic digestion and agricultural application of organic wastes. *Advances in Environmental Research*, *7*(2), pp. 73–85. http://www.techno-press.org/content/?page=article&journal=aer&volume=7&num=2&ordernum=1.

Leslie, B., 2022. *Sea Greens 101: Types, Benefits, Where to Buy Them and More*. Available from: https://www.everydayhealth.com/diet-nutrition/sea-greens-types-benefits-where-to-buy-them-and-more/ (Accessed:12th December, 2022).

Li, H., Zhang, Z., Xiong, T., Tang, K., He, C., Shi, Q., Jiao, N. and Zhang, Y., 2022. Carbon sequestration in the form of recalcitrant dissolved organic carbon in a seaweed (kelp) farming environment. *Environmental Science and Technology*, *56*(12), pp. 9112–9122.

Li, S., Luo, S. and Guo, R., 2013. Efficiency of CO_2 fixation by microalgae in a closed raceway pond. *Bioresource Technology*, *136*, pp. 267–272.

Mashoreng, S., La Nafie, Y.A. and Isyrini, R., 2019, November. Cultivated seaweed carbon sequestration capacity. In *IOP Conference Series: Earth and Environmental Science*, *370*(1), p. 012017. IOP Publishing.

Moreira, D. and Pires, J.C., 2016. Atmospheric CO_2 capture by algae: Negative carbon dioxide emission path. *Bioresource Technology*, *215*, pp. 371–379.

Mustafa, S., Bhatti, H.N., Maqbool, M. and Iqbal, M., 2021. Microalgae biosorption, bioaccumulation and biodegradation efficiency for the remediation of wastewater and carbon dioxide mitigation: Prospects, challenges and opportunities. *Journal of Water Process Engineering*, *41*, 102009.

Onyeaka, H., Miri, T., Obileke, K., Hart, A., Anumudu, C. and Al-Sharify, Z.T., 2021. Minimizing carbon footprint via microalgae as a biological capture. *Carbon Capture Science & Technology*, *1*, 100007.

Ould, E. and Caldwell, G.S., 2022. The potential of seaweed for carbon capture. *CABI Reviews*, https://doi.org/10.1079/cabireviews202217009.

Roberts, C.M., O'Leary, B.C., McCauley, D.J., Cury, P.M., Duarte, C.M., Lubchenco, J., Pauly, D., Sáenz-Arroyo, A., Sumaila, U.R., Wilson, R.W., Worm, B. and Castilla, J.C., 2017. Marine reserves can mitigate and promote adaptation to climate change. *Proceedings of the National Academy of Sciences of the United States of America*, *114*(24), pp. 6167–6175.

Saravanan, A., Deivayanai, V.C., Kumar, P.S., Rangasamy, G. and Varjani, S., 2022. CO_2 bio-mitigation using genetically modified algae and biofuel production towards a carbon net-zero society. *Bioresource Technology*, *363*, 127982.

Singh, H.M., Kothari, R., Gupta, R. and Tyagi, V.V., 2019. Bio-fixation of flue gas from thermal power plants with algal biomass: Overview and research perspectives. *Journal of Environmental Management*, *245*, pp. 519–539.

Sultana, F., Wahab, M.A., Nahiduzzaman, M., Mohiuddin, M., Iqbal, M.Z., Shakil, A., Mamun, A.A., Khan, M.S.R., Wong, L. and Asaduzzaman, M., 2022. Seaweed farming for food and nutritional security, climate change mitigation and adaptation, and women empowerment: A review. *Aquaculture and Fisheries*, *8*(5), pp. 463–480.

Troell, M., Henriksson, P.J., Buschmann, A.H., Chopin, T. and Quahe, S., 2022. Farming the ocean–seaweeds as a quick fix for the climate? *Reviews in Fisheries Science and Aquaculture*, *31*(3), pp. 285–295.

Wood, D., Capuzzo, E., Kirby, D., Mooney-McAuley, K. and Kerrison, P., 2017. UK macroalgae aquaculture: What are the key environmental and licensing considerations? *Marine Policy*, *83*, pp. 29–39.

Wu, J., Keller, D.P. and Oschlies, A., 2023. Carbon dioxide removal via macroalgae open-ocean mariculture and sinking: An Earth system modeling study. *Earth System Dynamics*, *14*(1), pp. 185–221.

Yan, C., McWilliams, J.C. and Chamecki, M., 2021. Generation of attached Langmuir circulations by a suspended macroalgal farm. *Journal of Fluid Mechanics*, *915*, p. 76.

Yen, H.W., Ho, S.H., Chen, C.Y. and Chang, J.S., 2015. CO_2, NOx and SOx removal from flue gas via microalgae cultivation: A critical review. *Biotechnology Journal*, *10*(6), pp. 829–839.

Yun, H.S., Ji, M.K., Park, Y.T., Salama, E.S. and Choi, J., 2016. Microalga, *Acutodesmus obliquus* KGE 30 as a potential candidate for CO_2 mitigation and biodiesel production. *Environmental Science and Pollution Research International*, *23*(17), pp. 17831–17839.

Zhang, Y., Zhang, J., Liang, Y., Li, H., Li, G., Chen, X., Zhao, P., Jiang, Z., Zou, D., Liu, X. and Liu, J., 2017. Carbon sequestration processes and mechanisms in coastal mariculture environments in China. *Science in China (Earth Sciences)*, *60*(12), pp. 2097–2107.

5 Marine Green Nanotechnology for Remediation of Wastewater

O.D. Ogundele, J.O. Jayeola, O.S. Ilesanmi, D.A. Oyegoke, O.F. Adedugbe, and V.A. Olagunju

5.1 INTRODUCTION

Marine green nanotechnology (MGN) refers to the application of nanoscience principles and techniques to harness the potential of marine-derived biomolecules and microorganisms, particularly marine algae (both microalgae and macroalgae), for the development of novel nanomaterials and nanoscale devices (Abd-Elnaby *et al.*, 2018; Duraia *et al.*, 2016). This interdisciplinary field has gained significant attention in recent years, as it offers sustainable and eco-friendly alternatives to conventional chemical and physical synthesis methods for producing nanomaterials, which often involve high energy consumption, hazardous reagents, and toxic waste generation (Phang *et al.*, 2015).

Marine algae are considered a promising source of bioactive compounds and functional biomolecules, such as proteins, polysaccharides, lipids, pigments, and phenolic compounds, which exhibit diverse biological activities and potential applications in various industries, including pharmaceuticals, nutraceuticals, cosmetics, and biotechnology (Kim *et al.*, 2014; Rodrigues *et al.*, 2015). These marine-derived biomolecules can serve as natural templates, stabilizers, and reducing agents for the synthesis of nanoparticles, such as metal nanoparticles (e.g., gold, silver, and platinum) and metal oxide nanoparticles (e.g., zinc oxide and titanium dioxide), with controlled size, shape, and surface properties (Abd-Elnaby *et al.*, 2018; Duraia *et al.*, 2016).

For example, phlorotannins, which are phenolic compounds found in brown macroalgae, have been shown to reduce metal ions and stabilize the formation of gold and silver nanoparticles, while sulfated polysaccharides, such as fucoidan and carrageenan, have been utilized for the green synthesis of metal oxide nanoparticles, such as zinc oxide and iron oxide (Jung *et al.*, 2013; Kathiresan *et al.*, 2009). These biogenic nanoparticles have demonstrated various biological activities, such as antimicrobial, antioxidant, anticancer, and anti-inflammatory properties, and have been proposed for applications in drug delivery, tissue engineering, biosensing, and environmental remediation (Abd-Elnaby *et al.*, 2018; Kim *et al.*, 2014).

MGN represents an emerging and promising field, which seeks to exploit the unique biochemical and structural properties of marine algae and microorganisms for the eco-friendly synthesis and development of innovative nanomaterials and nanodevices with diverse applications across various industries (Abd-Elnaby *et al.*, 2018; Duraia *et al.*, 2016). Further research and interdisciplinary collaboration between nanoscience, marine biology, and materials science are being explored for the full potential of MGN and to address the challenges associated with the scalability, reproducibility, and stability of the biogenic nanoparticles and nanoscale structures, as well as their integration into functional devices and systems (Phang *et al.*, 2015; Rodrigues *et al.*, 2015).

DOI: 10.1201/9781003369738-6

5.2 IMPORTANCE OF WASTEWATER REMEDIATION

Wastewater remediation is the process of removing pollutants and contaminants from wastewater, which can originate from various sources, such as industrial processes, agricultural activities, and domestic sewage (EPA, 2021). The importance of wastewater remediation cannot be overstated, as untreated or inadequately treated wastewater can have severe consequences on public health, the environment, and socio-economic development (UN-Water, 2020).

One of the primary concerns related to untreated wastewater is the potential for the spread of waterborne diseases, which can result from the discharge of pathogens, such as bacteria, viruses, and parasites, into water bodies used for drinking, bathing, and recreational activities (Snyder *et al.*, 2003). Wastewater remediation helps mitigate this risk by employing physical, chemical, and biological treatment processes to eliminate or reduce the levels of pathogens and other harmful substances, such as heavy metals, organic pollutants, and nutrients, in wastewater (EPA, 2021).

Another critical aspect of wastewater remediation is the prevention of environmental pollution and degradation (Aransiola *et al.*, 2021). The discharge of untreated wastewater into rivers, lakes, and coastal areas can lead to eutrophication, which is characterized by excessive growth of algae and other aquatic plants due to high concentrations of nutrients, such as nitrogen and phosphorus (Conley *et al.*, 2009). This can cause oxygen depletion, fish kills, and loss of biodiversity, as well as negatively affecting the aesthetic and recreational value of water bodies (Smith *et al.*, 1999).

Furthermore, investments in wastewater remediation infrastructure and technologies can create employment opportunities, stimulate innovation, and foster economic growth in the water and sanitation sector, as well as related industries, such as agriculture, energy, and manufacturing (UN-Water, 2018). Wastewater remediation can also contribute to climate change mitigation and adaptation by reducing greenhouse gas emissions from untreated wastewater and sludge decomposition, as well as promoting the use of renewable energy sources, such as biogas, and enhancing water security and resilience to water scarcity and pollution (Metcalf & Eddy, 2014).

5.3 MARINE GREEN NANOTECHNOLOGY

MGN is an emerging interdisciplinary field that combines the principles of green chemistry, nanoscience, and marine biology to harness the unique properties and capabilities of marine organisms, such as algae and microorganisms, for the synthesis and assembly of nanoscale materials and structures (Azizi *et al.*, 2018). This approach offers a sustainable and eco-friendly alternative to conventional methods of nanomaterial production, which often rely on toxic chemicals, high energy consumption, and non-renewable resources (Rai *et al.*, 2016).

Marine algae, also known as marine greens, are particularly attractive candidates for the development of green nanotechnology due to their abundance, diversity, and ability to produce a wide range of biologically active compounds and biomolecules, such as polysaccharides, proteins, and lipids, that can serve as templates, stabilizers, and reducing agents for the synthesis of nanoparticles (Pankhurst *et al.*, 2017). Moreover, marine algae exhibit unique features, such as high surface area, porosity, and functional groups, which can facilitate the formation of nanoscale structures with controlled size, shape, and surface properties (Nagarajan *et al.*, 2018).

Various types of nanoparticles, such as metal, metal oxide, and semiconductor nanoparticles, can be synthesized using MGN (Siddiqi & Husen, 2020). For instance, gold and silver nanoparticles have been successfully produced using extracts from marine algae, such as *Sargassum muticum*, *Ulva lactuca*, and *Fucus vesiculosus*, which act as both reducing and capping agents, resulting in stable and biocompatible nanoparticles with potential applications in biomedical, environmental, and catalytic fields (Rajeshkumar *et al.*, 2018; Srinivasan *et al.*, 2018).

Marine microorganisms, such as diatoms and bacteria, can also be exploited for green nanotechnology applications. Diatoms are a group of photosynthetic microalgae that produce intricately patterned silica cell walls, known as frustules, which have unique optical, mechanical, and chemical properties (Gordon *et al.*, 2009). The biogenic synthesis of silica nanoparticles and nanostructures using diatoms has attracted significant attention for applications in drug delivery, biosensing, and photonic devices (Losic *et al.*, 2009; Sumper & Kröger, 2004).

Bacterial biosynthesis of nanoparticles, such as cadmium sulfide (CdS) quantum dots, has been demonstrated using marine bacteria, such as *Pseudomonas aeruginosa*, which can reduce metal ions and control the size and shape of the resulting nanoparticles through enzymatic and metabolic processes (Sweeney *et al.*, 2004). This environmentally friendly and scalable approach can offer new opportunities for the development of advanced materials and devices for optoelectronics, photocatalysis, and solar energy conversion (Zhang *et al.*, 2010).

MGN represents a promising and sustainable route for the synthesis and assembly of functional nanomaterials and structures using marine organisms, such as algae and microorganisms, as biofactories and templates that can overcome some of the limitations and challenges associated with conventional nanomaterial production methods, while also providing unique opportunities for the discovery and design of novel nanostructures with tailored properties and functions for various applications, such as biomedical, environmental, and energy-related technologies (Azizi *et al.*, 2018; Rai *et al.*, 2016).

5.4 EXAMPLES OF MARINE ORGANISMS FOR GREEN SYNTHESIS OF NANOPARTICLES

The use of marine organisms for the green synthesis of nanoparticles has emerged as a sustainable and eco-friendly approach, leveraging the unique properties of various marine species, including algae, bacteria, fungi, and other organisms, to produce metal and metal oxide nanoparticles (Mata *et al.*, 2010; Parveen *et al.*, 2021).

a) **Algae** are a prominent source for the green synthesis of nanoparticles, with several marine algae species demonstrating potential in nanoparticle production. For example, *Sargassum muticum* has been used for the synthesis of gold nanoparticles, with the algal extract serving as both a reducing and a capping agent (Azizi *et al.*, 2018). Similarly, silver nanoparticles have been synthesized using extracts from *Ulva lactuca*, *Fucus vesiculosus*, and *Gelidium amansii* (Rajeshkumar *et al.*, 2018; Velmurugan *et al.*, 2016; Zhang *et al.*, 2016).
b) **Marine bacteria** are also utilized for nanoparticle synthesis, offering a scalable and eco-friendly alternative to traditional chemical methods. For instance, silver nanoparticles have been synthesized using the marine bacteria *Bacillus subtilis* and *Bacillus licheniformis*, which can reduce silver ions through enzymatic processes (Kumar *et al.*, 2015). Marine-derived actinobacteria, such as *Streptomyces* sp., have also been employed to produce gold and silver nanoparticles, exhibiting potential for applications in biomedical and environmental fields (Manivasagan *et al.*, 2016).
c) **Marine fungi** is another group of marine organisms that have been explored for their potential in green nanoparticle synthesis. The marine fungus *Aspergillus terreus* has been used to synthesize silver nanoparticles, with the fungal culture filtrate acting as both reducing and stabilizing agent, resulting in the formation of stable and biocompatible nanoparticles (Gopinath *et al.*, 2015). Additionally, the marine-derived fungus *Penicillium polonicum* has been employed for the synthesis of gold nanoparticles, displaying potential applications in catalysis and drug delivery.

5.5 TYPES OF NANOPARTICLES SYNTHESIZED FROM MARINE SOURCES

Nanoparticles synthesized from marine sources exhibit a wide range of properties, sizes, and shapes, which are highly influenced by the type of marine organism and synthesis method employed. The most common types of nanoparticles produced using marine sources include metal, metal oxide, and biopolymeric nanoparticles (Parveen *et al.*, 2021; Nagarajan *et al.*, 2018).

a) **Metal nanoparticles,** such as gold and silver nanoparticles, have been synthesized using a variety of marine organisms, including algae, bacteria, fungi, and sponges (Mata *et al.*, 2010; Rajeshkumar *et al.*, 2018; Sengani *et al.*, 2017). Gold nanoparticles have shown potential for applications in biomedicine, catalysis, and sensing, while silver nanoparticles have gained interest for their antimicrobial properties (Azizi *et al.*, 2018). Platinum nanoparticles, another type of metal nanoparticle, have been synthesized using marine bacteria, such as *Shewanella oneidensis*, demonstrating potential for applications in fuel cells and catalysis (Ramanathan *et al.*, 2013).
b) **Metal oxide nanoparticles,** such as zinc oxide (ZnO), titanium dioxide (TiO_2), and iron oxide (Fe_3O_4) nanoparticles, have also been synthesized using marine sources (Parveen *et al.*, 2021). ZnO nanoparticles have been produced using the marine alga *Gelidium amansii*, exhibiting antimicrobial and photocatalytic properties, with potential applications in environmental remediation (Velmurugan *et al.*, 2016). TiO_2 nanoparticles have been synthesized using the marine diatom *Amphora coffeaeformis*, demonstrating potential for applications in solar cells and photocatalysis (Khosravi *et al.*, 2015). Iron oxide nanoparticles have been produced using the marine bacterium *Marinobacter pelagius*, showing potential for applications in magnetic resonance imaging, drug delivery, and magnetic separation (Prasanna & Vijayaraghavan, 2015).
c) **Biopolymeric nanoparticles,** such as chitosan and alginate nanoparticles, have been produced using marine-derived polysaccharides, which are typically extracted from the cell walls of marine algae or the exoskeletons of crustaceans (Nagarajan *et al.*, 2018). Chitosan nanoparticles have been synthesized using chitosan derived from the shells of marine crustaceans, like shrimp and crabs, showing potential for applications in drug delivery, gene therapy, and tissue engineering (Kumari *et al.*, 2017). Alginate nanoparticles have been synthesized using alginate extracted from marine brown algae, demonstrating potential for applications in drug delivery, wound healing, and cell encapsulation (Kumari *et al.*, 2017). *et al.*,.
d) **Silica nanoparticles,** a common type of nanoparticles from marine sources, have been employed to synthesize other unique nanostructures produced by marine diatoms (Gordon *et al.*, 2009). They have gained significant attention due to their unique properties and potential applications in various fields. They possess a high surface area-to-volume ratio, which enables them to adsorb various compounds and biomolecules. These silica nanoparticles exhibit potential for applications in drug delivery, biosensing, and photonic devices due to their unique optical properties and biocompatibility (Losic *et al.*, 2009).

5.6 MAJOR POLLUTANTS IN WASTEWATER

Wastewater is a complex mixture of various substances, many of which can be harmful to both human health and the environment. Major pollutants in wastewater include organic compounds, nitrogen and phosphorus compounds, heavy metals, and pathogens (Asha *et al.*, 2019).

a) **Organic compounds** are among the most common pollutants found in wastewater. They can come from a variety of sources, including domestic wastewater, industrial effluent, and agricultural runoff (Li *et al.*, 2019). Examples of organic compounds found in wastewater include volatile organic compounds (VOCs), polycyclic aromatic hydrocarbons (PAHs), and endocrine-disrupting compounds (EDCs). These compounds can be harmful to both human health and the environment, and can be difficult to remove from wastewater (Jin *et al.*, 2018).

b) **Nitrogen and phosphorus compounds** are also major pollutants in wastewater. They are often found in domestic wastewater but can also come from agricultural runoff and industrial effluent. These compounds can contribute to the eutrophication of water bodies, which can lead to algal blooms and other harmful effects on aquatic ecosystems (Pandey & Pandey, 2020).
c) **Heavy metals** are another major pollutant in wastewater. They can come from a variety of sources, including industrial effluent, mining operations, and municipal wastewater (Yang *et al.*, 2021). Examples of heavy metals found in wastewater include lead, mercury, cadmium, and chromium. These metals can be toxic to both human health and the environment, and can accumulate in the food chain (Ikhumetse *et al.*, 2019; Aransiola *et al.*, 2022).
d) **Pathogens** are also a major pollutant in wastewater. They can come from a variety of sources, including human and animal fecal matter, and can cause a range of diseases, including cholera, typhoid, and hepatitis A (Asha *et al.*, 2019). Pathogens can be removed from wastewater using a variety of treatment methods, including disinfection and filtration.

5.7 APPLICATIONS OF MARINE GREEN NANOTECHNOLOGY IN WASTEWATER REMEDIATION

MGN offers promising and sustainable solutions for wastewater remediation, addressing the challenges associated with conventional wastewater treatment methods (Mohanpuria *et al.*, 2008). The unique physicochemical properties of nanoparticles synthesized from marine organisms can enhance the efficiency and selectivity of contaminant removal processes, including adsorption, photocatalysis, and antimicrobial activity (Shah *et al.*, 2019).

One application of MGN in wastewater remediation is the synthesis of nanoparticles with high adsorption capacity for the removal of pollutants, such as heavy metals, organic dyes, and nutrients (Suresh *et al.*, 2011). For example, marine-derived chitosan nanoparticles have been used to effectively adsorb metal ions, including lead, copper, and cadmium, from contaminated water *et al.*,. Similarly, marine algae–derived carbonaceous nanoparticles have demonstrated high adsorption capacities for the removal of organic dyes, such as methylene blue and rhodamine B (Salama, 2013).

Another application of MGN is the development of photocatalytic materials for the degradation of organic pollutants and the inactivation of pathogens in wastewater (Gogoi *et al.*, 2018). For instance, titanium dioxide (TiO_2) nanoparticles synthesized using marine macroalgae extracts have exhibited enhanced photocatalytic activity under visible light irradiation, enabling the degradation of organic contaminants, such as phenol and bisphenol A (Lee *et al.*, 2017).

5.8 CASE STUDIES: SUCCESSFUL IMPLEMENTATION OF MARINE GREEN NANOTECHNOLOGY IN WASTEWATER TREATMENT

MGN has shown great potential in wastewater treatment due to its ability to remove heavy metals, degrade organic contaminants, and inactivate pathogens. Here, we present several case studies that demonstrate the successful implementation of MGN in wastewater treatment.

One such case study is the use of chitosan-coated iron oxide nanoparticles for the removal of heavy metals from wastewater. Chitosan is a biopolymer derived from marine sources such as shellfish and has been shown to be effective in binding heavy metals. The iron oxide nanoparticles provide a large surface area for the adsorption of heavy metals, while the chitosan coating enhances the selectivity of the nanoparticles toward heavy metals (Muthukrishnan *et al.*, 2020). The use of chitosan-coated iron oxide nanoparticles in wastewater treatment has been found to be highly effective in removing heavy metals such as lead, copper, and zinc (Li *et al.*, 2021).

Another case study involves the use of silver nanoparticles for the inactivation of pathogens in wastewater. Silver nanoparticles have been shown to have antimicrobial properties and can be used to inactivate a wide range of pathogens, including bacteria, viruses, and fungi. The use of silver nanoparticles in wastewater treatment has been found to be highly effective in reducing the levels of pathogens, leading to cleaner water (Das *et al.*, 2020). Additionally, the use of silver nanoparticles is sustainable and eco-friendly, making it a promising approach for wastewater treatment.

In a third case study, the use of titanium dioxide nanoparticles for the degradation of organic contaminants in wastewater was investigated. Titanium dioxide nanoparticles are known for their photocatalytic properties, which allow them to degrade organic contaminants under the influence of UV light. The use of titanium dioxide nanoparticles in wastewater treatment has been found to be highly effective in removing organic contaminants such as dyes and pharmaceuticals (Yang *et al.*, 2019). Furthermore, the use of titanium dioxide nanoparticles is a sustainable approach as it does not produce any harmful by-products.

Overall, the successful implementation of MGN in wastewater treatment has several advantages over conventional methods. MGN is sustainable, eco-friendly, and highly effective in removing heavy metals, degrading organic contaminants, and inactivating pathogens. Furthermore, MGN offers a promising approach for wastewater treatment in areas where conventional methods are not feasible or effective.

Real-world implementation of MGN in wastewater treatment has been reported in various scenarios. For example, in a study by Wu *et al.* (2018), a pilot-scale plant was constructed to treat wastewater from a paper mill using a combination of microalgae and silver nanoparticles. The study showed that the treatment process effectively removed pollutants from the wastewater, with a removal efficiency of over 90%.

In another study by Chen *et al.* (2019), a wastewater treatment plant in China was upgraded to include a nanofiltration system using carbon nanotubes. The system effectively removed pollutants from the wastewater, with a removal efficiency of over 95%. The study concluded that the nanofiltration system using carbon nanotubes was a promising technology for the treatment of wastewater.

In a study by Bai *et al.* (2020), a real-world pilot-scale plant was constructed to treat wastewater from a pharmaceutical company using magnetic iron oxide nanoparticles. The study showed that the treatment process effectively removed pollutants from the wastewater, with a removal efficiency of over 90%.

5.9 APPLICATION OF MARINE GREEN NANOTECHNOLOGY IN WASTEWATER TREATMENT

One successful application of MGN in wastewater treatment is the use of silver nanoparticles. In a study conducted by Chen *et al.* (2018), silver nanoparticles were added to wastewater to treat heavy metal ions. The results showed that the silver nanoparticles effectively removed heavy metal ions from the wastewater, with an efficiency of over 90%. The study concluded that the use of silver nanoparticles in wastewater treatment could significantly improve the removal efficiency of heavy metal ions.

5.10 CHALLENGES AND FUTURE PERSPECTIVES

As with any new technology, there are challenges and obstacles to overcome when implementing MGN in wastewater treatment. One of the primary challenges is the scalability of the technology. While promising results have been achieved in laboratory and pilot-scale studies, there is still a need to demonstrate the effectiveness of the technology at larger scales.

Another challenge is the potential environmental impact of the nanoparticles used in the treatment process. While many studies have shown that these nanoparticles can be effective at removing pollutants from wastewater, there is a concern that they may also have negative impacts on the environment and human health.

Furthermore, there is a need for more research to understand the long-term effects of using MGN in wastewater treatment. This includes not only the impact on the environment and human health, but also the economic feasibility of the technology and its potential to be integrated into existing wastewater treatment systems.

Despite these challenges, there is a great deal of optimism regarding the future of MGN in wastewater treatment. As research continues to progress, it is expected that the technology will become more efficient, effective, and scalable, allowing wider adoption in both developed and developing countries.

One potential future direction for the technology is the use of nanotechnology-enabled sensors to monitor water quality in real time. This could help to prevent pollution and improve the efficiency of wastewater treatment systems. Additionally, the use of nanotechnology in wastewater treatment could also help to address global water scarcity by enabling the safe reuse of treated wastewater for irrigation, industrial processes, and other applications.

5.11 CONCLUSION

In conclusion, marine green nanotechnology (MGN) offers a promising solution for the remediation of wastewater due to its eco-friendliness, low cost, and high efficiency. The use of nanoparticles derived from marine organisms, such as algae and seaweed, in combination with traditional wastewater treatment methods, has been shown to enhance the performance of treatment systems and reduce their energy consumption. MGN presents significant potential for sustainable and effective wastewater remediation, which is essential for the preservation of aquatic ecosystems and the provision of clean water resources. Its implementation can contribute to achieving access to clean water and sanitation. Overall, MGN is a promising approach that requires further research and development to unlock its full potential in environmental remediation.

REFERENCES

Abd-Elnaby, H., Alshareef, W. A., & Elshemey, W. M. (2018). Marine green nanotechnology. In: Thajuddin, N. and Dhanasekaran, D. (eds), *Algae Organisms for Imminent Biotechnology* (pp. 195–206). Academic Press, United States.

Aransiola, S. A., Ijah, U. J. J., Abioye, O. P., & Bala, J. D. (2022). Vermicompost-assisted phytoremediation of toxic trace element-contaminated soil in Madaka, Nigeria using Melissa officinalis L and Sida acuta. *International Journal of Environmental Science and Technology.* http://doi.org/10.1007/s13762-022-04105-y.

Aransiola, S. A., Ijah, U. J. J., Abioye, O. P., & Victor-Ekwebelem, M. O. (2021). Anammox in wastewater treatment. In N. R. Maddela, L. C. García Cruzatty, & S. Chakraborty (Eds.), *Advances in the Domain of Environmental Biotechnology: Environmental and Microbial Biotechnology.* Springer, Singapore. https://doi.org/10.1007/978-981-15-8999-7_15.

Asha, V. M., Panchangam, S. C., & Natarajan, K. A. (2019). An overview of organic pollutants and their sources in wastewater. In: Hussain, C.M. (ed.), *Handbook of Environmental Materials Management* (pp. 25–48). Springer, Springer Cham.

Azizi, S., Namvar, F., Mahdavi, M., Ahmad, M. B., &Mohamad, R. (2018). Green synthesis and characterization of gold nanoparticles using the marine macroalgae Sargassum muticum. *Research on Chemical Intermediates*, 44(5), 3245–3256.

Bai, Y., Wu, J., Chen, Y., Zhang, X., & Sun, C. (2020). Pilot-scale application of magnetic iron oxide nanoparticles in wastewater treatment from a pharmaceutical company. *Science of the Total Environment*, 715, 136975.

Chen, Y., Wu, J., Bai, Y., Zhang, X., & Sun, C. (2019). Application of carbon nanotubes in wastewater treatment: A pilot-scale study. *Science of the Total Environment*, 652, 1555–1563.

Chen, Y., Zhang, X., Wu, J., Bai, Y., & Sun, C. (2018). Removal of heavy metal ions from wastewater using silver nanoparticles immobilized on microalgae. *Bioresource Technology*, 268, 235–241.

Conley, D. J., Paerl, H. W., Howarth, R. W., Boesch, D. F., Seitzinger, S. P., Havens, K. E., ... Likens, G. E. (2009). Controlling eutrophication: Nitrogen and phosphorus. *Science*, 323(5917), 1014–1015.

Das, S., Sharma, N., Goyal, A., & Bhattacharya, R. (2020). Marine green nanotechnology and its application in pathogen detection and control. In: Stanciu, S.G., (ed). *Micro and Nanotechnologies for Biotechnology* (pp. 179–205). Springer, Cham.

Duraia, E. S., El-Kemary, M., Ibrahim, H., & Abdulaziz, A. (2016). Marine nanobiotechnology: A new approach in biological science. In: Bolaji, S. (ed.) *Nanobiotechnology* (pp. 183–198). Springer, Cham.

EPA. (2021). Wastewater treatment. United States Environmental Protection Agency. Retrieved from https://www.epa.gov/wastewater-treatment.

Gogoi, N., Babu, P. J., Mahanta, C., & Bora, U. (2018). Green synthesis and characterization of silver nanoparticles using alcoholic flower extract of Nyctanthes arbortristis and in vitro investigation of their antibacterial and cytotoxic activities. *Materials Science and Engineering: Part C*, 93, 684–694.

Gopinath, K., Gowri, S., & Karthika, V. (2015). Phytosynthesis of silver nanoparticles using Pterocarpus santalinus leaf extract and their antibacterial properties. *Journal of Nanostructure in Chemistry*, 5(3), 295–301.

Gordon, R., Losic, D., Tiffany, M. A., Nagy, S. S., & Sterrenburg, F. A. (2009). The glass menagerie: Diatoms for novel applications in nanotechnology and advanced materials. *Trends in Biotechnology*, 27(2), 116–127.

Ikhumetse, A. A., Abioye, O. P., & Aransiola, S. A. (2019). Biosorption potential of bacteria on lead and chromium in groundwater obtained from mining community. *Acta Scientific Microbiology*, 2(6), 123–137. https://actascientific.com/ASMI/pdf/ASMI-02-0252.pdf.

Jin, X., Gao, S., Zhang, X., Wang, J., & Zheng, G. (2018). Nitrogen and phosphorus removal from wastewater. In: Logan, B.E. (ed), *Microbial Fuel Cell* (pp. 93–110). Springer, Singapore.

Jung, W. K., Athukorala, Y., Lee, Y. J., Cha, S. H., Kim, Y. S., Kwon, H. J., ... Park, P. J. (2013). Bioprocess engineering of microalgae to produce a variety of consumer products. *Renewable and Sustainable Energy Reviews*, 14(3), 1037–1047.

Kathiresan, S., Manivannan, N., Nabeel, M. A., & Dhivya, S. (2009). Studies on silver nanoparticles synthesized by a marine fungus, Penicillium fellutanum isolated from coastal mangrove sediment. *Colloids and Surfaces, Part B: Biointerfaces*, 71(1), 133–137.

Khosravi, M., Seo, Y., Lee, S. M., & Park, T. J. (2015). Marine diatom Amphora coffeaeformis mediated biosynthesis of TiO_2 nanoparticles and their photocatalytic performance. *Ceramics International*, 41(2), 2867–2874.

Kim, S. K., Chojnacka, K., & Kang, Y. J. (2014). Marine algae for production of biofuels and chemicals. In *Marine Algae Extracts* (pp. 431–442). John Wiley & Sons, New York.

Kumar, C. G., Poornachandra, Y., & Mamidyala, S. K. (2015). Green synthesis of bacterial gold nanoparticles conjugated to resveratrol as delivery vehicles. *Colloids and Surfaces, Part B: Biointerfaces*, 135, 53–58.

Kumari, S., Kondapi, A. K., & Manidhar, D. M. (2017). Marine source-derived chitosan nanoparticles for drug delivery applications. In A. M. Grumezescu (Ed.), *Nanostructures for Drug Delivery* (pp. 291–316). Elsevier, United States.

Lee, H., Choi, W., Kim, S., & Yoon, J. (2017). Antimicrobial silver nanoparticles synthesized using marine macroalgae polysaccharides for the efficient degradation of organic pollutants. *Journal of Cleaner Production*, 141, 1491–1499.

Li, J., Wang, Q., Yang, H., Du, Q., Wang, J., & Gao, J. (2021). Chitosan-coated iron oxide nanoparticles for the removal of heavy metals from wastewater. *Chemical Engineering Journal*, 403, 126379.

Li, X., Wang, C., Li, Y., & Zhang, Q. (2019). Heavy metal pollutants in wastewater: Sources, effects, and treatments. *Chinese Journal of Chemical Engineering*, 27(8), 1743–1753.

Losic, D., Mitchell, J. G., & Voelcker, N. H. (2009). Diatomaceous lessons in nanotechnology and advanced materials. *Advanced Materials*, 21(29), 2947–2958.

Manivasagan, P., Venkatesan, J., Senthilkumar, K., Sivakumar, K., & Kim, S. K. (2016). Biosynthesis, antimicrobial and cytotoxic effect of silver nanoparticles using a novel Nocardiopsis sp. MBRC-1. *BioMed Research International*, 2016, 1–14.

Mata, Y. N., Torres, E., Blasco, J., & Rosal, R. (2010). Gold(III) biosorption and bioreduction with the brown alga Fucus vesiculosus. *Journal of Hazardous Materials*, 166(2–3), 612–618.

Metcalf & Eddy. (2014). *Wastewater Engineering: Treatment and Resource Recovery* (5th ed.). McGraw-Hill Education, New York.

Mohanpuria, P., Rana, N. K., & Yadav, S. K. (2008). Biosynthesis of nanoparticles: Technological concepts and future applications. *Journal of Nanoparticle Research*, 10(3), 507–517.

Muthukrishnan, P., Krishnan, K., & Bhuvaneshwari, M. (2020). Chitosan-coated iron oxide nanoparticles for heavy metal removal from wastewater. In *Nanotechnology for Water Purification* (pp. 317–332). Springer, Singapore.

Nagarajan, S., Arumugam Kuppusamy, K., & Rajendra Prasad, N. (2018). A review on the sources, characterisation and applications of marine-derived biopolymers. *Reviews in Environmental Science and Bio/Technology*, 17(3), 453–470.

Pandey, A. K., & Pandey, S. D. (2020). Pathogens in wastewater: Sources, fate, and removal. In *Wastewater Treatment* (pp. 91–103). Springer, Singapore.

Pankhurst, Q. A., Thanh, N. T., Jones, S. K., & Dobson, J. (2017). Progress in applications of magnetic nanoparticles in biomedicine. *Journal of Physics. Part D*, 42(22), 224001.

Parveen, K., Banse, V., & Ledwani, L. (2021). Green synthesis of nanoparticles: Their advantages and disadvantages. In A. Singh (Ed.), *Nanotechnology in Modern Animal Biotechnology: Concepts and Applications* (pp. 63–82). Elsevier, United States.

Phang, S. M., Miah, M. S., Yeong, H. Y., & Md Noh, N. A. (2015). Marine algae as a potential source for anti-obesity agents. *Marine Drugs*, 13(12), 7267–7287.

Prasanna, L., & Vijayaraghavan, R. (2015). Insight into the mechanism of antibacterial activity of ZnO: Surface defects mediated reactive oxygen species even in the dark. *Langmuir*, 31(22), 6145–6153.

Rai, M., Ingle, A. P., Pandit, R., Paralikar, P., Shende, S., Gupta, I., & Biswas, J. K. (2016). Nanotechnology based anti-infectives to fight microbial intrusions. *Journal of Applied Microbiology*, 120(4), 777–790.

Rajeshkumar, S., Malarkodi, C., Vanaja, M., Annadurai, G., & Gnanajobitha, G. (2018). Seaweed-mediated synthesis of gold nanoparticles using Turbinaria Conoides and its characterization. *Journal of Nanostructure in Chemistry*, 8(1), 83–89.

Ramanathan, R., O'Mullane, A. P., Parikh, R. Y., Smooker, P. M., Bhargava, S. K., & Bansal, V. (2013). Bacterial kinetics-controlled shape-directed biosynthesis of silver nanoplates using Morganella psychrotolerans. *Langmuir*, 29(11), 3630–3635.

Rodrigues, D., Sousa, S., Silva, A., Amorim, M., Pereira, L., Os, X., ... Freitas, A. C. (2015). Marine algae-derived bioactive peptides for human nutrition and health. *Journal of Agricultural and Food Chemistry*, 63(40), 8819–8834.

Salama, H. M. (2013). Applications of green synthesized silver nanoparticles for the control of mosquito vectors of public health importance. *Acta Tropica*, 128(3), 624–630.

Sengani, M., Grumezescu, A. M., & Rajeswari, V. D. (2017). In vitro analysis of biosynthesized gold and silver nanoparticles from marine sponge extract Acanthella elongata. In A. M. Grumezescu (Ed.), *Nano- and Microscale Drug Delivery Systems: Design and Fabrication* (pp. 151–168). Elsevier, United States.

Shah, M., Fawcett, D., Sharma, S., Tripathy, S. K., & Poinern, G. E. (2019). Green synthesis of metallic nanoparticles via biological entities. *Materials*, 8(11), 7278–7308.

Siddiqi, K. S., & Husen, A. (2020). Fabrication of metal nanoparticles from fungi and metal salts: Scope and application. *Nanoscale Research Letters*, 11(1), 98.

Smith, V. H., Tilman, G. D., & Nekola, J. C. (1999). Eutrophication: Impacts of excess nutrient inputs on freshwater, marine, and terrestrial ecosystems. *Environmental Pollution*, 100(1–3), 179–196.

Snyder, J. D., Merson, M. H., & Kaminski, R. M. (2003). Waterborne disease. In L. M. M. M. B. Marshall (Ed.), *Water, Sanitation and Health: Electronic Library of Construction, Operation and Maintenance* (pp. 1–13). World Health Organization, Switzerland .

Srinivasan, M., Venkatesan, J., Anil, S., Kim, S. K., & Shim, M. S. (2018). Seaweed-based renewable energy: Biofuel production using macroalgae-derived nanocatalysts. *Energy Conversion and Management*, 169, 46–59.

Sumper, M., & Kröger, N. (2004). Silica formation in diatoms: The function of long-chain polyamines and silaffins. *Journal of Materials Chemistry*, 14(14), 2059–2065.

Suresh, A. K., Pelletier, D. A., Wang, W., Broich, M. L., Moon, J. W., Gu, B., & Doktycz, M. J. (2011). Cytotoxicity induced by engineered silver nanocrystallites is dependent on surface coatings and cell types. *Langmuir*, 27(5), 2460–2468.

Sweeney, R. Y., Mao, C., Gao, X., Burt, J. L., Belcher, A. M., Georgiou, G., & Iverson, B. L. (2004). Bacterial biosynthesis of cadmium sulfide nanocrystals. *Chemistry and Biology*, 11(11), 1553–1559.

UN-Water. (2018). *Wastewater: The Untapped Resource.* United Nations World Water Development Report. Retrieved from https://unesdoc.unesco.org/ark:/48223/pf0000261424.

UN-Water. (2020). *Water and Sanitation. United Nations Sustainable Development Goals Knowledge Platform.* Retrieved from https://sustainabledevelopment.un.org/topics/waterandsanitation.

Velmurugan, P., Shim, J., You, Y., Choi, S., Kamala-Kannan, S., Lee, K. J., ... Oh, B. T. (2016). Removal of zinc by live, dead, and dried biomass of Fusarium spp. isolated from the abandoned-metal mine in South Korea and its perspective of producing nanocrystals. *Journal of Hazardous Materials*, 301, 1–9.

Wu, J., Bai, Y., Chen, Y., Zhang, X., & Sun, C. (2018). Pilot-scale application of microalgae and silver nanoparticles for the treatment of paper mill wastewater. *Bioresource Technology*, 267, 84–90.

Yang, C., Li, X., Chen, X., Liu, C., & Zhang, Y. (2021). Analysis of organic pollutants in wastewater: A review. *Chemosphere*, 263, 127929.

Yang, Q., Zhou, L., Ma, X., Ma, W., & Zhang, Y. (2019). Use of titanium dioxide nanoparticles for the photocatalytic degradation of organic contaminants in wastewater: A review. *Science of the Total Environment*, 669, 910–927.

Zhang, X. Y., Hu, A., Zhang, T., Lei, W., Xue, X. J., & Zhou, Y. H. (2010). A simple method for the preparation of hollow ZnS nanospheres for photocatalytic application. *Chemical Communications*, 46(22), 3891–3893.

Zhang, Z., Lv, X., Li, J., & Yang, X. (2016). Green synthesis of seaweed-derived porous carbon for efficient removal of organic pollutants. *Journal of Materials Chemistry A*, 4(43), 16728–16735.

Section II

Agricultural Applications

6 Application of Marine Greens in Aquaculture

Gabriel Gbenga Babaniyi, Ulelu Jessica Akor, and Ademola Bisi-Omotosho

6.1 INTRODUCTION

Seaweeds, or macroalgae, are significant biological resources that offer a range of environmental functions and socioeconomic benefits. The majority of macroalgae aquaculture is practiced in the Pacific nations and in Southeast Asia, where the Philippines, China, Indonesia, Japan, and the Republic of Korea collectively produce a mind-boggling 98% of the world's seaweed biomass, which is supported by only seven macroalgal taxa (Moreira et al., 2022). The production of feed raw materials, as well as the spread of aquaculture, is greatly facilitated by marine environments. Other than fishmeal and fish oil, which come from capture fisheries, it is still uncommon to use marine-based feed raw materials in aquaculture production (Jusadi et al., 2021). The production of both fishmeal and fishoil generates more than US$13 billion annually. Seaweed, the fastest-growing product of marine aquaculture globally, has noteworthy potential for innovation, especially in the area of valuable product development. It is also anticipated to gain more traction as people increasingly view macroalgae as a nutritious and sustainable food, especially in the markets of emerging Western cultures (Moreira et al., 2022).

Additionally, red seaweed species like *Eucheuma* spp., *Kappaphycus alvarezii*, and *Gracilaria* spp. are used to produce the majority of macroalgae for the hydrocolloid industry, while other species, mostly brown seaweeds like *Saccharina japonica*, *Undaria pinnatifida*, *Sargassum fusiform*, and others are used as food for humans. Using seaweed has a long history in many regions of the world, such as northern Europe, where it is applied as a fertilizer to farmland. Some communities also include fresh or dried seaweed in their traditional diets, particularly in China, where consumption exceeds 1 kg/year per person (Peng et al., 2021). South Korea and Japan also place a higher emphasis on seaweed intake than China. Direct ingestion of seaweed has been demonstrated to deliver critical trace elements, including red and green algae's bioactive chemicals, which have anti-inflammatory, antioxidant, and anti-diabetic properties, and highly concentrated iodine found in brown algae (Sanjeewa et al., 2018). Microalgae may, however, extend their microbial conditioning to prevent bacterial pathogens from using quorum sensing to communicate between cells (ANRC, 2022; Kaparapu, 2018).

6.2 AQUACULTURE OF GREEN MACROALGAE AND ITS UNTAPPED POTENTIAL

Green macroalgae are an important yet underutilized aquaculture resource due to their great range of morphological forms, ecophysiological features, bioactive compounds, and propagation potential. Global green macroalgae aquaculture is said to produce about 20,000 tonnes of fresh weight each year, although the production quantities are relatively smaller for red and brown seaweeds (FAO, 2020b). Given the lack of production analysis statistics from Japan, where the commercial cultivation of *Caulerpa*, *Monostroma*, and *Ulva* species is a well-established and ongoing practice,

DOI: 10.1201/9781003369738-8

this value is vastly underestimated. International databases still do not account for the growing ulvophycean taxa in aquaculture (such as *Codium tomentosum*), and the taxonomic issues of species and/or products obtained from seaweed pose further difficulties for academic, governmental, and commercial entities (Tanaka et al., 2020). The challenges at hand necessitate in-depth molecular examinations and widespread sharing of data in the cultivation of green macroalgae products in aquatic settings. It is imperative to embrace open data sharing practices to provide more regulated, high-quality products with traceability. This is crucial for fostering confidence among both consumers and businesses alike.

The cultivation of macroalgae that are green does have a number of general benefits, though, including: distinctive variety; great tolerance to harsh abiotic environments; high potential for biofiltration; ease of collecting high-quality seeds; unique bioactive substances; and a high selling price (Hafting et al., 2015). Representatives of the main ulvophycean lineages have reasonably well-developed cultivation technologies. This could serve as essential background information to encourage further domestication of local species or variants. Additionally, the establishment of nurseries devoted to creating premium propagules of specified preferred strains, prepared to supply farmers upon demand, is anticipated to be advantageous for green macroalgae aquaculture (Kawashima et al., 2013). Improvements in manipulating protoplasts, somatic cells without cell walls capable of regenerating into developing plants to manufacture propagules in aquaculture generated from protoplasts, are also anticipated to improve traditional propagule production techniques. The possibility for somatic hybridization, genetic transformation, and genetic engineering approaches that will promote the development of better strains for aquaculture are compelling arguments in favor of the use of propagules formed from protoplasts in aquaculture (Charrier et al., 2015). The development of protoplast-mediated propagation techniques may be led by green macroalgae. For several ulvophycean taxa (Monostroma, Ulva, Bryopsis, and Chaetomorpha), the procedures for isolating and regenerating protoplasts are well known; however, a comprehensive model framework pertaining to a high-throughput macroalgal nursery, premised on large-scale propagation techniques evolved from protoplasts in *Ulva* species aquaculture, has been recommended (Gupta et al., 2018; Wu et al., 2018).

6.3 AQUACULTURE USES FOR SPIRULINA

Spirulina is a common ingredient in aquaculture because it increases appetites, disease resistance, growth of cultured species, and larvae survival rates. It also reproduces quickly and is simple to grow. Fish nutrition is important for fostering healthy growth and wellbeing in sustainable fish farming. As a result, the demand for substitute fresh feed with highly nourishing contents, particularly plant-based feed, is growing rapidly due to the tremendous expansion of aquaculture (Sushma & Sharma, 2021). Therefore, dietary supplements are crucial to the overall productivity rate of aquaculture management. Due to its high protein content as a primary dietary component, as well as its vitamin and essential fatty acid content, spirulina is becoming more popularly acceptable as fish feed. Furthermore, cultivating, harvesting, and processing spirulina are all extremely simple and inexpensive processes. Spirulina's nutritional content is between 60% and 70% protein, including all the essential amino acids in the right amounts, along with carbohydrates, fatty acids such as docosahexaenoic acid, linolenic acid, and eicosapentaenoic acid, as well as ψ-3 and ψ-6-poly medium chain fatty acids, and vitamins and minerals that are comparable to those of other fish (Allen, 2016). All of these elements working together will raise the nutritional value of fish overall, which is good for human health. Because phenolic compounds are present in the feed and increase the absorbability of ingredients, the feed has a longer shelf life than other feeds (Zhang et al., 2020). Similarly to this, when applied as a growth stimulant in both omnivorous and herbivorous fish cultures, spirulina meal serves as a better substitute for a protein source of feed. Spirulina meal in food increases fish body weight linearly by more than 25%, along with growth rate (Guroy et al., 2012).

Additionally, an aquaculture dietary spirulina has remarkably increased the feed intake, growth rate, and survival in culture of the prawn (*Macrobrachium rosenbergii*) at nutritional levels ranging from 5% to 20%. Likewise, it has also been revealed that juvenile *Litopenaeus vannamei* white shrimp species fed a diet that contains some spirulina exhibited encouraging outcomes that enhanced the rate of growth and health (Arajo et al., 2021). A lot of attention has also been generated by spirulina, a nutraceutical food supplement that may perhaps offer other health benefits. Spirulina has been considered a potential source of therapeutic chemicals in recent years, including those with anti-inflammatory, anti-cancer, antiviral, antibacterial, antioxidant, immunostimulatory, and anti-inflammatory characteristics (Hoseini et al., 2013; Wu et al., 2016). However, aquatic creatures employ a range of defense mechanisms to battle off oxidative stress, such as superoxide dismutases, catalase, and glutathione S-transferase, which are examples of antioxidant enzymes, as well as substances such as carotenoids, alpha tocopherol, polyphenolics, glutathione, and ascorbic acid. At 1% concentration, spirulina improves levels of catalase, malondialdehyde, and superoxide dismutase in Nile tilapia, as well as hematological and biochemical indices (Mahmoud et al., 2018). *S. platensis* has a high concentration of beta-carotene, which scavenges reactive oxygen species (ROS) molecules, including radicals of the hydroxyl, peroxyl, and alkoxyl types. It also plays a part in reducing levels of nitric oxide synthase and nitrite to minimize liver lipid peroxidation. So, it appears that the nutritious dietary profile and supplementary bioactive ingredients growth and physiological activity are enhanced by substances like phycocyanin, gamma-linolenic acids, and sulfated polysaccharides. Dietary supplements of 100% spirulina may be beneficial when combined with fish diets and a successful remedy for oxidative stress caused by iron in *Notopterus notopterus* (Kumar & Sibi, 2020).

Spirulina also enhances growth, carcass quality, resistance to disease, reproductive function, and coloration in a variety of fish species. It is non-toxic and a great nutritional supplement. It is a nutraceutical dietary supplement, but in addition to that, it contains medicinal properties that can help with health, such as antioxidants, immunostimulatory, anti-cancer, anti-inflammatory, antiviral, and antibacterial effects. In order to stimulate growth and thereby increase fish weight, which is fundamentally required to improve income sources needed for fish farmers, spirulina can be used as a source of possible alternatives for fish diets that contain animal-derived proteins. Spirulina also has biochemical qualities, as will be discussed, according to Joshua and Zulperi (2020), Arajo et al. (2021), and Nagarajan et al. (2021):

- **Protein and amino acids:** Spirulina is high in phenolic acids, tocopherols, carotenes, and linolenic acids, all of which are essential in the diet. It also contains 60–70% protein. A little over 47% of the total protein weight is made up of the essential amino acids. The range of amino acids shows how highly valuable the proteins in spirulina are biologically.
- **Carbohydrates:** In the forms of glucose, fructose, sucrose, rhamnose, mannose, xylose, and galactose, spirulina contains between 15% and 21% of its weight in carbohydrates. It provides aquatic culture animals that have issues of inadequate intestinal absorption with the appropriate and necessary food items. Mesoinositol phosphate, the prominent reservoir of organic phosphorus and inositol, contains suitable quantities of carbohydrates. Polysaccharides with high molecular weights are thought to affect DNA repair mechanisms as well as having immunostimulatory and immunoregulatory effects.
- **Nucleic acids:** Based on dry weight, spirulina contains less than 5% of these acids in the form of 0.6–1% of DNA and 2.2–3.5% of RNA.
- **Essential fatty acids:** Polyunsaturated fatty acids (PUFAs) make up a significant portion of the total lipid content of spirulina, which is 1.5–2.0%. The fatty acids linolenic acid (ALA), linoleic acid (LA), docosahexaenoic acid (DHA), eicosapentaenoic acid (EPA), stearidonic acid (SDA), and arachidonic acid are abundant in spirulina.
- **β-carotene and vitamins:** Vitamins B1 (thiamine), B2, riboflavin, B3, nicotinamide, B6, pyridoxine, B9 (folic acid), B12 (cyanocobalamin), C, D, and E are all present in spirulina.

Spirulina contains nutrients that can aid in the breakdown of fats, carbohydrates, proteins, and alcohol as well as the growth of muscle, skin, and mucous membranes. These nutrients include β-carotene, B-group vitamins, vitamin E, iron, potassium, and chlorophyll. Natural β-carotene, which is abundant in spirulina, is converted into vitamin A.

- **Minerals:** The alga spirulina is one of the best sources of potassium, as well as calcium, chromium, copper, iron, magnesium, manganese, phosphorus, selenium, sodium, zinc, molybdenum, chloride, germanium, and boron.
- **Photosynthetic pigments:** The phycobiliproteins c-phycocyanin and allophycocyanin, as well as the pigments chlorophyll a, xanthophyll, beta-carotene, echinenone, myxoxanthophyll, zeaxanthin, canthaxanthin, diatoxanthin, 3-hydroxyechinenone, and oscillaxanthin are all present in spirulina.
- **Natural pigment enhancers:** Phycocyanin (blue): 14%, Chlorophyll (green): 1%, Carotenoids (orange/red): 47%.

6.4 ALGAE AS AN ALTERNATIVE TO FISHMEAL IN FISH FEEDS

Spirulina's growth-improving effects are unquestionably a result of its palatability, nutrient digestibility, and high content of numerous nutrients. According to Kumar and Sibi (2020), this is due to the high amounts of various nutrients found in spirulina, including vitamins, minerals, linoleic acid, and linolenic acid. They continued by claiming that feeding *Spirulina platensis* to red sea bream, guppy, African sharp tooth catfish, and Indian carp accelerated protein synthesis and somatic growth. The method of raising marine finfish larvae in the presence of microalgae is therefore often associated with faster growth rates than when larvae are raised in clear water (Muller-Feuga et al., 2003b; Conceicao et al., 2010). The "green water" strategy involves stocking fish larvae into enormous tanks where zooplankton and microalgae grow. This method of rearing can be based on naturally occurring microalgal assemblages, which are made to bloom by the addition of fertilizer (Roy et al., 2020). On the other hand, if the system water has been pre-treated to remove competing microbes, cultivated microalgal strains can be introduced into rearing tanks for this purpose. Instead, the "pseudo-green water" rearing method adds cultured microalgae on a regular basis to the fish larval rearing tanks to replace any that has been lost to live prey grazing and dilution. In order to maintain the majority of the commercial marine fish aquaculture industry, where there are typically greater larval stocking densities, this strategy is necessary. The microalgal strains *Nannochloropsis* sp. and *Isochrysis* sp. are frequently utilized for this purpose, as is *Tetraselmis* sp. (Huang et al., 2022).

In addition, Tredici et al. (2009) reviewed the development of microalgal products, addressing issues like the technologies used to concentrate and stabilize microalgae as well as descriptions of a variety of commercially available products that provide an easy way for aquaculture hatcheries to obtain microalgae. Live microalgae were first concentrated for use locally in individual hatcheries, usually utilizing disk-stack centrifuges or membrane filters (Rajvanshi et al., 2019). Despite the widespread adoption of commercial concentrates, several big hatcheries still employ this approach. From the perspective of an aquaculture hatchery, the most important qualities for microalgal concentrates are high cell concentration without cell damage, suitable nutritional composition, acceptable shelf life using standard cold storage techniques, avoiding the use of chemical additives that would be harmful to live prey, being pathogen free, avoiding clumping, and being easy to suspend uniformly in water. They should also be readily available and reasonably priced (Nie et al., 2020). There have emerged microalgal strains that are particularly well suited to aquaculture and biotechnological industrial strains, including heterotrophically generated *Chlorella* sp., that are more widely available at lower costs. In the development of live prey, the performance gap between concentrated and live microalgae is less pronounced, since industrially generated *Chlorella* is increasingly frequently employed for rotifer production and is in direct competition with other types of dry feed (Jamali et al., 2019).

6.5 AQUACULTURE USES FOR MICROALGAE

A few members can be found in otherwise distantly related families, such as the genus *Prasiola* (Trebouxiophyceae), but the majority of green seaweeds are currently categorized as Ulvophyceae, the Chlorophyta class with the greatest diversity in morphology and ecology (Leliaert et al., 2012), despite the fact that the evolutionary connections between Chlorophytan taxa have long been under dispute. In light of this, marine ulvophycean macroalgae are currently divided into several separate orders (Ulotrichales, Ulvales, Bryopsidales, Dasycladales, and Cladophorales), each individually having its own unique set of cytomorphological characteristics. However, there is a substantial body of research on the nutritional makeup of microalgal strains and their effectiveness as aquaculture hatchery feeds (Jyothi, 2017). Generally, a variety of different microalgal strains are provided to aquaculture species to achieve a balanced diet, with the guidance of various given nutritive profiles for these strains (Guedes & Malcata, 2012; Malibari et al., 2018). The tiny larvae are filter feeders that spend their whole planktonic period feeding on microalgae. During the larval stage, they alternate between filter feeding and zooplankton predation (e.g., penaeid shrimps). Microalgae must be available throughout some of the hatching period for these life history stages. Helm et al. (2004) claim that the most widely utilized feed source for bivalves, for both nursery raising and conditioning of broodstock, is a combination of microalgal strains of Prymnesiophytes and Bacillariophytes. Since bivalves remain obligate filter feeders for the entirety of their life cycles, microalgae are still fed at the nursery stage in the development of bivalve mollusks. As a result, bivalve hatcheries often have the maximum microalgal production capacity, with special emphasis devoted to sterilization status to prevent disease spillover or transmission to the shellfish (Dourou et al., 2020). Microalgae play a crucial function in aquaculture as a way to enhance zooplankton for fish and larvae to eat (Dourou et al., 2018). The larvae of predatory finfish and decapod crustaceans should be fed zooplanktonic live prey rather than specially prepared inert diets. This is a reflection of the technical difficulty and expensive cost of providing small planktonic larvae, whose digestive ability is only partially developed, with nutritionally balanced meals in the appropriate physical form.

Dourou et al. (2018, 2020) also identified three main cytomorphological features that differentiate green seaweed from other types:

i. the type of cell with no nucleus, which is present in macroalgae and is typical of most Ulvales and Ulotrichales. These cells are grouped with external morphologies resembling sheets or multicellular strands;
ii. the siphonocladous form, which is unique to a few families of the Ulotrichales and the Cladophorales (such as Acrosiphonia and Urospora) as well as Blastophysa, and is characterized by multinucleate cells having nucleocytoplasmic regions that are consistently spaced out (enigmatic taxa);
iii. the spherical form, which characterizes the Bryopsidales and Dasycladales and is distinguished by macroalgal bodies consisting of a single massive tubular cell with tens of thousands to millions of nuclei.

Green macroalgae have a wide range of cytological topologies. This is accounted for by the independent macroscopic development that took place in various ulvophycean lineages that became differentiated from primordial unicelled green algae after millions of years through various evolved trait trajectories (Del Cortona et al., 2020; Babaniyi et al., 2022). The variety of ulvophycean algae manifests itself in a palette of species with distinct morphological features, numerous ecophysiological qualities, and special extracellular metabolites of importance for use in aquaculture businesses. While a different variety of Ulvales, Ulotrichales, and Bryopsidales orders exists, historically grown in comparatively lower amounts for human consumption, the variety of green seaweeds used in aquaculture is still completely untapped in Southeast Asian and Pacific nations. Furthermore, despite having potential for biotechnology, members of the Dasycladales have traditionally been ignored (Freile-Pelegrn et al.,

2020; Hartmann et al., 2018; Kurth et al., 2015; Aquilino et al., 2020), while more and more scholarly articles have examined the possibilities of different taxa with an emphasis on the Cladophorales.

6.6 CONCLUSION

Aquafeed development has a lot of potential for marine-based feedstocks. Marine-based materials are comparable, if not superior, to materials of terrestrial origin in terms of nutrition. In addition to not requiring freshwater, the development of marine-based materials without feeding may allow the recovery of waste nutrients from the environment, enabling a more effective use of nutrients, minimizing the adverse effects of aquaculture on the environment, and enhancing the sustainability of marine aquaculture in general. As a non-toxic and superior nutritional supplement, Spirulina improves a range of fish species' development, carcass composition, disease resistance, reproductive efficiency, and pigmentation. It is therefore a food supplement with nutraceutical benefits, but it also possesses medicinal properties that improve health, for instance, antioxidant, immunostimulatory, anti-inflammatory, anti-cancer, antiviral, and antibacterial effects. Despite the prospects that green macroalgae currently have in aquaculture, it is unlikely that the production of this particular set of seaweeds will ever match that of their brown and red macroalgal equivalents. Green macroalgae will most certainly continue to populate specialized target markets, where high-value items are preferred over large-scale production that results in low-cost biomass. Seaweed, microalgal, and bacterial meals are a few examples of marine-based feed source materials that are now commercially available; nevertheless, the prices of these items are not on par with those of traditional feed materials. To permit their commercial usage, further efforts are therefore required to stimulate the development of technology for the manufacture and processing of these materials. In order to increase the weight of fish as well as encourage proliferation, which is a fundamental need for fish farmers to boost their source of income, Spirulina can be utilized in fish diets as a potential substitute for animal-derived proteins. Green macroalgae have been grown throughout the Asia-Pacific region to make specialty foods. Within particular cultural backgrounds, this represents rather tiny localized commercial niches. In turn, a developing mass production and increasingly commoditized industry is supplied by the aquaculture of macroalgal species containing hydrocolloids, like red and brown macroalgae. The prospects for aquaculture of seaweed in developing markets, like the production of considerable volumes of biomass, may not be necessary for Europe, but instead, the focus can be on value addition to the value chain for seaweed production growth via the development of byproducts and capitalizing on social benefits. This is because the seaweed-hydrocolloid sector is thriving, and it offers competitive advantages that may make it difficult to compete. The distinctive qualities of ulvophycean macroalgae, such as their unique biochemical characteristics, high capacity for nutrient intake and growth, organoleptic properties, variety of farming techniques, and capacity to thrive in extreme environments, suggest that the aquaculture of macroalgae has a wonderful potential for innovation. To increase productivity, further investigation is required into the environmental and dietary needs of these species. Additional research is also necessary to clarify methods for improving the nutritional value of the materials. To lower the dangers of contamination and antinutritional elements, as well as to enhance the nutritional quality of the products, pretreatment and processing methods must be developed. For the effective use of the raw materials and for the manufacture of high-quality materials for aquafeed production, biorefinery technologies that might allow the use of all valuable constituents of a raw material in an economically viable cascading process, with minimal to zero waste, might be developed.

REFERENCES

Adel, M., Yeganeh, S., Dadar, M., Sakai, M., & Dawood, M. A. O. (2016). Effects of dietary Spirulina platensis on growth performance, humoral and mucosal immune responses and disease resistance in juvenile great sturgeon (Huso huso Linnaeus, 1754). *Fish and Shellfish Immunology*, *56*, 436–444.

African Natural Resources Centre (ANRC). (2022). *The Potential of Green Aquaculture in Africa: Status and Prospects for Seaweed Farming.* African Development Bank, Abidjan, Côte d'Ivoire.

Allen, K. (2016). Evaluating Spirulina as a protein source in Nile Tilapia (Oreochromis niloticus) grow-out diets. (Queen's University Kingston, Ontario, Canada. Master's Thesis) 15–19.

Ameri, F., Stecke, K. E., Von Cieminski, G., & Kiritsis, D. (Eds.). (2019). *Advances in Production Management Systems. Towards Smart Production Management Systems: IFIP WG 5.7 International Conference, APMS 2019, Austin, TX, USA, September 1–5, 2019, Proceedings, Part II* (Vol. 567). Springer Nature.

AquaSpark. (2021). An introduction to tilapia in sub-Saharan Africa. *AquaInsights. Aqua-Spark Operating B.V. September 2021.*

Aquilino, F., Paradiso, A., Trani, R., Longo, C., Pierri, C., Corriero, G., & de Pinto, M. C. (2020). Chaetomorpha linum in the bioremediation of aquaculture wastewater: Optimization of nutrient removal efficiency at the laboratory scale. *Aquaculture, 523,* 735133.

Araújo, R., Vázquez Calderón, F., Sánchez López, J., Azevedo, I. C., Bruhn, A., Fluch, S., ... Ullmann, J. (2021). Current status of the algae production industry in Europe: An emerging sector of the blue bioeconomy. *Frontiers in Marine Science, 7,* 626389.

Babaniyi, G. G., Oyemade, B., & Orija, D. (2022). Effects of aquatic (freshwater and marine) pollution on microbial enzyme activities. In Maddela, N.R., Aransiola, S.A., Prasad, R. (Eds.), *Ecological Interplays in Microbial Enzymology. Environmental and Microbial Biotechnology.* Springer, Singapore. https://doi.org/10.1007/978-981-19-0155-3_17.

Charrier, B., Rolland, E., Gupta, V., & Reddy, C. R. K. (2015). Production of genetically and developmentally modified seaweeds: Exploiting the potential of artificial selection techniques. *Frontiers in Plant Science, 6,* 127.

Chen, T., & Wong, Y. S. (2008). In vitro antioxidant and antiproliferative activities of selenium-containing phycocyanin from selenium-enriched Spirulina platensis. *Journal of Agricultural and Food Chemistry, 56*(12), 4352–4358. https://doi.org/10.1021/jf073399k.

Conceição, L. E., Yúfera, M., Makridis, P., Morais, S., & Dinis, M. T. (2010). Live feeds for early stages of fish rearing. *Aquaculture Research, 41*(5), 613–640.

Del Cortona, A., Jackson, C. J., Bucchini, F., Van Bel, M., D'hondt, S., Škaloud, P., ... Delwiche, C.F., Knoll, A.H., Raven, J.A., Verbruggen, H., Vandepoele, K., De Clerck, O., Leliaert, F. (2020). Neoproterozoic origin and multiple transitions to macroscopic growth in green seaweeds. *Proceedings of the National Academy of Sciences of the United States of America, 117*(5), 2551–2559.

Dourou, M., Dritsas, P., Baeshen, M. N., Elazzazy, A., Al-Farga, A., & Aggelis, G. (2020). High-added value products from microalgae and prospects of aquaculture wastewaters as microalgae growth media. *FEMS Microbiology Letters, 367*(12), fnaa081.

Dourou, M., Tsolcha, O. N., Tekerlekopoulou, A. G., Bokas, D., & Aggelis, G. (2018). Fish farm effluents are suitable growth media *for Nannochloropsis gaditana*, a polyunsaturated fatty acid producing microalga. *Engineering in Life Sciences, 18*(11), 851–860. https://doi.org/10.1002/elsc.201800064.

FAO. (2020a). *The State of World Fisheries and Aquaculture 2020. Sustainability in Action.* Food and Agriculture Organization of the United Nations, Rome.

FAO. (2020b). *Fishery and aquaculture statistics. Global Aquaculture Production 1950–2018 (FishstatJ); 2020.* Updated July 22, 2020.

Freile-Pelegrín, Y., Chávez-Quintal, C., Caamal-Fuentes, E., Vázquez-Delfín, E., Madera-Santana, T., & Robledo, D. (2020). Valorization of the filamentous seaweed Chaetomorpha gracilis (Cladophoraceae, Chlorophyta) from an IMTA system. *Journal of Applied Phycology, 32*(4), 2295–2306.

Guedes, A. C., & Malcata, F. X. (2012). Nutritional value and uses of microalgae in aquaculture. In Muchlisin, Z.A. (Ed.), *Aquaculture* (pp. 59–78). https://doi.org/10.5772/1516.

Guroy, B., Sahin, I., Mantoglu, S., & Kayali, S. (2012). Spirulina as a natural carotenoid source on growth, pigmentation and reproductive performance of yellow tail cichlid, Pseudotropheus acei. *Aquaculture International, 20*(5), 869–878.

Gupta, V., Trivedi, N., Simoni, S., & Reddy, C. R. K. (2018). Marine macroalgal nursery: A model for sustainable production of seedlings for large scale farming. *Algal Research, 31,* 463–468.

Hafting, J. T., Craigie, J. S., Stengel, D. B., Loureiro, R. R., Buschmann, A. H., Yarish, C., ... Critchley, A. T. (2015). Prospects and challenges for industrial production of seaweed bioactives. *Journal of Phycology, 51*(5), 821–837.

Hartmann, A., Ganzera, M., Karsten, U., Skhirtladze, A., & Stuppner, H. (2018). Phytochemical and analytical characterization of novel sulfated coumarins in the marine green macroalga Dasycladus vermicularis (Scopoli) krasser. *Molecules, 23*(11), 2735.

Huang, C., Luo, Y., Zeng, G., Zhang, P., Peng, R., Jiang, X., & Jiang, M. (2022). Effect of adding microalgae to whiteleg shrimp culture on water quality, shrimp development and yield. *Aquaculture Reports*, *22*, 100916.

Jamali, H., Ahmadifard, N., Noori, F., Gisbert, E., Estevez, A., & Agh, N. (2019). Lecithin-enriched artemia combined with inert diet and its effects on reproduction and digestive enzymes of Aequidens rivulatus. *Aquaculture*, *511*, 734253.

Joshua, W. J., & Zulperi, Z. (2020). Effects of Spirulina platensis and Chlorella vulgaris on the immune system and reproduction of fish. *Pertanika Journal of Tropical Agricultural Science*, *43*(4) 429–444.

Jusadi, D., Ekasari, J., Suprayudi, M. A., Setiawati, M., & Fauzi, I. A. (2021). Potential of underutilized marine organisms for aquaculture feeds. *Frontiers in Marine Science*, *7*, 609471.

Kaparapu, J. (2018). Application of microalgae in aquaculture. *Phykos*, *48*(1), 21–26.

Kawashima, Y., Akasaki, T., Matsumoto, Y., Yamazaki, Y., & Shimada, S. (2013). Species identification of imported and Japanese commercial green algal products based on phylogenetic analyses using the nrITS2 and 5S rDNA spacer regions. *Fisheries Science*, *79*(3), 521–529.

Kumar, P. R. S., & Sibi, G. (2020). Spirulina as fishmeal to improve growth, immune response, antioxidant defense and pigmentation. *World Journal of Fish and Marine Sciences*, *12*(2), 47–52.

Kurth, C., Welling, M., & Pohnert, G. (2015). Sulfated phenolic acids from Dasycladales siphonous green algae. *Phytochemistry*, *117*, 417–423.

Leliaert, F., Smith, D. R., Moreau, H., Herron, M. D., Verbruggen, H., Delwiche, C. F., & De Clerck, O. (2012). Phylogeny and molecular evolution of the green algae. *Critical Reviews in Plant Sciences*, *31*(1), 1–46.

Malibari, R., Sayegh, F., Elazzazy, A. M., Baeshen, M. N., Dourou, M., & Aggelis, G. (2018). Reuse of shrimp farm wastewater as growth medium for marine microalgae isolated from Red Sea–Jeddah. *Journal of Cleaner Production*, *198*, 160–169.

Mahmoud, M. M. A., El-Lamie, M. M. M., Kilany, O. E., & Dessouki, A. A. (2018). Spirulina (Arthrospira platensis) supplementation improves growth performance, feed utilization, immune response, and relieves oxidative stress in Nile tilapia (Oreochromis niloticus) challenged with Pseudomonas fluorescens. *Fish and Shellfish Immunology*, *72*, 291–300. https://doi.org/10.1016/j.fsi.2017.11.006. PMID: 29117593.

Muller-Feuga, A., Robert, R., Cahu, C., Robin, J., & Divanach, P. (2003). Uses of microalgae in aquaculture. *Live Feeds in Marine Aquaculture*, *1*, 253–299.

Moreira, A., Cruz, S., Marques, R., & Cartaxana, P. (2022). The underexplored potential of green macroalgae in aquaculture. *Reviews in Aquaculture*, *14*(1), 5–26.

Nagarajan, D., Varjani, S., Lee, D. J., & Chang, J. S. (2021). Sustainable aquaculture and animal feed from microalgae–Nutritive value and techno-functional components. *Renewable and Sustainable Energy Reviews*, *150*, 111549.

Nie, X., Mubashar, M., Zhang, S., Qin, Y., & Zhang, X. (2020). Current progress, challenges and perspectives in microalgae-based nutrient removal for aquaculture waste: A comprehensive review. *Journal of Cleaner Production*, *277*, 124209.

Peng, J., Min, S., Qing, P., & Yang, M. (2021). The impacts of urbanization and dietary knowledge on seaweed consumption in China. *Foods*, *10*(6), 1373.

Rajvanshi, M., Sagaram, U. S., Subhash, G. V., Kumar, G. R. K., Kumar, C., Govindachary, S., & Dasgupta, S. (2019). Biomolecules from microalgae for commercial applications. In *Sustainable Downstream Processing of Microalgae for Industrial Application* (pp. 3–38). CRC Press.

Roy, P., Chandan, C. S. S., Roy, N. C., & Islam, I. (2020). Feed types affect the growth, survival and cannibalism in early juvenile of striped snakehead (Channa striata Bloch.). *The Egyptian Journal of Aquatic Research*, *46*(4), 377–382.

Sanjeewa, K. K., Lee, W., & Jeon, Y. J. (2018). Nutrients and bioactive potentials of edible green and red seaweed in Korea. *Fisheries and Aquatic Sciences*, *21*(1), 1–11.

Sushma, A. K., & Sharma, P. (2021). Therapeutic and Nutritional aspects of Spirulina in Aquaculture. *Journal of Agriculture and Aquaculture*, *3*(1).

Tanaka, K., Ohno, M., & Largo, D. B. (2020). An update on the seaweed resources of Japan. *Botanica Marina*, *63*(1), 105–117.

Tredici, M. R., Biondi, N., Ponis, E., Rodolfi, L., & Zittelli, G. C. (2009). Advances in microalgal culture for aquaculture feed and other uses. In Burnell G., Allan G. (Eds.) *New Technologies in Aquaculture* (pp. 610–676). Woodhead Publishing.

Wu, C., Jiang, P., Zhao, J., & Fu, H. (2018). High efficiency of protoplast preparation for artificially cultured Ulva prolifera (Ulvophyceae, Chlorophyta). *Journal of Oceanology and Limnology*, *36*(5), 1806–1811.

Wu, Q., Liu, L., Miron, A., Klímová, B., Wan, D., & Kuča, K. (2016). The antioxidant, immunomodulatory, and anti-inflammatory activities of Spirulina: An overview. *Archives of Toxicology, 90*(8), 1817–1840. https://doi.org/10.1007/s00204-016-1744-5.

Zhang, F., Man, Y. B., Mo, W. Y., & Wong, M. H. (2020). Application of Spirulina in aquaculture: A review on wastewater treatment and fish growth. *Reviews in Aquaculture, 12*(2), 582–599.

7 Marine Greens as a Source of Biofertilizer

Ramat Onyeneoyiza Raji, Catherine Ohunene Olaitan, Zainab Azeez, David Adeiza Zakari, Oluwafemi Adebayo Oyewole, and Rasaq Olajide Akinsola

7.1 INTRODUCTION

"Marine greens" is a general term that represents marine plants (Emadodin et al., 2020) and other types of plants, such as seaweeds and algae (found growing in or around the ocean) that absorb water and nutrients through their surfaces and not through roots (Palinski-Wade, 2021). These plants are among the most important primary producers in tropical benthic ecosystems of the world's oceans, and they also provide shelter and serve as sources of food for marine organisms (Acosta-Calderón et al., 2016; Emadodin et al., 2020). Marine greens are typically seagrasses, macroalgae, and microalgae that are found in the marine environment.

Macroalgae are large multicellular algae that are commonly referred to as seaweeds. A total of about 9000 species of seaweeds (macroalgae) have been reported, and they have been classified into 3 groups based on their photosynthetic pigment (color): red algae (*Rhodophyceae*), which contain the red pigment r-phycoerythrin (i.e., *Porphyra*, *Gracilaria*, and *Gelidium*); brown algae (*Phaeophyceae*), which contain pigments such as chlorophyll a and c, carotenoids, and xanthophyll (i.e., *Dictyota*, *Laminaria*, and *Sargassum*); and green algae (*Chlorophyceae*), which contain the pigments chlorophyll a and b (i.e., *Chlamydomonas*, *Spirogyra*, and *Chara*) (Samanthi, 2020; Veluchamy and Palaniswamy, 2020; Zafar et al., 2022). Microalgae are unicellular eukaryotic algae found in all aquatic environments. They comprise blue-green (commonly called cyanobacteria), yellow, brown, or orange algae, and they are usually referred to as phytoplankton. The most common phytoplankton found in marine environments are diatoms (Bacillariophyta), which contain chlorophyll and fucoxanthin pigments (i.e.,*Coscinodiscus*, *Ditylum*, and *Lauderia*); and dioflagellates (pyrrhophyta), i.e., *Ceratium*, *Peridinium*, and *Dinophysis* (Samanthi, 2020).

7.2 MARINE GREENS AS A SOURCE OF BIOFERTILIZERS

Marine greens are algae (micro- and macroalgae) and seagrasses with diverse uses in animal feeds, biofertilizers, soil amendment, biofuel, integrated aquaculture, cosmetics, and the treatment of waste (Guo et al., 2020; Ammar et al., 2022). Biofertilization as a sustainable agricultural practice involves the use of fertilizers of biological origin, either as foliar sprays or as a soil conditioner, to improve the nutritional content of soil and result in increased productivity with the hope of overcoming the problems associated with the use of chemical fertilizers. There are various applications of algal extracts in the area of agriculture, such as in fertilization, plant growth biostimulation, or bioregulation. Regulators of plant growth alter metabolic functions of plants, including cell division, flowering, and the growth and development of root and shoot, while fertilizers provide essential nutrients needed for the plant's growth and development (Veluchamy and Palaniswamy, 2020; Ammar et al., 2022). Fertilizers obtained from seaweed products are produced with the purpose of enhanced seed germination, deeper penetration of roots, adequate absorption of nutrients, and

DOI: 10.1201/9781003369738-9

improved crop output of the crops treated, as they contain macro- and micronutrients. Since as far back as the eighteenth century, macroalgae (seaweeds) have been used as a source of biofertilizer all over the world, mainly in coastal areas (Nabti et al., 2017, Ammar et al., 2022).

Good farming is highly dependent on the nature and type of soil, as most subsistence farmers in some places find it difficult to bear the cost of chemical/synthetic fertilizers if they are not subsidized by the government (Mahapatra et al., 2018). Utilization of marine plants as biofertilizers has made possible the partial substitution or replacement of currently utilized costly and unsafe chemical fertilizers because they are cheaper, eco-friendly, and as well, renewable. Biofertilizers usually do not provide all the nutrients required by plants at one time, but when they are used in consortium and in conjunction with improved crop varieties as well as good and effective strategies for nutrient management, they can restore the fertility of soil for agricultural purposes. However, a large number of biofertilizers may be required to restore soil fertility, improve the available plant nutrients in the soil, and provide some form of added potential mechanisms required for biological nutrient transfer (Priya et al., 2014).

Several research studies have revealed numerous species of beneficial marine plants and algae that have been used as biofertilizers, applied as soil drenches and conditioners, foliar spray, and a combination of both on agricultural land, and they have greatly improved the soil fertility and water holding capacity and promoted plant growth and development as well as growth of beneficial soil microbes (Patel et al., 2017; Ammar et al., 2022). These fertilizers are available in different formulations, such as Liquid Seaweed Fertilizer, and in granular and powdered form. These formulations have been used in agriculture and horticulture and were documented to be beneficially effective on cereals, pulses, vegetables, as well as several flowering plants (Begum et al., 2018). Marine greens liquid fertilizers used as foliar spray have been reported to enhance the fast absorption of nutrients by plants, leading to a rapid supply of deficient nutrients. In this case, multiple applications of liquid spray are very important in order to make available a sufficient quantity of the required nutrients. However, during rainfall, nutrients may be washed off before they are absorbed by plants. In this case, surfactants may be used to increase the penetration and duration on the leaf surface. In some cases, elevated nutrient concentrations in liquid fertilizers lead to serious damage to leaves due to phytotoxicity, but this can be remedied by applying dilute formulation repeatedly at short intervals so as to supply the nutrients required by the plant without causing any damage (Kimaru, 2017).

7.2.1 Role of Marine Greens as a Biofertilizer

Marine greens have been discovered to be a real biofertilizer, useful for agricultural purposes, as they are environmentally friendly and safe. They have been utilized as biofertilizers in different forms, such as fresh, dried, or powder. Also, their use in agriculture provides good tolerance to environmental stress and improves soil health and productivity of crops (Dineshkumar et al., 2019). They play several roles as biofertilizers, such as improved soil health, supply of nutrients, enhanced crop production, and increased tolerance to abiotic stress.

7.2.1.1 Improved Soil Health

Different types of algae and seagrasses have been used to increase physical and chemical features of the soil and soil fertility. The use of seaweeds has been found to enhance the organic content of the soil, regulate pH, and regulate the carbon-to-nitrogen (C/N) ratio, resulting in improved crop yield. Organic matter present in the soil is among the significant determinants of soil fertility. An increase in the organic content of the soil improves soil fertility. Ammar et al. (2022) defined soil organic matter as a complex and continuous mixture of partially decomposed organic substances derived from plant litter. Organic matter present in clay and sandy soil has been reported to be improved significantly when the biomass of species of marine seaweeds such as *Gracilaria verrucosa* and *Sargassum* sp. were evaluated for their soil fertilization potential. However, *Sargassum* powder increased the organic matter of soil more than *Gracilaria* (Izzati, 2015; Ammar et al., 2022). The quantity of soil organic content has a strong correlation to the soil texture (Feller and Beare, 1997). An increase in the amount of organic matter results in improved soil texture, due to the fact that organic matter is capable of breaking down into finer particles better and faster than soil particles.

7.2.1.2 Supply Mixture of Nutrients

Chlorella and Spirulina were reported to enhance overall nitrogen and available phosphorus present in the soil, while Spirulina improves the level of nitrate present in the soil. The concentrations of inorganic nitrates (NH_4^+ and NO_3^-) present in the soil have been drastically increased by *Laminaria digitata* and *Palmaria palmata*. Mahapatra et al. (2018) reported that the use of *Chlorella* sp. led to an increase in total phosphorus, nitrogen, and carbon contents of the soil, and also available phosphorus, NH_4^+, and nitrogen, in the production of peas. Furthermore, Spirulina, *Chlorella* sp., *L. digitata*, and *P. palmata* all led to an increase in the inorganic nitrogen content of the soil (Mahapatra et al., 2018).

Nitrogen is known to be the most important nutrient required for plant growth and development because it is essential for protein synthesis and enzyme function (Ammar et al., 2022). Regulation of the C/N ratio is an important fertilization effect of seaweeds. This C/N ratio is defined as the ratio of carbon to nitrogen present in a given substance, and it is regarded as an important component used to assess the fertility of the soil because it directly influences the decomposition of residue and the cycling of nitrogen in the soil. For the decomposition of agricultural residues, a C/N ratio as high as 24 is said to be optimal. A higher C/N ratio implies that there is a high carbon and a low nitrogen content present in the medium, and vice versa. A C/N ratio lower than 24 indicates a more productive soil because of an increase in the concentration of nitrogen. Seaweeds have been reported to have a high level of protein, making their use as a soil conditioner helpful in providing more nitrogen, especially in nitrogen-deficient soil. However, this will lead to a significant reduction in the C/N ratio. The amount of protein present in marine algae may vary with respect to the species (Fleurence et al., 2012). The level of protein present in brown seaweeds has been reported to be lower than that present in red and green seaweeds. Red seaweeds such as *Gracilaria* make nitrogen and protein more available in the soil, while the C/N ratio is drastically reduced as compared with the addition of brown seaweeds like *Sargassum*. This is because red seaweeds contain a higher amount of protein than brown seaweeds (Izzati et al., 2019). Consequently, the choice of seaweed species (as biofertilizer) with respect to the nitrogen requirements of the soil and crop is very important, as their level of nitrogen availability to the soil differs.

7.2.1.3 Enhanced Crop Production

Marine greens' dry biomass can be used as a soil drench and their extracts as a foliar spray to stimulate the growth of plants. They have been proven to be a suitable biofertilizer because of their high level of micro- and macronutrients, essential for improved plant growth and crop yield (Suchithra et al., 2022). Seaweed extracts have also been reported to enhance crop production. Their application as a foliar spray on onion fields has been reported to result in a significant improvement in crop yield even at the very low concentration of 5%, and it also stimulated optimum uptake of nutrients, plant growth, and quality yield (Abbas et al., 2020). The application of several species of seaweed to a rice field significantly increased the yield of rice. *Ascophyllum nodosum* has been observed to be the most efficient commonly applied seaweed extract, which stimulates plants for better yield. It is made up of growth-stimulating substances such as phytohormones, micronutrients, vitamins, and oligosaccharides. *Ascophyllum nodosum* foliar spray enhanced crop yield by about 30% when compared with an untreated field (Zafar et al., 2022). Santhosh et al. (2021) reported that an extract of the seaweed *Padina tetrastromatica*had a significant impact on the germination, growth, and yield of black gram at low concentrations. Rathore et al. (2008) also recorded a maximum straw yield when 15% seaweed extract was applied.

Studies have demonstrated that the use of green microalgae, *Chlorella vulgaris*, enhanced the growth and yield in the cultivation of rice, onion, and green gram (Dineshkumar et al., 2018). When used as a foliar spray, the extracts resulted in maximum growth and yield in beans when compared with the control. Its effect on seed germination was also enhanced in cucumber. It also improves the soil nutrients, quality, and productivity of okra, paddy, maize, onion, and black gram. The effect of microalgae on the growth of crops is attributed to plant growth regulators such as auxins, cytokinins, and gibberellins, and also increased levels of vital macronutrients and micronutrients. *Chlorella vulgaris* as a foliar spray and a soil drench enhanced the growth, quality, and yield of tomato plants (Suchithra et al., 2022).

7.2.2 Effect of Marine Green Biofertilizers as Foliar Spray

Marine plant extracts in foliar spray have been known to increase yield in many agricultural crops, as they stimulate the rate of growth and increase the yield of vegetables, cereals, fruits, and horticultural crops more than the powdered form. This approach supplies nutrients to plants faster than the soil application approach (Elansary et al., 2016). This could be attributed to the absorption of active nutrients through the pores of the stomata instead of the cuticles. Recently, varieties of seaweed extracts have been made available commercially under different names, such as Algifert (marinure), Maxicrop (Sea born), Kelpak 66, Goemar GA14, Seaspray, Seasol, Cytex, SM3, and Seacrop 16 (Begum et al., 2018)

Sridhar and Rengasamy (2010) tested the significance of green seaweed (*Ulva lactuca* L) and brown seaweed (*Sargassum wightii*) on ground nut yield. They reported that plants treated with liquid fertilizer (LF) showed a decrease in yield as compared with those treated with chemical fertilizers at 100% recommended level, which indicated that the level of macronutrients in the LFs may not be able to meet the crop's demand. But when the plants were treated with 1.0% LFs plus chemical fertilizers at a 50% recommended rate, there was a significant difference in the yield compared with when chemical fertilizers at a 100% recommended rate were used. This indicates that the combined effect of both gave a better yield. However, LF obtained from *S.wightii* gave a better yield of ground nut than *U. lactuca*. Therefore, seaweed fertilizer could help reduce the cost of chemical fertilizer by 50%. Arumugam and Anantharaman (2009)investigated the effect of brown seaweed LF(*Rosenvigea intricate*)on seed germination of okra. It was discovered that 20% of the LF had maximum effect on seed germination whether or not the recommended dose of chemical fertilizer was utilized. Maximum effect on growth parameters (such as shoot and root length, number of lateral roots and leaves) and yield (such as length, number, and weight of the vegetables) were achieved when a 20% concentration of the LF with the recommended dose of chemical fertilizer was used.

Uthirapandi et al. (2018) investigated the green alga*Caulerpa racemosa* and the brown algae*Sargassum wightii and Turbinaria ornata*on the growth of *Ocimum sanctum*(holy basil). Foliar application of the individual seaweeds improved the entire physiological processes and growth of *Ocimum sanctum*. Furthermore, there was a reasonable increase in total plant growth, such as total plant height, shoot and root length, and number of leaves when a mixture of the seaweeds was used compared with individual treatment and control. This resulted from the synergistic effect of active ingredients like vitamins, micro- and macroelements, as well as phytohormones that are present in the liquid extracts of the seaweed. Therefore, a combination of liquid extracts of different seaweeds could be a promising and efficient biofertilizer that may replace chemical fertilizers for sustainable agriculture.

El-Din (2015) studied the effect of seaweed LFs, the green alga*Codium tomentosum*, and the brown alga Sargassum *vulgare* on the germination and growth of wheat seedlings. The seed germination percentage of the seedlings was increased by over 90% when the seedlings were soaked in a 20% concentration of both LFs. Several research studies revealed that higher concentrations of marine green biofertilizers showed a decreasing trend with respect to germination, overall plant growth, and yield, and increased plant tolerance to environmental stresses (El-Din, 2015; Uthirapandi et al., 2018; Zafar et al., 2022). Layek et al. (2018) reported an increase in the rate of germination, length of shoot and root, and vigor index of seedlings when a lower concentration of seaweed extracts was used in soaking rice seeds. An increased rate of germination was also obtained when maize seeds were soaked in low concentrations (5%) of *Kappaphycus* and *Gracilaria* species seaweed fertilizers, while the germination was inhibited when a higher concentration of the extracts (15%) was used. Patel et al. (2018)also reported improved germination of fenugreek, spinach, and coriander vegetable seeds when they were soaked in *Ulva lactuca* LF at lower concentrations of 6% and 8%. The increase in the rate of germination and seedling vigor when seaweed extracts were used at low concentrations was reported to be a result of the presence of substances that promote growth, such as gibberellins, auxins, phenylacetic acid, and other macro- and micronutrients (Begum et al., 2018).

Abbas et al. (2020) also reported that a 0.5% extract of the seaweed *Ascophyllum nodosum* significantly increased the yield, nutritional contents (such as nitrogen, phosphorus, and potassium contents of bulbs), and total soluble solids (TSS) of four onion cultivars investigated. They also reported a significant increase in the length and width of leaf, length and weight of root, bulb diameter, and the weight of the onion cultivars as well as the yield. Hussain et al. (2021) reported that seaweed extracts obtained from the brown algae*Ascophyllum nodosum* and *Durvillaea potatorum* had a significant effect on the growth of tomato plants (such as number of flowers and clusters, number of fruits, length of root, and dry weight of root and shoot) and plant productivity (yield and quality), and also improved soil health and microbial diversity and population. Field studies have revealed that marine algal extracts used at slightly low concentrations, either as foliar sprays or as a soil amendment, increased the yield of several crops, including vegetables, wine grapes, strawberries, and sugarcane (Shukla et al., 2019). Research on plants such as tomato, *Arabidopsis*, and soybean has revealed that the use of these biofertilizers may enhance the ability of some plants to tolerate some abiotic stresses (Jithesh et al., 2019; Hussain et al., 2021).

7.3 PREPARATION OF MARINE GREEN BIOFERTILIZERS

Since marine greens constitute microalgae, macroalgae (seaweeds), and seagrasses, their preparation as biofertilizers is slightly different. For this reason, the preparation of seaweeds and seagrasses is similar, while that of microalgae is different.

7.3.1 Preparation of Sea Plant (Seaweeds/Seagrasses) Biofertilizers

Seaweeds have been reported to be crucial marine sources, available worldwide. Their extracts are usually available in the form of fertilizers of varying forms, such as the liquid and powdered form of the fertilizers. These seaweed biofertilizers in powdered form are usually used as manure (Patel et al., 2017). Seaweeds are harvested and washed thoroughly to eliminate all epiphytes and particles of sand, and then sun-dried in order to minimize the water content. Dried seaweeds are thereafter soaked in freshwater to get rid of salt and reduce the level of salinity. The process is repeated until the level of salt in the water is zero (0 ppt). When this is achieved, the seaweed is dried and milled into powdered form. This powder can then be applied to treat the soil at varying concentrations (Ammar et al., 2022). The LF is prepared by mixing the powder in distilled water in a ratio of 1:20 (w/v) and boiling for 45 to 60 minutes. It is then filtered using muslin cloth to obtain the extract. The extract is considered to be the 100% concentration, from which different concentrations of the LF can be made for use as a foliar spray (Patel et al., 2017). This method of preparation is referred to as the boiling method. Other methods include the following.

7.3.1.1 Low-Temperature Heating Method

This involves heating the mixture to 60°C and maintaining this for a period of 24 hours in a hot air oven. The extract is filtered using a muslin cloth.

7.3.1.2 Autoclave Method

Seaweed or seagrass powder is weighed and then soaked in distilled water in a ratio of 1:10. The mixture is then autoclaved for a period of 1 hour at 15 lb pressure. The extract is filtered and kept inside a refrigerator until needed for use.

7.3.1.3 Alcohol Extraction Method

Seaweed or seagrass powder is weighed and then mixed with alcohol in a ratio of 1:1, stirred properly, and then filtered. This filtrate is regarded as containing a 100% concentrated extract.

7.3.1.4 Alcoholic-Aqueous Method

Seaweed or seagrass powder is weighed and then mixed with alcohol in a ratio of 2:1 for a period of 12 hours. The mixture is vigorously shaken to dissolve the constituents that are soluble in alcohol. The supernatant is retained, while the residue is boiled in distilled water of the same volume as the alcohol for 30 minutes, cooled, and then filtered. The supernatants obtained from alcohol and water are mixed, and then distilled water is added to make up the volume to 100 mL in order to get a 100% concentrated extract.

7.3.2 Preparation of Microalgae Biofertilizers

A pure culture of an efficient strain of a microalga, such as nitrogen-fixing cyanobacteria grown on a slant using the proper growth medium, is obtained. This is known as the inoculum. A loopful of the inoculum is collected and placed in a 250-mL capacity conical flask containing a liquid medium. The flask is placed in an incubator or incubator with a rotary shaker (if the organism is a slow grower) for 3–7 days. After the incubation period, the cells in the flask should contain 10^5–10^6 cells/mL, and this is known as the starter or mother culture. This mother culture is reproduced in a bigger-volume conical flask, using a rotary shaker for about 96–120 hours until the cell count per mL is up to 10^9–10^{10} cells. This culture is not to be stored for a period exceeding 24 hours or at a temperature above 4°C, since the broth thickens and becomes inconsistent. For large-scale production, fermenters are used to produce this biofertilizer for commercial purposes. The broth is thereafter transferred to an automated filling machine and packaged in 250-mL, 500-mL, or 1-L bottles having 0.5 mm thickness, leaving two-thirds of the space for the purpose of aeration (Ritika and Utpal, 2014). To obtain the powdered form, centrifugation is utilized to harvest the microalgal biomass from the broth culture. The biomass is then spray dried to obtain the powdered form (Chen et al., 2016). A microalgal liquid spray is obtained by boiling about 500 mg of dried algal powder in 200 mL of double distilled water and then boiling in a water bath at a temperature of 100°C for 15 minutes. The mixture is then cooled and filtered with a cheesecloth. The filtrate is regarded as a standard solution, which will be used in preparing different concentrations of the biofertilizer (Deepika and MubarakAli, 2020).

7.4 ADVANTAGES OF MARINE GREENS AS A BIOFERTILIZER

Marine greens can be cultivated on a large scale in the sea to provide biomass for soil additives and biofertilizers. The use of this green technology will help reduce the effect of chemicals present in chemical fertilizers on plants, animals, and humans as well as surface and groundwater pollution (Ammar et al., 2022).

The use of sea greens as biofertilizers helps in the reduction of phosphorus and nitrogen runoff into water bodies, which eventually results in an increase in the quality of the water that flows in water bodies (Zafar et al., 2022).

Several varieties of marine greens have been prepared as liquid biofertilizers, and sometimes either whole or finely powdered algal forms are used as manures. Research has shown that the use of marine greens as biofertilizers enhances plants' tolerance to abiotic stresses, including salinity, drought, ion toxicity, and high temperature. Advanced methods of extraction such as microwave-assisted extraction, enzyme-assisted extraction, supercritical fluid extraction, and pressurized liquid extraction have been discovered for better extraction without causing deterioration.

Macroalgal extracts contain a diverse array of phytohormones such as gibberellins, abscisic acid, betaine, polyamines, and auxins. Micronutrients and trace elements present are active as plant growth regulators and increase crop yields when they are applied externally. The application of seaweed biofertilizer to crops showed a significant effect on crop production (Nabti et al., 2017; Zafar et al., 2022).

Biofertilizers obtained from marine greens such as *Sargassum*, *Fucus*, *Ascophyllum*, *Laminaria*, etc. are biodegradable, non-hazardous, and eco-friendly to humans and animals as well as surface and underground waters (Zafar et al., 2022).

7.5 DISADVANTAGES OF ORGANIC FERTILIZER

Biofertilizers must undergo decomposition before their nutrients are released into the soil, which eventually improves the fertility of the soil. Factors such as temperature and soil moisture greatly influence the rate at which organic matter in biofertilizers decompose; therefore, available nutrients may be released at a time when the plants do not need them. Furthermore, biofertilizers do not provide all the required nutrients needed by plants, and biofertilizers are not readily available or obtainable in some places; hence, it becomes very difficult to rely on them to supply the required nutrients for crop production. Biofertilizers are needed in large amounts and are not readily available to subsistence farmers (Guo et al., 2020; Ammar et al., 2022).

The use of marine greens as biofertilizers allows nutrient transfer from the marine environment to the soil and could replace chemical fertilizers, especially in soil with poor levels of nutrients, thus reducing water pollution (Emadodin et al., 2020).

7.6 LIMITATIONS

Some of the limitations of biofertilizers are lack of rules and standards for biofertilizer production, limited availability of transportation means and storage space, inadequate labeling of the products, inadequate exposure of farmers to the products, scarcity of suitable and efficient strains, mutation of strains as they age, soil physicochemical properties, and biotic and abiotic factors that may affect the efficiency and potency of the biofertilizer (Coppens et al., 2016).

Fertilizers are known to be composed of several groups of supplements. Producing and implementing biofertilizers is laborious and expensive when compared with the use of compost. Due to the low amounts of nutrients when compared with complex manures, large quantities are usually required for effective land application and for an extended period of time or repeatedly. This may result in the accumulation of nutrients and salts, which may negatively affect the development of plants as well as soil microbes, water quality, and human health. Essential micro- and macronutrients may not be readily available in adequate amounts for proper plant growth and development. This eventually results in nutritional deficiencies in plants (Coppens et al., 2016).

REFERENCES

Abbas, M., Anwar, J., Zafar-ul-Hye, M., Iqbal Khan, R., Saleem, M., Rahi, A. A., Danish, S., &Datta, R. (2020). Effect of seaweed extract on productivity and quality attributes of four onion cultivars. *Horticulture*, 6(2), 28.

Acosta-Calderón, J. A., Mateo-Cid, L. E., &Mendoza-González, A. C. (2016). An updated list of marine green algae (Chlorophyta, Ulvophyceae) from the Biosphere Reserve of Sian Ka'an, Quintana Roo, Mexico. *Check List*, 12(3), 1886–1902.

Ammar, E. E., Aioub, A. A., Elesawy, A. E., Karkour, A. M., Mouhamed, M. S., Amer, A. A., &El-Shershaby, N. A. (2022). Algae as bio-fertilizers: Between current situation and future prospective. *Saudi Journal of Biological Sciences*, 29(5), 3083–3096.

Arumugam, R., &Anantharaman, P. (2009). Effect of seaweed liquid fertilizer on growth and pigment concentration of Abelmoschus esculentus (l) Medikus. *Journal of Agronomy*, 2, 57–66.

Begum, M., Bordoloi, B. C., Singha, D. D., &Ojha, N. J. (2018). Role of seaweed extract on growth, yield and quality of some agricultural crops: A review. *Agricultural Reviews*, 39(4), 321–326.

Chen, J., Wang, Y., Benemann, J. R., Zhang, X., Hu, H., &Qin, S. (2016). Microalgal industry in China: Challenges and prospects. *Journal of Applied Phycology*, 28(2), 715–725.

Coppens, J., Grunert, O., Van Den Hende, S., Vanhoutte, I., Boon, N., Haesaert, G., &De Gelder, L. (2016). The use of microalgae as a high-value organic slow-release fertilizer results in tomatoes with increased carotenoid and sugar levels. *Journal of Applied Phycology*, 28(4), 2367–2377.

Deepika, P., &MubarakAli, D. (2020). Production and assessment of microalgal liquid fertilizer for the enhanced growth of four crop plants. *Biocatalysis and Agricultural Biotechnology*, 28, 101701.

Dineshkumar, R., Ahamed Rasheeq, A., Arumugam, A., Nathiga Nambi, K. S., &Sampathkumar, P. (2019). Marine microalgal extracts on cultivable crops as a considerable biofertilizer: A review. *Indian Journal of Traditional Knowledge*, 18(4), 849–854.

Dineshkumar, R., Kumaravel, R., Gopalsamy, J., Sikder, M. N. A., &Sampathkumar, P. (2018). Microalgae as bio-fertilizers for rice growth and seed yield productivity. *Waste and Biomass Valorization*, 9(5), 793–800.

Elansary, H. O., Skalicka-Wozniak, K., &King, I. W. (2016). Enhancing stress growth traits as well as phytochemical and antioxidant contents of Spiraea and Pittosporum under seaweed extract treatments. *Plant Physiology and Biochemistry*, 105, 310–320.

El-Din, S. M. (2015). Utilization of seaweed extracts as bio-fertilizers to stimulate the growth of wheat seedlings. *The Egyptian Journal of Experimental Biology*, 11, 31–39.

Emadodin, I., Reinsch, T., Rotter, A., Orlando-Bonaca, M., Taube, F., &Javidpour, J. (2020). A perspective on the potential of using marine organic fertilizers for the sustainable management of coastal ecosystem services. *Environmental Sustainability*, 3(1), 105–115.

Feller, C., &Beare, M. H. (1997). Physical control of soil organic matter dynamics in the tropics. *Geoderma*, 79(1–4), 69–116. https://doi.org/10.1016/S0016-7061(97)00039-6.

Fleurence, J., Morançais, M., Dumay, J., Decottignies, P., Turpin, V., Munier, M., Garcia-Bueno, N., &Jaouen, P. (2012). What are the prospects for using seaweed in human nutrition and for marine animals raised through aquaculture?*Trends in Food Science and Technology*, 27(1), 57–61. https://doi.org/10.1016/J.TIFS.2012.03.004.

Guo, S., Wang, P., Wang, X., Zou, M., Liu, C., &Hao, J. (2020). Microalgae as biofertilizer in modern agriculture. In *Microalgae Biotechnology for Food, Health and High Value Products* (pp. 397–411). Springer, Singapore. https://doi.org/10.1007/978-981-15-0169-2_12.

Hussain, H. I., Kasinadhuni, N., &Arioli, T. (2021). The effect of seaweed extract on tomato plant growth, productivity and soil. *Journal of Applied Phycology*, 33(2), 1305–1314.

Izzati, M. (2015). The use of seaweeds *Sargassum* sp and *Gracilaria verrucosa* as soil conditioneer to enhance the growth of Vigna radiata in sandy and clay soil. Proceeding of 5th International Seminar on New Paradigm and Innovation on Natural Science and its Application (pp. 86–89).

Izzati, M., Haryanti, S., &Setiari, N. (2019). The use of macroalga *Sargassum* sp. and *Gracilaria verrucosa* in improving sandy and clay soil fertility. *Journal of Physics: Conference Series*, 1217(1), 012179. IOP Publishing. https://doi.org/10.1088/1742-6596/1217/1/012179.

Jithesh, M., Shukla, S., Kant, P., Joshi, J., Critchley, T., &Prithiviraj, B. (2019). Physiological and transcriptomics analyses reveal that *Ascophyllum nodosum* extracts induce salinity tolerance in Arabidopsis by regulating the expression of stress responsive genes. *Journal of Plant Growth Regulution*, 38(2), 463–478.

Kimaru, A. W. (2017). *The Potential of Seagrass Litter as Organic Fertilizer in Kilifi Kenya* (Doctoral Dissertation, Pwani University).

Layek, J., Das, A., Idapuganti, R. G., Sarkar, D., Ghosh, A., Zodape, S. T., Lal, R., Yadav, G. S., Panwar, A. S., Ngachan, S., &Meena, R. S. (2018). Seaweed extract as organic bio-stimulant improves productivity and quality of rice in eastern Himalayas. *Journal of Applied Phycology*, 30(1), 547–558.

Mahapatra, D. M., Chanakya, H. N., Joshi, N. V., Ramachandra, T. V., &Murthy, G. S. (2018). Algae-based biofertilizers: A biorefinery approach. In *Microorganisms for Green Revolution* (pp. 177–196). Springer, Singapore. https://doi.org/10.1007/978-981-10-7146-1_10.

Nabti, E., Jha, B., &Hartmann, A. (2017). Impact of seaweeds on agricultural crop production as biofertilizer. *International Journal of Environmental Science and Technology*, 145(14), 1119–1134.

Palinski-Wade, E. (2021). What are sea geeens and should you be eating them? Available from http://erinpalinski.com/what-are-sea-greens/.

Patel, R. V., Pandya, K. Y., Jasrai, R. T., &Brahmbhatt, N. (2018). Significance of green and brown seaweed liquid fertilizer on seed germination of Solanum melongena, Solanum Lycopersicum and Capsicum annum by paper towel and pot method. *International Journal of Recent Scientific Research*, 9, 24065–24072.

Patel, R. V., Pandya, K. Y., Jasral, R. T., &Brahmbhatt, N. (2017). A review: Scope of utilizing seaweed as a biofertilzer in agriculture. *International Journal of Advanced Research*, 5(7), 2046–2054.

Priya, M., Gurung, N., Mukherjee, K., &Bose, S. (2014). Microalgae in removal of heavy metal and organic pollutants from soil. In *Microbial Biodegradation and Bioremediation* (pp. 519–537). Elsevier. https://doi.org/10.1016/B978-0-12-800021-2.00023-6.

Rathore, S., Chaundary, D., Boricha, G., Gosh, A., Bhat, B., Zodape, S., &Patolia, J. (2008). Effect of seagrass extract on the growth, yield and nutrient uptake of soybean (glycine max) under rainfed conditions. *Journal of Botany*, 75, 351–355.

Samanthi. (2020). Differences between macroalgae and microalgae. Available at http://www.differncebetween.com/difference-between-macroalgae-and microalgae/amp/.

Santhosh, C., Vino, P., &Maruthupandian, A. (2021). Effect of Padina tetrastromatica and Cymodocea serrulata using different methods extract on growth and pigments of black gram (Vigna mungo L.). *Indian Journal of Natural Products and Resources*, 12(2), 300–306.

Shukla, P., Mantin, E., Adil, M., Bajpai, S., Critchley, A., &Prithiviraj, B. (2019). Ascophyllum nodosum-based biostimulants: Sustainable applications in agriculture for the stimulation of plant growth, stress tolerance, and disease management. *Frontiers in Plant Science*, 10, 655.

Sridhar, S., &Rengasamy, R. (2010). Significance of seaweed liquid fertilizers for minimizing chemical fertilizers and improving yield of Arachis hypogaea under field trial. *Recent Research in Science and Technology*, 2(6), 61–68.

Suchithra, M. R., Muniswami, D. M., Sri, M. S., Usha, R., Rasheeq, A. A., Preethi, B. A., &Dineshkumar, R. (2022). Effectiveness of green microalgae as biostimulants and biofertilizer through foliar spray and soil drench method for tomato cultivation. *South African Journal of Botany*, 146, 740–750.

Uthirapandi, V., Suriya, S., Boomibalagan, P., Eswaran, S., Ramya, S. S., Vijayanand, N., &Kathiresan, D. (2018). Bio-fertilizer potential of seaweed liquid extracts of marine macro algae on growth and biochemical parameters of Ocimum sanctum. *Journal of Pharmacognosy and Phytochemistry*, 7(3), 3528–3532.

Veluchamy, C., &Palaniswamy, R. (2020). A review on marine algae and its applications. *Asian Journal of Pharmaceutical and Climical Resources*, 13(3), 21–27.

Zafar, A., Ali, I., &Rahayu, F. (2022). Marine seaweeds (biofertilizer) significance in sustainable agricultural activities: A review. *IOP Conference Series: Earth and Environmental Science*, 974(1), 012080. IOP Publishing.

8 Marine Green Macroalgae Biomass Production and Application

Srinivasan Kameswaran, Ch. Manjula, Sudhakara Gujjala, and Bellamkonda Ramesh

8.1 INTRODUCTION

A varied range of photosynthetic, multicellular, eukaryotic organisms classified as either Plantae or Chromista make up macroalgae (also known as seaweeds). The majority are marine, where they are the primary producers in most of the world's coastal waters (Levine 2016). Some macroalgae resemble plants in appearance despite being vascular and lacking real roots, stems, leaves, and complicated reproductive structures because they have distinct thalli with attachment organs (holdfasts), stem-like structures (stipes), and photosynthetic blades (fronds) (Dawes 2016). Marine seaweeds can be categorized into three phyla, Chlorophyta (green), Rhodophyta (red), and Ochrophyta (brown), based on the amount of photosynthetic pigment, the amount of stored carbohydrates, the components of the cell wall, and the structure and orientation of the flagella (Levine 2016). The variety of this group as a whole surpasses 10,000 species, and new taxa are described every year. Chlorophyta currently includes more than 1500 macroalgal species, Ochrophyta about 2000, and Rhodophyta about 7300 (Guiry 2021). Together, macroalgae make up a significant biological resource that offers a range of socioeconomic benefits and ecosystem services (Buchholz et al. 2012; Rebours et al. 2014).

The majority of macroalgae cultivation is concentrated in South-East Asian and Pacific nations, where China, the Philippines, Indonesia, Republic of Korea, and Japan produce an astounding 98% of the world's seaweed biomass. However, only seven macroalgal genera, all of which are classified as Rhodophyta or Ochrophyta, are responsible for supporting this industry (FAO 2020). *Kappaphycusalvarezii*, *Eucheuma* spp., and *Gracilaria* spp. are a few of these macroalgae that are utilized in the hydrocolloid business, while Saccharina japonica, Undaria pinnatifida, Sargassum fusiform, and Pyropia spp. are taxa that are used directly as human food (FAO 2020; Buschmann et al. 2017). The output of green macroalgae (Chlorophyta) cultivation, on the other hand, now accounts for only a small percentage of global landings (1%), far below the most pertinent taxa in terms of production metrics. Despite showing similarly lower output figures, green seaweed cultivation has seen an increase in productivity and economic diversification over the past few decades (FAO 2020). However, inequalities resulting from the dominance of a small number of species in macroalgal farming exacerbate the need for increased innovation and industry sustainability. The commoditization of the sector is regarded as being particularly incompatible with the need for high-value seaweed goods in the future, since it does not produce standardized, high-quality, traceable items (Hafting et al. 2015). By diversifying species and/or cultivars, standardizing farming practices, and raising understanding of regional genetic and environmental heterogeneity, advancements are therefore needed to meet the difficulties of a changing sector (Valero et al. 2017; Buschmann and Camus 2019; Hafting et al. 2015; Winberg et al. 2014). The goal of the current chapter is to give a thorough overview of the current situation, ongoing advancements, and potential future uses of green macroalgae in cultivation. Together

DOI: 10.1201/9781003369738-10

with known applications, estimations of regional relevance and worldwide production, and current cultivation methods, key taxa are provided in detail, along with descriptions of their diversity and potential. We analyze the present research on the search for bioactive substances in more detail, and we wrap up by describing the potential uses of this mostly unstudied class of macroalgaein cultivation.

8.2 METHODS FOR CULTIVATING GREEN MARINE MACROALGAE

Green macroalgae are currently being cultivated on a commercial basis using a number of technical methods. Depending on the target species' biology and ecophysiology, as well as the particulars of the site location, different stages of the production chain (propagule production, nursery, growout) require distinct procedures (Figure 8.1).

The ability to produce seed stock of the target species in enough quantity and quality to easily meet demand for biomass production is the primary factor that heavily influences the production chain (Radulovich et al. 2015). In commercial-scale green seaweed aquaculture, several techniques are being used to produce viable propagules, including vegetative propagation, asexual propagation by parthenogenesis, and sexual reproduction through gamete conjugation. In taxa with particularly high vegetative thalli proliferation potential, such as *Caulerpa* (de Gaillande et al. 2017; Gao et al. 2019), *Chaetomorpha* (Odom and Walters 2014), and *Codium*, vegetative propagation has been successful; other taxa, however, are preferentially propagated via manipulation of reproductive propagules (zoospores or gametes), necessitating a deeper understanding of species life cycle patterns and developmental biology (Charrier et al. 2017). In these taxa, the industry has transitioned from basic natural seeding techniques—placing artificial substrata (ropes, nets) near wild populations of the target species to act as settling structures for naturally occurring seed—to more advanced artificial propagule production techniques used in land-based nursery facilities, such as *Ulva* (Dan et al. 2002), *Monostroma* (Kida 1990), and *Capsosiphon*(Hwang et al. 2003).

The majority of asexual propagation techniques used in commercial-scale aquaculture of Ulvophycean taxa are rather well developed and widely used (vegetative propagation or parthenogenesis). Green seaweeds (Caulerpa, Codium, Ulva, and Capsosiphon) can be grown more successfully using asexual propagation techniques than other significant taxa like kelps (Ochrophyta) and *Pyropia* spp. (Rhodophyta), which typically require altering the entire sexual life cycle. As a result, easier culture techniques are made possible by avoiding the manipulation of one of the sexual life cycle phases (sporophyte or gametophyte) (Charrier et al. 2017). Asexual reproduction through

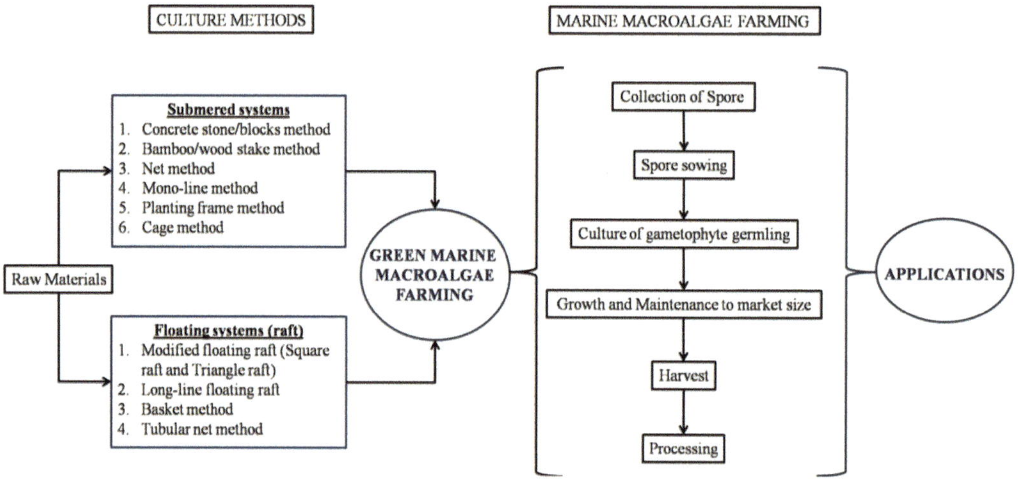

FIGURE 8.1 Different methods for growing and processing green marine macroalgae.

ongoing vegetative growth of macroalgae may lead to decreased productivity and heightened susceptibility to disease, as seen in red seaweeds; however, this has not been documented in green seaweeds (e.g. *Eucheuma*, *Kappaphycus*, and *Gracilaria*) (Loureiro et al. 2015; Massad et al. 2020).

The grow-out phase is when propagules are transferred to open-water culture grounds or to land-based systems once they have been collected and are of a size that is suitable for transplantation. Conventional cultivation methods use land-based tanks or ponds, or artificial substrata like ropes or nets in open water. The following categories apply to them. Floating raft cultivation methods use seeded ropes or nets held within a rigid floating frame at the surface. Tank or tumble culture cultivation methods grow plants in tanks under free-floating conditions. Other methods include planting directly on the ocean floor or attaching to artificial substrates close to the bottom (Radulovich et al. 2015). In Ulvophycean macroalgae, all traditional culture techniques are well represented. Novel strategies and iterations of these techniques have occasionally been experimentally validated. The procedures used in commercial production of Ulvophycean taxa during the grow-out stages are as follows.

8.2.1 Vegetative Propagation

The most distinctive genus of Ulvophyceans is *Caulerpa* (sea grapes), which is grown by conventional vegetative propagation techniques. Seed stock is obtained by simple thalli excision, which then regenerates the entire plant, according to Santelices' concept of "one-step technology" (Santelices 1999). Yet, the ability of removed fragments to survive and grow can vary depending on the organ and size. In *Caulerpa okamurae*, for instance, excised erect fronds and stolons (horizontal runners) can regrow from >3 cm and >5 cm fragments, respectively (Gao et al. 2019).

Beginning in the early 1950s, bottom-planting techniques were utilized in the Philippines to cultivate the *Caulerpa* species (*C. lentillifera* and *C. chemnitzia*) in clay fish ponds. This straightforward technique is still in use today in this South-East Asian nation. In order to enable multiple harvests, propagules (rhizoidal thalli) are buried on the surface sediment layer (2 cm) and allowed to grow (Horstmann 1983). The growth of this cultivation technique in the Philippines, where more sustainable ways at sea (cages or tubular nets placed off-bottom) have been tried, was, however, limited by a permanent ban on turning mangrove areas into earthen ponds (Tanduyan et al. 2013) (Figure 8.1).

Moreover, it has been claimed that *C. lentillifera* is being grown in land-based systems and in submerged cylindrical cages at sea as part of commercial-scale *Caulerpa* production (off-bottom). In concrete tanks (4 m^3) with two plastic mesh grids fastened to a floating PVC frame, *C. lentillifera* has been produced in Japan. Unfiltered seawater is circulated five times daily through the tanks, with intense aeration increasing water movement (de Gaillande et al. 2017). *Caulerpa lentillifera* has been raised in open water in cages that are 100 cm long and 60 cm in diameter, made of net stretched across a metal frame. At a depth of 0.5 m, buoys are used to hang the frames via ropes (Titlyanov and Titlyanova 2010). Many diverse culture techniques have been used in experimental *Caulerpa* species production around the world, including tying thalli to lines, ropes, nets (submerged at various water depths), cages, and trays (off-bottom); and employing plastic trays supported by bamboo frames in clay ponds (de Gaillande et al. 2017).

Several methods have been established in *Codium* (green sponge fingers, velvet horn) for the vegetative multiplication of two distinct species (*C. fragile* and *C. tomentosum*). In Korea, isolated utricles and medullary filaments are regenerated to produce *Codium fragile* propagules (Hwang et al. 2008; Hwang et al. 2005). These structures are created by hand chopping adult vegetative spongy thalli into tiny (5 mm) fragments. During a nursery cultivation stage (1–3 months) in land-based culture tanks, the resultant fragments are allowed to connect directly to a framed seed fiber (100 m) and regenerate from a filamentous morphology to juvenile spongy thalli. The seeded fiber is then transported offshore to a long-line horizontal cultivation system after being wound on culture ropes (100 m). Propagules go through a quiescent pre-main cultivation stage of growth at a depth of

FIGURE 8.2 Green seaweeds are frequently found along the Indian coast.

2 meters (5 months) before moving on to a fast growth main culture stage at a depth of 1 meter (3 months) before harvest (Hwang et al. 2008).

8.3 GREEN ALGAE AS SOURCES OF BIOLOGICAL COMPOUNDS

Biased screening and bioprospecting efforts have been made in the main categories of seaweeds, according to a review of the published literature on natural chemicals isolated from macroalgae. Just 8% of marine natural compounds extracted from macroalgae between 1965 and 2012 in the MarinLit database were green macroalgae–derived, while red and brown macroalgae accounted for 53% and 39%, respectively, according to Leal et al. (2013). The Bryopsidales order of green macroalgae was responsible for almost two-thirds of these substances (Leal et al. 2013). Sulfated polysaccharides, lipids, photosynthetic pigments, and different secondary metabolites are just a few of the naturally occurring compounds that may be isolated from green macroalgae and have been shown to have strong bioactive capabilities (de Souza et al. 2009; Pangestuti and Kim 2011; Patel 2012; Pereira et al. 2012) (Table 8.1).

8.3.1 Sulfated Polysaccharides

Macroalgae are known to be potential sources of sulfated polysaccharides, vital parts of cell walls(Patel 2012). Fucans, carrageenans, and ulvans are the three main forms of macroalgal

TABLE 8.1
Biological Active Compounds Derived from Marine Green Macroalgae and Reported Applications

S. No.	Green Macroalgae Species	Byproducts	Applications	References
1	*Cladophora rupestris*	Cellulose nanocrystals (CNCs)	Strengthening of starch-based films' mechanical properties	Sucaldito and Camacho 2017
2	*Cladophora sericioides*	Modified composite form	Extraction of copper from aqueous solution	Ebrahimi et al. 2016
3	*Ulva lactuca*	Activated carbon	Elimination of harmful hexavalent chromium ions from wastewater, saline water, and aqueous solutions	El-Sikaily et al. 2007
4	*Cladophora glomerata, Ulva intestinalis,*and *Microsporaamoena*	Dry solid material	Elimination of hexavalent chromium (Cr(VI)) from aqueous solution	Al-Homaidan et al. 2018
5	*Ulva lactuca*	Cellulose nanocrystals (CNCs)	Development of polymer materials' mechanical characteristics for food packaging	Doh et al. 2020
6	*Ulva lactuca*	Biochar	Remazol Brilliant Orange 3R remediation in an up-flow fixed column	Gokulan et al. 2019
7	*Ulva lactuca*	Sap, lipids, ulvan, and protein	Uses in cosmetics, medicine, food, and biofuels	Gajaria et al. 2017
8	*Ulva lactuca*	Biogas, sap, ulvan, and protein	High-value chemicals, biofuels	Mhatre et al. 2019
9	*Ulva fasciata*	Biodiesel	Biofuel	Khan et al. 2016
10	*Ulva fasciata*	Dry solid material	Taking copper out of its aqueous solution	Kumar et al. 2006
11	*Ulva* sp. mixed with cow dung	Biogas and bio-fertilizer	Mung bean growth using organic fertilizer, biogas, as biofuel	Akila et al. 2019
12	*Ulva lactuca*	Acetone, butanol, ethanol, 1,2-propanediol, and organic acid	Potential for feeding 1,2-propanediol synthesis using seaweeds rich in rhamnose	Van der Wal et al. 2013
13	*Ulva* sp.	Bio-hydrogen	Bioenergy	Margareta et al. 2020
14	*Ulva lactuca*	Biodiesel	Biofuel	Kalavathy and Baskar 2019
15	*Ulva lactuca*	Butanol	Biofuels	Potts et al. 2012
16	*Ulva lactuca*	Ulvan (hydrogel)	Transport and delivery of bioactive chemicals that are not water soluble	Bang et al. 2019
17	*Ulva* spp.	Ulvan (hydrogel)	–	Lahaye and Axelos 1993

(*Continued*)

TABLE 8.1 (CONTINUED)
Biological Active Compounds Derived from Marine Green Macroalgae and Reported Applications

S. No.	Green Macroalgae Species	Byproducts	Applications	References
18	*Ulva* spp.	Ulvan (hydrogel)	Applications of in situ gelling systems in biomedicine	Morelli et al. 2016
19	*Ulva armoricana*	Ulvan (hydrogel)	Encapsulation of cell scaffolds that can support cells	Morelli and Chiellini 2010
20	*Ulva lactuca*	Ulvan (hydrogel)	Transport and distribution of water-insoluble biologically active substances	Bang et al. 2019
21	*Ulva* spp.	Ulvan (hydrogel)	Vehicle for viable cells in the use of injectable cell delivery systems	Morelli et al. 2016
22	*Ulva lactuca*	Ulvan (film)	Packaging material	Guidara et al. 2019
23	*Ulva fasciata*	Ulvan (film)	Food packaging	Ganesan et al. 2018
24	*Ulva fasciata*/ polycaprolactone (PCL) or ulvan	Ulvan (fiber)	Drug release over time and scaffolding for tissue engineering	Kikionis et al. 2015
25	*Ulva rigida*	Ulvan (fiber)	Drug delivery methods	Toskas et al. 2011
26	*Ulva armoricana*	Ulvan (scaffolds)	Tissue engineering	Dash et al. 2018
27	*Ulva lactuca*	Ulvan (scaffolds)	Applications of bone tissue engineering	Alves et al. 2012
28	*Ulva armoricana*	Ulvan (scaffolds)	Alternatives to resorbable bone grafts	Dash et al. 2014

sulfated polysaccharides, which are derived from brown, red, and green seaweeds, respectively. Ulvans, which make up 8–29% of the dry weight of algae, are water-soluble sulfated polysaccharides made of disaccharide repeating moieties built of sulfated rhamnose linked to glucuronic acid, iduronic acid, or xylose (Lahaye and Robic 2007). In addition to *Ulva* species, ulvans can be found in other Ulvophycean genera, including *Monostroma*, *Caulerpa*, *Codium*, or *Gayralia* (Patel 2012). Green macroalgal sulfated polysaccharides have been shown to have antioxidant, anti-inflammatory, antitumoral, immunomodulatory, anticoagulant, and antiviral properties. Desulfation greatly reduces the bioactivity of ulvans, demonstrating the significance of the sulfateresidues for the stimulatory properties of these molecules (Qi et al. 2005; Leiro et al. 2007; Ciancia et al. 2007). Green macroalgae extracts that contain glucuronic acid have been linked to enhanced skin hydration and defense. In the cosmetics business, extracts from the green macroalgae *Codium tomentosum* are being employed as a moisturizing ingredient (Wang et al. 2015; Pereira 2018) (Table 8.2).

Moreover, the sulfated polysaccharides isolated from Ulvophycean taxa have bioactive characteristics that are highly valued for many biomedical applications. Sulfated polysaccharides, for instance, were isolated from a number of green macroalgae belonging to the genera *Codium* and *Monostroma*, and they showed potent blood anticoagulant action (Ciancia et al. 2007; Siddhanta et al. 1999; Li et al. 2011; Liu et al. 2018). Qi et al. (2005)discovered high scavenging activity of superoxide and hydroxyl radicals in ulvans derived from *Ulva australis*. It has been demonstrated that *Ulva lactuca* polysaccharides have an antitumoral effect against cancer cell lines by inducing

TABLE 8.2
Natural Substances Taken from Marine Green Macroalgae and Their Reported Bioactivities

S. No.	Chemical substances	Bioactivity	Green Macroalgae	References
1	Sulfated polysaccharides	Anticoagulant	*Monostromaangicava*	Liu et al. 2018
			Codium vermilara	Ciancia et al. 2007
			Codium dwarkense	Siddhanta et al. 1999
			Monostromalatissimum	Li et al. 2011
		Antioxidant	*Ulva australis*	Qi et al. 2005
		Anti-inflammatory	*Ulva rigida*	Leiro et al. 2007
		Antiviral	*Gayraliaoxysperma*	Cassolata et al. 2008
			Codium fragile	Ohta et al. 2009
			Ulva lactuca	Ivanova et al. 1994
			Monostromanitidum	Kazlowski et al. 2012
		Antitumoral	*Ulva lactuca*	Ahmed and Ahmed 2014
		Immunomodulatory	*Ulva rigida*	Leiro et al. 2007
			Ulva prolifera	Kim et al. 2011
2	Lipids	Antioxidant	*Codium tomentosum*	Rey et al. 2020
3	Sulfolipids	Antibacterial Antiviral Antitumoral	*Ulva fasciata*	El Baz et al. 2013
4	Clerosterol	Antitumoral	*Codium fragile*	Kim et al. 2013
		Anti-inflammatory	*Codium fragile*	Lee et al. 2013
5	Glycolipids	Antitumoral	*Ulva armoricana*	Kendel et al. 2015
6	Siphonaxanthin	Antitumoral	*Codium fragile*	Ganesan et al. 2011
		Antiangiogenic	*Codium fragile*	Ganesan et al. 2010
7	Squalene	Anti-inflammatory Antioxidant	*Caulerpa racemosa*	Fernando et al. 2018
8	Terpenoids	Neuroprotective	*Caulerpa racemosa*	Yang et al. 2017
9	Caulerprenylols (para-xylene)	Antifungal	*Caulerpa racemosa*	Liu et al. 2013
10	Caulerpin (alkaloid)	Anti-inflammatory	*Caulerpa racemosa*	De Souza et al. 2009
11	Racemosin A (alkaloid)	Neuroprotective	*Caulerpa racemosa*	Liu et al. 2013
12	Cladophorols (phenol)	Antibacterial	*Cladophora socialis*	Lavoie et al. 2019
13	Kahalalide F (depsipeptide)	Antitumoral Antiviral Antimalarial	*Bryopsis*sp.	Hamann et al. 1996
14	Pheophytin *a*	Anti-inflammatory	*Ulva prolifera*	Okai and Higashi-Okai 1997
		Antitumoral	*Ulva prolifera*	Hiqashi-Okaj et al. 1999
15	Sesquiterpenes	Antitumoral	*Ulva fasciata*	Lakshmi et al. 2018
16	Pheophorbide *a*	Antioxidant	*Ulva prolifera*	Cho et al. 2011

apoptosis and suppressing cell division (Ahmed and Ahmed 2014). Sulfated polysaccharides that were isolated from various green macroalgal species had antiviral and immunostimulatory properties (Leiro et al. 2007; Ivanova et al. 1994; Cassolato et al. 2008; Kim et al. 2011; Kazlowski et al. 2012; Ohta et al. 2009).

8.3.2 Lipids

Lipids are nonpolar solvent-soluble molecules that serve as the primary structural elements of cell membranes in addition to playing a role in energy storage and cell signaling pathways. Mass spectrometry has made it possible to precisely characterize the polar lipid profile of macroalgae (glycolipids, phospholipids, and betaine lipids), some of which have been shown to have beneficial effects on one's diet and overall health (Maciel et al. 2016). The high concentration of polyunsaturated fatty acids (PUFA), which are important for both human and animal health and nutrition, is particularly important for lipid bioactivity (Newton 1996). Macroalgae have the ability to produce critical PUFA on a big scale, which has numerous uses in the pharmaceutical and nutraceutical industries (Pereira et al. 2012). Codium galeatum, Codium tomentosum, and Ulva armoricana were shown to have high relative abundances of long-chain PUFA, specifically omega-3 fatty acids (Skrzypezyk et al. 2019; Rey et al. 2020; Kendel et al. 2015) (Table 8.2).

Research on the biological functions of lipids from the Ulvophyceans in a few representative species has revealed some intriguing biomedical prospects. For instance, *sulfolipidsfromUlva fasciata* and other algae species showed potent antiviral, antitumoral, and antibacterial properties. In cancer cell lines (Kendel et al. 2015), glycolipids from *Ulva armoricana* demonstrated encouraging antiproliferative properties (El Baz et al. 2013). Human melanoma cells were demonstrated to undergo apoptosis by clerosterol derived from *Codium fragile* (Kim et al. 2013). Moreover, it has been demonstrated that this sterol lowers the expression of proinflammatory proteins, suggesting the potential for its usage as a treatment for UVB-induced inflammatory and oxidative skin damage (Lee et al. 2013). Squalene derived from *Caulerpa racemosa* showed notable anti-inflammatory and antioxidant properties (Fernando et al. 2018).

8.3.3 Three-Dimensional Porous Scaffolds

For applications involving tissue engineering, scaffold design is essential. The perfect scaffold should have the right amount of porosity and connectedness, as well as mechanical qualities, high biocompatibility, and biodegradability (Alves et al. 2012; Alves et al. 2013). Dinoro et al. (2019) examined sulfonated polysaccharide-based scaffolds for orthopedic tissue engineering.

The performance of these materials can be improved, and their range of applications can be increased, by combining both synthetic and natural materials. Ulvan and poly-D, L-lactic acid (PDLLA) were thus coupled by Alves et al. (2012) to create a new scaffold for bone tissue engineering applications. Subcritical fluid sintering with carbon dioxide was used to create the scaffolds for PDLLA filled with ulvan particles at 40°C and 50 bar. The PDLLA matrix was filled with both pure ulvan particles and ulvan particles that had been loaded with dexamethasone. The constructed scaffold was discovered to have the proper morphologic characteristics, mechanical performance, and cytocompatibility. Moreover, the created scaffold can be utilized as a means of medication delivery (Alves et al. 2012). A method for creating photo-crosslinked polymeric ulvan scaffolds that are then subjected to enzymatic treatment for calcium phosphate deposition was suggested by Dash et al. (2014). Ulvan was first altered with methacrylic anhydride to produce photoreactive groups; this altered ulvan was then utilized to create photo-crosslinked scaffolds under the influence of UV light. Alkaline phosphatase (ALP) was used to treat the crosslinked ulvan scaffolds in order to encourage mineralization. The ulvan scaffolds were discovered to have undergone uniform mineralization at room temperature, and the resulting minerals contained apatite. The produced minerals enhanced the osteogenic cell activity on the mineralized scaffolds, which were non-toxic. The biofunctionalized scaffolds might be utilized as an alternative to resorbable bone grafts (Dash et al. 2014). When mineralized with apatite, the polymer scaffolds exhibit strong osteoconductivity and osteoinductivity, which is crucial for their medicinal uses. In order to study how alkaline phosphatase impacts the production of apatitic minerals, Dash et al. (2018) developed an ulvan-based scaffold employing polyelectrolyte complexes of chitosan (40 wt%) and ulvan (60 wt%). On the ALP-treated scaffold, it was discovered that the calcium phosphate minerals had been successfully deposited. The extracellular matrix development, cell proliferation, and cell adhesion are all facilitated by the globular nature of the deposited minerals. The mineralized scaffolds are also non-toxic. Hence, creating scaffolds out of polyelectrolyte complexes is an environmentally friendly method. For tissue engineering, the produced scaffold can be employed as a resorbable material (Dash et al. 2018). Because it is edible and has an amphiphilic nature, ulvan has also been studied as a stabilizing and emulsifying agent in colloidal formulations including functional ingredients for food and cosmetic applications (Morelli et al. 2019). In mild conditions, ulvan was also utilized as a reducing and stabilizing agent for the manufacture of silver nanoparticles (Massironi et al. 2019). Moreover, lysozyme/ulvan complexes were created using ulvan to boost the antibacterial activity of lysozyme. A potential nanocarrier for positively charged bioactive compounds can be made from the produced lysozyme/ulvan (Tziveleka et al. 2018). Its putative antioxidant action has also generated a lot of interest in its beauty sector uses (Kidgell et al. 2019). HeLa cells have been demonstrated to be protected in vitro from oxidative stress caused by hydrogen peroxide by extracts from *Ulva rigida* that contain sulfated polysaccharides (Mezghani et al. 2013). Moreover, ulvans include a lot of rhamnosyl residues, which are thought to encourage collagen formation and cell proliferation (Alves et al. 2013). Additionally, glucuronic acids are present, which have moisturizing qualities that protect skin from environmental dryness. Ulvan's distinctive chemical makeup gives its suspensions distinct rheological properties and gelling characteristics at a range of temperatures, pH levels, and cations, making it useful as an emulsifier and stabilizer in a variety of applications, including the food and cosmetics industries (Lahaye and Robic 2007; Morelli et al. 2019). Ulvan is a special biopolymer for the creation of functional materials due to its distinctive structure, high bioactivity, and presence of both hydrophilic (hydroxyl, carboxyl, and sulfate) and hydrophobic (methyl) groups (Table 8.1).

8.3.4 Photosynthetic Pigments

Research on the biological functions of lipids from the Ulvophyceans in a few representative species has revealed some intriguing biomedical prospects. Light-harvesting substances called photosynthetic pigments are found in the chloroplasts of plants and algae. Their primary job is to take in visible light energy, or the 400–700 nm range, which is transformed into chemical energy during photosynthesis. Yet, some pigments have potent antioxidant properties, and their primary job in the cell is to prevent light damage due to heat dissipation or as a component of systems that scavenge reactive oxygen species. Chlorophylls and carotenoids are the two primary types of photosynthetic pigments found in green macroalgae.

Green macroalgae's photosynthetic pigments have reportedly been shown to have antioxidant, anti-inflammatory, anticancer, and antiangiogenic properties. The xanthophyll siphonaxanthin isolated from *Codium fragile* showed strong antiangiogenic properties (Ganesan et al. 2010). The ability of this green algae–specific keto-carotenoid to induce apoptosis in human leukemia cells has also been demonstrated (Ganesan et al. 2011). Siphonaxanthin is a more effective antitumoral drug than fucoxanthin, a carotenoid that has previously been demonstrated to prevent the growth of cancer cells by triggering apoptosis (Hosokawa et al. 1999; Sugawara et al. 2014). It has also been demonstrated that pigments produced from chlorophyll a, like pheophytin and pheophorbide a, have bioactive qualities, while pheophorbide has been found to have antioxidant characteristics (Cho et al. 2011), pheophytin a, which is produced from *Ulva prolifera*, has been reported to exhibit substantial anti-inflammatory and anticarcinogenic activities (Okai and Higashi-Okai 1997; Hiqashi-Okaj et al. 1999). Although natural pigments have less stability than their synthetic equivalents due to technological limitations, many techniques and substances have been successfully used to increase the stability of pigment characteristics and bioactivities (Cortez et al. 2017). Pigment stability and bioavailability can also be enhanced by creating effective delivery systems (Liang et al. 2019).

8.3.5 Secondary Metabolites

Green macroalgae have been found to include a variety of different bioactive substances, mostly secondary metabolites that frequently play a crucial role in defense against herbivory. Particularly, diversified secondary metabolites from *Caulerpa* species have been reported to have neuroprotective, anti-inflammatory, anticancer, and antifungal properties (de Sonza et al. 2009; Yang et al. 2017; Liu et al. 2013; Liu et al. 2013; Yu et al. 2017). The alkaloid caulerpin, one of these metabolites, may have therapeutic use in the management of colorectal cancer (Yu et al. 2017). A new sesquiterpene derivative isolated from *Ulva fasciata* was discovered to have selective cytotoxicity against breast cancer cells (Lakshmi et al. 2018). Cladophorols isolated from *Cladophora socialis* demonstrated strong and selective antibacterial action (Lavoie et al. 2019).

A powerful cytotoxic substance known as kahalalide F was derived from *Bryopsis* sp. and has been shown to have antitumoral, antiviral, and antimalarial properties (Hamann et al. 1996; Suárez et al. 2003). Kahalalides, particularly kahalalide F, are certainly produced by an obligate bacterial symbiont of the marine alga as defense chemicals to protect the host from herbivory, according to research by Zan et al. (2019) (Table 8.2). *Elysia rufescens*, a mollusk that consumes *Bryopsis*, is tolerant to kahalalides and uses these chemicals to protect itself from predators (Zan et al. 2019).

8.5 A FUTURE POINT OF VIEW

The current chapter describes the possibility of further investigating green macroalgae in aquaculture and makes the case that this category of macroalgae presents itself as a compelling choice under the current search for commercial diversification of goods and expansion of the industry. Green macroalgae are a significant, but underutilized, aquaculture resource because of their great range of morphological forms, eco-physiological features, bioactive compounds, and propagation potential. Global green macroalgae aquaculture apparently produces about 20,000 tonnes of fresh weight each year, while having comparably smaller production volumes than red and brown seaweeds (FAO 2020). The present chapter draws the conclusion that this value is significantly undervalued because there are no production statistics from Japan, where commercial cultivation of Ulva, Monostroma, and Caulerpa species has been going on for a while (Tanaka et al. 2020). Moreover, new Ulvophycean taxa in aquaculture are still not included in international databases (such as *Codium tomentosum*), and taxonomic uncertainty of species and/or products made from seaweed presents extra difficulties for academic, societal, and commercial participants. These problems demand more thorough molecular analyses and open data sharing of green macroalgae aquaculture products in order to provide more standardized, better-quality products that are traceable—important requirements to boost consumer and industrial confidence (Hafting et al. 2015; Kawashima et al. 2013).

Still, there are a number of general benefits to growing green macroalgal taxa that can be emphasized, including: distinctive diversity; high tolerance to extreme abiotic conditions; high potential for biofiltration; ease of obtaining high-quality seed; unique bioactive compounds; and high market value. For representatives of the main Ulvophycean lineages, cultivation technologies are comparatively well developed. This could serve as crucial background information to encourage further domestication of indigenous species or variants. The establishment of nurseries devoted to creating high-quality propagules of specified preferred strains, prepared to serve farmers upon demand, is also anticipated to be advantageous for green macroalgae aquaculture. For instance, there are currently established commercial nurseries across Europe (e.g. Hortimare). Advances in the manipulation of protoplasts—somatic cells lacking cell walls, capable of regenerating into de novo plants—to manufacture protoplast-derived propagules in aquaculture are also anticipated to improve traditional propagule production techniques. The potential for genetic engineering, genetic transformation, and somatic hybridization processes that will aid the development of improved strains in aquaculture are compelling grounds for the use of propagules formed from protoplasts in aquaculture (Reddy et al. 2008; Charrier et al. 2015). The development of protoplast-mediated propagation techniques may be led by green macroalgae. For several Ulvophyceantaxa (Ulva, Monostroma, Chaetomorpha, and Bryopsis), protoplast isolation and regeneration protocols are well established; however, a detailed model system for high-throughput macroalgal nursery, based on protoplast-derived propagation technologies for large-scale *Ulva* species aquaculture, has recently been proposed (Reddy et al. 2008; Beer and Björk 1994; Reddy et al. 2006; Wu et al. 2018; Gupta et al. 2018).

Macroalgal aquaculture is anticipated to take off in developing markets, particularly in Europe, where large-scale consortium projects and networks have been working together to promote the industry (for example, GENIALG and PHYCOMORPH). These projects and networks have also been assisting in the creation and publication of guidelines for the sustainability of the macroalgal aquaculture industry (Barbier et al. 2020). The following are some of the main forces driving the movement: the desire for alternate sustainable protein sources, the increased perception of macroalgae as sustainable nutritious meals, and growing industrial interest are all contributing factors. Positive signs are emerging in the field of "phycogastronomy," (Mouritsen et al. 2019) using the example of high-end restaurants that currently offer a variety of dishes featuring green macroalgae as the main delicacy ingredient. While changing Western cultures' dietary regimens to include macroalgae in the menu is still a work in progress (van den Burg et al. 2019), these restaurants are demonstrating the potential of this emerging field (e.g. Codium tomentosum, Caulerpa lentillifera) (Pérez-Lloréns et al. 2018). Yet, given the increased interest in employing biorefineries to manufacture a variety of products, it is anticipated that establishing industrial operations using green seaweeds as raw materials will enhance demand (Zollmann et al.2019). In the end, each operation's viability will be determined by the sustainability of the entire production chain.

In order to fully realize the potential of green macroalgae (and other seaweeds), sustainability studies are currently rare but essential. Three factors make a product sustainable: economic, social, and environmental factors (Kloepffer 2008). While designing new products or making existing ones better, these three factors must be accurately evaluated and balanced. An organized, thorough, and internationally standardized method for assessing the environmental effects of the bioeconomy is lifecycle assessment (LCA) (Giuntoli et al. 2019). The LCA seeks to evaluate potential environmental effects throughout the life cycle of a system, process, or product. The idea of life cycle sustainability assessment (LCSA) has been presented as sustainability and green markets gain popularity worldwide (Kloepffer 2008; Zamagni et al. 2013). The inclusion of fairness in the environmental costs makes LCSA innovative. To arrive at a single common life cycle sustainability outcome, conclusions are formed from a variety of weighing factors, including financial profitability, market demand, social acceptability, willingness to pay, and bioeconomic modeling. Although there are now no directly applicable methods for LCSA applied to macroalgae aquaculture (Banach et al. 2020), emerging LCSA case studies illustrate the potential advantages of carrying out such analysis. Improvements can be made to boost sustainability by locating hotspots within the aquaculture value chain. For instance, LCA studies on the large-scale cultivation of Saccharinalatissima, a brown alga, suggested that seaweed farming has the potential to develop into a lucrative industry in north-west Europe (Taelman et al. 2015; Hasselström et al. 2020), with great promise to reduce the resource footprint of seaweed cultivation once energetic efficiency is improved (electricity and transport), biomass productivity increases (Taelman et al. 2015; Halfdanarson et al. 2019), and/or improvements are made in infrastructural component materials (e.g. stainless steel) (van Oirschot et al. 2017). Similarly to this, a preliminary LCA study conducted to assess the environmental effects of producing bioplastics from green macroalgae (*Ulva* spp.) allowed the identification of the main impacting hotspots in the production chain, those related to high energetic costs of cultivation in land-based systems and material efficiency (Helmes et al. 2018).Due to the fact that the suggested design changes can be implemented without making sizable financial investments, conducting LCAs at the beginning of the implementation of green macroalgal aquaculture and conversion will become attractive in this context (Seghetta and Goglio 2020). Overall, these findings indicate that macroalgae aquaculture sustainability will be significantly increased by an effective transition to renewable energies and more environmentally acceptable materials. Collaborations between the academic and business worlds will make it possible to establish diverse approaches for evaluating macroalgae aquaculture activity and win over the public.

Notwithstanding the prospects that green macroalgae currently have in aquaculture, it is unlikely that the production of this particular subset of seaweeds will ever match that of their red and brown macroalgal counterparts. It is more likely that green macroalgae will continue to fill specialized market niches, where high-value products are preferred over large-scale production that results in low-cost biomass. This has been the case in the Asia-Pacific region, where green macroalgae have been raised to generate specialist foods that fill relatively narrow, localized market niches within certain cultural backgrounds. In turn, a developing, large-scale, and increasingly commoditized business is supplied by aquaculture of hydrocolloid-bearing macroalgal species (red and brown) (Porse and Rudolph 2017). The future of seaweed aquaculture in developing markets (like Europe) may not necessarily depend on producing large amounts of biomass but instead focus on adding value to the seaweed production chain, through by-product development (such as fodder, fertilizers, and bioactive compounds), and capitalizing on societal benefits. This will help to overcome the competitive advantages of the seaweed-hydrocolloid industry thriving in Asia (e.g. economic, environmental) (van den Burg et al. 2019). Various biochemical profiles and organoleptic properties, high nutrient uptake and propagation capacity, a variety of cultivation techniques, and tolerance to extreme conditions are just a few of the distinctive qualities Ulvophycean macroalgae offer. These characteristics represent a remarkable potential for innovation in macroalgal aquaculture.

REFERENCES

Ahmed, O.M. and Ahmed, R.R. 2014. Anti-proliferative and apoptotic efficacies of ulvan polysaccharides against different types of carcinoma cells *in vitro* and *in vivo*. *J. Cancer Sci. Ther.*6(6):202–208.

Akila, V., Manikandan, A., SahayaSukeetha, D., Balakrishnan, S., Ayyasamy, P.M. and Rajakumar, S. 2019. Biogas and biofertilizer production of marine macroalgae: An effective anaerobic digestion of *Ulva* sp. *Biocatal. Agric. Biotechnol.*18:101035.

Al-Homaidan, A.A., Al-Qahtani, H.S., Al-Ghanayem, A.A., Ameen, F. and Ibraheem, I.B.M. 2018. Potential use of green algae as a biosorbent for hexavalent chromium removal from aqueous solutions. *Saudi J. Biol. Sci.*25(8):1733–1738.

Alves, A., Duarte, A.R.C., Mano, J.F., Sousa, R.A. and Reis, R.L. 2012. PDLLA enriched with ulvan particles as a novel 3D porous scaffold targeted for bone engineering. *J. Supercrit. Fluids* 65: 32–38.

Alves, A., Pinho, E.D., Neves, N.M., Sousa, R.A. and Reis, R.L. 2012. Processing ulvan into 2D structures: Cross-linked ulvan membranes as new biomaterials for drug delivery applications. *Int. J. Pharm.*426(1–2):76–81.

Alves, A., Sousa, R.A. and Reis, R.L. 2013. A practical perspective on ulvan extracted from green algae. *J. Appl. Phycol.*25(2):407–424.

Banach, J.L., den Hil, E.F.H. and van der Fels-Klerx, H.J. 2020. Food safety hazards in the European seaweed chain. *Compr. Rev. Food Sci. Food Saf.*19(2):332–364.

Bang, T.H., Van, T.T.T., Hung, L.X., Ly, B.M., Nhut, N.D., Thuy, T.T.T. and Huy, B.T. 2019. Nanogels of acetylated ulvan enhance the solubility of hydrophobic drug curcumin. *Bull. Mater. Sci.*42(1):1.

Barbier, M., Araújo, R., Rebours, C., Jacquemin, B., Holdt, S.L. and Charrier, B. 2020. Development and objectives of the Phycomorph European Guidelines for the Sustainable Aquaculture of Seaweeds (PEGASUS). *Bot Mar.*63(1):5–16.

Beer, S. and Björk, M. 1994. Photosynthetic properties of protoplasts, as compared with thalli, of *Ulva fasciata* (Chlorophyta). *J. Phycol.*30(4):633–637.

Buchholz, C.M., Krause, G. and Buck, B.H. 2012. Seaweed and man. In Wiencke, C. and Bischof, K. (Eds.), *Seaweed Biology: Novel Insights into Ecophysiology, Ecology and Utilization*. Berlin Heidelberg: Springer-Verlag, pp. 471–493.

Buschmann, A.H. and Camus, C. 2019. An introduction to farming and biomass utilisation of marine macroalgae. *Phycologia*58(5):443–445.

Buschmann, A.H., Camus, C., Infante, J., Neori, A., Israel, Á., Hernández-González, M.C., Pereda, S.V., Gomez-Pinchetti, J.L., Golberg, A., Tadmor-Shalev, N. and Critchley, A.T. 2017. Seaweed production:overview of the global state of exploitation, farming and emerging research activity. *Eur. J. Phycol.* 52(4):391–406.

Cassolato, J.E.F., Noseda, M.D., Pujol, C.A., Pellizzari, F.M., Damonte, E.B. and Duarte, M.E.R. 2008. Chemical structure and antiviral activity of the sulfated heterorhamnan isolated from the green seaweed *Gayraliaoxysperma*. *Carbohydr. Res.*343(18):3085–3095.

Charrier, B., Abreu, M.H., Araujo, R., Bruhn, A., Coates, J.C., De Clerck, O., Katsaros, C., Robaina, R.R. and Wichard, T. 2017. Furthering knowledge of seaweed growth and development to facilitate sustainable aquaculture. *New Phytol.* 216(4):967–975.

Charrier, B., Rolland, E., Gupta, V. and Reddy, C.R.K. 2015. Production of genetically and developmentally modified seaweeds: Exploiting the potential of artificial selection techniques. *Front. Plant Sci.*6:127.

Cho, M., Lee, H.S., Kang, I.J., Won, M.H. and You, S. 2011. Antioxidant properties of extract and fractions from *Enteromorpha prolifera*, a type of green seaweed. *Food Chem.*127(3):999–1006.

Ciancia, M., Quintana, I., Vizcargüénaga, M.I., Kasulin, L., de Dios, A., Estevez, J.M. and Cerezo, A.S. 2007. Polysaccharides from the green seaweeds *Codium fragile* and *C. vermilara* with controversial effects on hemostasis. *Int. J. Biol. Macromol.*41(5):641–649.

Cortez, R., Luna-Vital, D.A., Margulis, D. and de Mejia, E.G. 2017. Natural pigments: Stabilization methods of anthocyanins for food applications. *Compr. Rev. Food Sci. Food Saf.*16(1):180–198.

Dan, A., Hiraoka, M., Ohno, M. and Critchley, A.T. 2002. Observations on the effect of salinity and photon fluence rate on the induction of sporulation and rhizoid formation in the green alga *Enteromorpha prolifera* (Müller). J. Agardh (Chlorophyta, Ulvales). *Fish. Sci.*68(6):1182–1188.

Dash, M., Samal, S.K., Bartoli, C., Morelli, A., Smet, P.F., Dubruel, P. and Chiellini, F. 2014. Biofunctionalization of ulvan scaffolds for bone tissue engineering. *ACS Appl. Mater. Interfaces*6(5):3211–3218.

Dash, M., Samal, S.K., Morelli, A., Bartoli, C., Declercq, H.A., Douglas, T.E., Dubruel, P. and Chiellini, F. 2018. Ulvan-chitosan polyelectrolyte complexes as matrices for enzyme induced biomimetic mineralization. *Carbohydr. Polym.*182:254–264.

Dawes, C. 2016. Macroalgae systematics. In Fleurence, J. and Levine, I. (Eds.), *Seaweed in Health and Disease Prevention*. San Diego, CA: Academic Press, pp. 107–148.

de Gaillande, C., Payri, C., Remoissenet, G. and Zubia, M. 2017. Caulerpa consumption, nutritional value and farming in the Indo-Pacific region. *J. Appl. Phycol.*29(5):2249–2266.

de Souza, É.T., Pereira de Lira, D., Cavalcanti de Queiroz, A., et al. 2009. The antinociceptive and anti-inflammatory activities of caulerpin, a bisindole alkaloid isolated from seaweeds of the genus *Caulerpa*. *Mar. Drugs*7(4):689–704.

Dinoro, J., Maher, M., Talebian, S., et al. 2019. Sulfated polysaccharide-based scaffolds for orthopaedic tissue engineering. *Biomaterials*214:119214.

Doh, H., Lee, M.H. and Whiteside, W.S. 2020. Physicochemical characteristics of cellulose nanocrystals isolated from seaweed biomass. *Food Hydrocoll.*102:105542.

Ebrahimi, A., Hashemi, S., Akbarzadeh, S. and Ramavandi, B. 2016. Modification of green algae harvested from the Persian Gulf by L-cysteine for enhancing copper adsorption from wastewater: Experimental data. *Chem. Data Collect.*2:36–42.

El Baz, F.K., Baroty, G.S.E., Baky, H.H.A.E., El-Salam, O.I.A. and Ibrahim, E.A. 2013. Structural characterization and biological activity of sulfolipids from selected marine algae. *Grasas Aceites*64(5):561–571.

El-Sikaily, A., El Nemr, A., Khaled, A. and Abdelwehab, O. 2007. Removal of toxic chromium from wastewater using green alga *Ulva lactuca* and its activated carbon. *J. Hazard. Mater.*148(1–2):216–228.

FAO. 2020. Fishery and aquaculture statistics. Global Aquaculture Production 1950–2018 (FishstatJ); Updated July 22, 2020.

Fernando, I.P.S., Sanjeewa, K.K.A., Samarakoon, K.W., Lee, W.W., Kim, H.S. and Jeon, Y.J. 2018. Squalene isolated from marine macroalgae *Caulerpa racemosa* and its potent antioxidant and anti-inflammatory activities. *J. Food Biochem.*42(5):e12628.

Gajaria, T.K., Suthar, P., Baghel, R.S., Balar, N.B., Sharnagat, P., Mantri, V.A. and Reddy, C. 2017. Integration of protein extraction with a stream of byproducts from marine macroalgae: A model forms the basis for marine bioeconomy. *Bioresour. Technol.*243:867–873.

Ganesan, A.R., Shanmugam, M. and Bhat, R. 2018. Producing novel edible films from semi refined carrageenan (SRC) and ulvan polysaccharides for potential food applications. *Int. J. Biol. Macromol.*112:1164–1170.

Ganesan, P., Matsubara, K., Ohkubo, T., Tanaka, Y., Noda, K., Sugawara, T. and Hirata, T. 2010. Anti-angiogenic effect of siphonaxanthin from green alga. *Codium fragile*. *Phytomedicine*17(14):1140–1144.

Ganesan, P., Noda, K., Manabe, Y., Ohkubo, T., Tanaka, Y., Maoka, T., Sugawara, T. and Hirata, T. 2011. Siphonaxanthin, a marine carotenoid from green algae, effectively induces apoptosis in human leukemia (HL-60) cells. *Biochim. Biophys. Acta - Gen. Subj.* 1810(5):497–503.

Gao, X., Choi, H.G., Park, S.K., Sun, Z.M. and Nam, K.W. 2019. Assessment of optimal growth conditions for cultivation of the edible *Caulerpa okamurae* (Caulerpales, Chlorophyta) from Korea. *J. Appl. Phycol.*31(3):1855–1862.

Giuntoli, J., Bulgheroni, C., Marelli, L., Sala, S. and Pant, R. 2019. Brief on the use of Life Cycle Assessment (LCA) to evaluate environmental impacts of the bioeconomy. In Lusser, M. and Avraamides, M. (Eds). Luxembourg: Publications Office of the European Union, pp. 1–8.

Gokulan, R., Ganesh Prabhu, G. and Jegan, J. 2019. A novel sorbent *Ulvalactuca* derived biochar for remediation of Remazol brilliant Orange 3R in packed column. *Water Environ. Res.*91(7):642–649.

Guidara, M., Yaich, H., Richel, A., Blecker, C., Boufi, S., Attia, H. and Garna, H. 2019. Effects of extraction procedures and plasticizer concentration on the optical, thermal, structural and antioxidant properties of novel ulvan films. *Int. J. Biol. Macromol.*135:647–658.

Guiry, M.D. and Guiry, G.M. 2021. *AlgaeBase, World-Wide Electronic Publication*, National University of Ireland, Galway, Ireland, 2013.

Gupta, V., Trivedi, N., Simoni, S. and Reddy, C.R.K. 2018. Marine macroalgal nursery: A model for sustainable production of seedlings for large scale farming. *Algal Res.*31:463–468.

Hafting, J.T., Craigie, J.S., Stengel, D.B., Loureiro, R.R., Buschmann, A.H., Yarish, C., Edwards, M.D. and Critchley, A.T. 2015. Prospects and challenges for industrial production of seaweed bioactives. *J. Phycol.* 51(5):821–837.

Hamann, M.T., Otto, C.S., Scheuer, P.J. and Dunbar, D.C. 1996. Kahalalides: Bioactive peptides from a marine mollusk *Elysia rufescens* and its algal diet *Bryopsis* sp. *J. Org. Chem.*61(19):6594–6600.

Hasselström, L., Thomas, J.B., Nordström, J., Cervin, G., Nylund, G.M., Pavia, H. and Gröndahl, F. 2020. Socioeconomic prospects of a seaweed bioeconomy in Sweden. *Sci. Rep.*10(1):1610.

Helmes, R.J.K., López-Contreras, A.M., Benoit, M., Abreu, H., Maguire, J., Moejes, F. and Burg, S. 2018. Environmental impacts of experimental production of lactic acid for Bioplastics from *Ulva* spp. *Sustainability*10(7):2462.

Hiqashi-Okaj, K., Otani, S. and Okai, Y. 1999. Potent suppressive effect of a Japanese edible seaweed, *Enteromorpha prolifera* (Sujiao-nori) on initiation and promotion phases of chemically induced mouse skin tumorigenesis. *Cancer Lett.*140(1–2):21–25.

Horstmann, U. 1983. Cultivation of the green alga, *Caulerpa racemosa*, in tropical waters and some aspects of its physiological ecology. *Aquaculture*32(3–4):361–371.

Hosokawa, M., Wanezaki, S., Miyauchi, K., et al. 1999. Apoptosis-inducing effect of fucoxanthin on human leukemia cell line HL-60. *Food Sci. Technol. Res.*5(3):243–246.

Hwang, E.K., Amano, H. and Park, C.S. 2008. Assessment of the nutritional value of *Capsosiphonfulvescens* (Chlorophyta): Developing a new species of marine macroalgae for cultivation in Korea. *J. Appl. Phycol.*20(2):147–151.

Hwang, E.K., Baek, J.M. and Park, C.S. 2005. Artificial seed production and nursery culture conditions using regeneration of isolated utricles and medullary filaments of *Codium fragile* (Suringar) Hariot. *Korean J. Fish Aquat. Sci.*38(6):393–398.

Hwang, E.K., Yi, Y.H., Shin, W.J., and Sohn, C.H. (2003). Growth and maturation of a green alga, Capsosiphon fulvescens, as a new candidate for seaweed cultivation in Korea. In *Proceedings of the 17th International Seaweed Symposium, Cape Town, South Africa, 28 January-2 February 2001* (pp. 59–64). Oxford University Press, Oxford.

Ivanova, V., Rouseva, R., Kolarova, M., Serkedjieva, J., Rachev, R. and Manolova, N. 1994. Isolation of a polysaccharide with antiviral effect from *Ulva lactuca. Prep. Biochem.*24(2):83–97.

Kalavathy, G. and Baskar, G. 2019. Synergism of clay with zinc oxide as nanocatalyst for production of biodiesel from marine *Ulva lactuca. Bioresour. Technol.*281:234–238.

Kawashima, Y., Akasaki, T., Matsumoto, Y., Yamazaki, Y. and Shimada, S. 2013. Species identification of imported and Japanese commercial green algal products based on phylogenetic analyses using the nrITS2 and 5S rDNA spacer regions. *Fish. Sci.*79(3):521–529.

Kazłowski, B., Chiu, Y.H., Kazłowska, K., Pan, C.L. and Wu, C.J. 2012. Prevention of Japanese encephalitis virus infections by low-degree- polymerisation sulfated saccharides from *Gracilaria* sp. and *Monostromanitidum. Food Chem.*133(3):866–874.

Kendel, M., Wielgosz-Collin, G., Bertrand, S., Roussakis, C., Bourgougnon, N. and Bedoux, G. 2015. Lipid composition, fatty acids and sterols in the seaweeds *Ulva armoricana*, and *Solieriachordalis* from Brittany (France): An analysis from nutritional, chemotaxonomic, and antiproliferative activity perspectives. *Mar. Drugs*13(9):5606–5628.

Khan, A.M., Fatima, N., Hussain, M.S. and Yasmeen, K. 2016. Biodiesel production from green seaweed *Ulva fasciata* catalyzed by novel waste catalysts from Pakistan steel Industry. *Chin. J. Chem. Eng.*24(8):1080–1086.

Kida, W. 1990. Culture of seaweeds *Monostroma. Mar. Behav. Physiol.*16(2):109–131.

Kidgell, J.T., Magnusson, M., de Nys, R. and Glasson, C.R.K. 2019. Ulvan: A systematic review of extraction, composition and function. *Algal Res.*39:101422.

Kikionis, S., Ioannou, E., Toskas, G. and Roussis, V. 2015. Electrospunbiocomposite nanofibers of ulvan/PCL and ulvan/PEO. *J. Appl. Polym. Sci.*132(26):42153.

Kim, A.D., Lee, Y., Kang, S.H., Kim, G.Y., Kim, H.S. and Hyun, J.W. 2013. Cytotoxic effect of clerosterol isolated from *Codium fragile* on A2058 human melanoma cells. *Mar. Drugs*11(2):418–430.

Kim, J.K., Cho, M.L., Karnjanapratum, S., Shin, I.S. and You, S.G. 2011. *In vitro* and *in vivo* immunomodulatory activity of sulfated polysaccharides from *Enteromorpha prolifera. Int. J. Biol. Macromol.*49(5):1051–1058.

Kloepffer, W. 2008. Life cycle sustainability assessment of products. *Int. J. Life Cycle Assess.*13(2):89.

Kumar, Y.P., King, P. and Prasad, V. 2006. Removal of copper from aqueous solution using *Ulva fasciata* sp. – A marine green algae. *J. Hazard. Mater.*137(1):367–373.

Lahaye, M. and Axelos, M.A.V. 1993. Gelling properties of water-soluble polysaccharides from proliferating marine green seaweeds (*Ulva* spp.). *Carbohydr. Polym.*22(4):261–265.

Lahaye, M. and Robic, A. 2007. Structure and functional properties of ulvan, a polysaccharide from green seaweeds. *Biomacromolecules*8(6):1765–1774.

Lakshmi, T.P., Vajravijayan, S., Moumita, M., Sakthivel, N., Gunasekaran, K. and Krishna, R. 2018. A novel guaiane sesquiterpene derivative, guai-2-en-10α-ol, from *Ulvafasciata* Delile inhibits EGFR/PI3K/Akt signaling and induces cytotoxicity in triple-negative breast cancer cells. *Mol. Cell. Biochem.*438(1–2):123–139.

Lavoie, S., Sweeney-Jones, A.M., Mojib, N., et al. 2019. Antibacterial oligomeric polyphenols from the green alga *Cladophorasocialis. J. Org. Chem.*84(9):5035–5045.

Leal, M.C., Munro, M.H., Blunt, J.W., Puga, J., Jesus, B., Calado, R., Rosa, R. and Madeira, C. 2013. Biogeography and biodiscovery hotspots of macroalgal marine natural products. *Nat. Prod. Rep.* 30(11):1380–1390.

Lee, C., Park, G.H., Ahn, E.M., Kim, B.A., Park, C.I. and Jang, J.H. 2013. Protective effect of *Codium fragile* against UVB-induced proinflammatory and oxidative damages in HaCaT cells and BALB/c mice. *Fitoterapia*86:54–63.

Leiro, J.M., Castro, R., Arranz, J.A. and Lamas, J. 2007. Immunomodulating activities of acidic sulphated polysaccharides obtained from the seaweed *Ulva rigida* C, Agardh. *Int. Immunopharmacol.*7(7):879–888.

Levine, I. 2016. Algae: A way of life and health. In Fleurence, J. and Levine, I. (Eds.), *Seaweed in Health and Disease Prevention.* San Diego, CA: Academic Press, pp. 1–5.

Li, H., Mao, W., Zhang, X., Qi, X., Chen, Y., Chen, Y., Xu, J., Zhao, C., Hou, Y., Yang, Y. and Li, N. 2011. Structural characterization of an anticoagulant-active sulfated polysaccharide isolated from green alga Monostroma latissimum. *Carbohydr. Polym.* 85(2):394–400.

Liang, X.P., Ma, C., Yan, X., Liu, X. and Liu, F. 2019. Advances in research on bioactivity, metabolism, stability and delivery systems of lycopene. *Trends Food Sci. Technol.*93:185–196.

Liu, A.H., Liu, D.Q., Liang, T.J., Yu, X.Q., Feng, M.T., Yao, L.G., Fang, Y., Wang, B., Feng, L.H., Zhang, M.X. and Mao, S.C. 2013. Caulerprenylols A and B, two rare antifungal prenylated para-xylenes from the green alga Caulerpa racemosa. *Bioorganic Med. Chem. Lett.* 23(9):2491–2494.

Liu, D.Q., Mao, S.C., Zhang, H.Y., Yu, X.Q., Feng, M.T., Wang, B., Feng, L.H. and Guo, Y.W. 2013. Racemosins A and B, two novel bisindole alkaloids from the green alga Caulerpa racemosa. *Fitoterapia*, 91:15–20.

Liu, X., Wang, S., Cao, S., He, X., Qin, L., He, M., Yang, Y., Hao, J. and Mao, W. 2018. Structural characteristics and anticoagulant property in vitro and in vivo of a seaweed sulfated rhamnan. *Mar. Drugs*, 16(7):243.

Loureiro, R., Gachon, C.M.M. and Rebours, C. 2015. Seaweed cultivation: Potential and challenges of crop domestication at an unprecedented pace. *New Phytol.*206(2):489–492.

Maciel, E., Leal, M.C., Lillebø, A.I., Domingues, P., Domingues, M.R. and Calado, R. 2016. Bioprospecting of marine macrophytes using MS-based lipidomics as a new approach. *Mar. Drugs*14(3):49.

Margareta, W., Nagarajan, D., Chang, J.S. and Lee, D.J. 2020. Dark fermentative hydrogen production using macroalgae (*Ulva* sp.) as the renewable feedstock. *Appl. Energy*262:114574.

Massad, I.P., Ávila, M., Contreras-Porcia, L. and Bulboa, C.C. 2020. Spores resuspending technology, a new system improving spore seeding for culture of commercial red seaweeds. *Aquaculture*526:735374.

Massironi, A., Morelli, A., Grassi, L., Puppi, D., Braccini, S., Maisetta, G., Esin, S., Batoni, G., Della Pina, C. and Chiellini, F. 2019. Ulvan as novel reducing and stabilizing agent from renewable algal biomass: Application to green synthesis of silver nanoparticles. *Carbohydr. Polym.* 203:310–321.

Mezghani, S., Bourguiba, I., Hfaiedh, I. and Amri, M. 2013. Antioxidant potential of *Ulva rigida* extracts: Protection of HeLa cells against H_2O_2 cytotoxicity. *Biol. Bull.*225(1):1–7.

Mhatre, A., Gore, S., Mhatre, A., Trivedi, N., Sharma, M., Pandit, R., Anil, A. and Lali, A. 2019. Effect of multiple product extractions on bio-methane potential of marine macrophytic green alga *Ulva lactuca. Renew. Energy*132:742–751.

Morelli, A., Betti, M., Puppi, D. and Chiellini, F. 2016. Design, preparation and characterization of ulvan based thermosensitive hydrogels. *Carbohydr. Polym.*136:1108–1117.

Morelli, A. and Chiellini, F. 2010. Ulvan as a new type of biomaterial from renewable resources: Functionalization and hydrogel preparation. *Macromol. Chem. Phys.*211(7):821–832.

Morelli, A., Massironi, A., Puppi, D., Creti, D., Domingo Martinez, E., Bonistalli, C., Fabroni, C., Morgenni, F. and Chiellini, F. 2019. Development of ulvan-based emulsions containing flavour and fragrances for food and cosmetic applications. *Flavour Fragr. J.* 34(6):411–425.

Mouritsen, O.G., Rhatigan, P. and Pérez-Lloréns, J.L. 2019. The rise of seaweed gastronomy: Phycogastronomy. *Bot Mar.*62(3):195–209.

Newton, I.S. 1996. Long chain fatty acids in health and nutrition. *J. Food Lipids*3(4):233–249.

Nisizawa, K., Noda, H., Kikuchi, R. and Watanabe, T. 1987. The main seaweed foods in Japan. *Hydrobiologia*151(1):5–29.

Odom, R.L. and Walters, L.J. 2014. A safe alternative to invasive *Caulerpa taxifolia* (Chlorophtya)? Assessing aquarium-release invasion potential of aquarium strains of the macroalgal genus *Chaetomorpha* (Chlorophyta). *Biol. Invas.*16(8):1589–1597.

Ohta, Y., Lee, J.B., Hayashi, K. and Hayashi, T. 2009. Isolation of sulfated galactan from *Codium fragile* and its antiviral effect. *Biol. Pharm. Bull.*32(5):892–898.

Okai, Y. and Higashi-Okai, K. 1997. Potent anti-inflammatory activity of pheophytin a derived from edible green alga, *Enteromorpha prolifera* (Sujiao-nori). *Int. J. Immunopharmacol.*19(6):355–358.

Pangestuti, R. and Kim, S.K. 2011. Biological activities and health benefit effects of natural pigments derived from marine algae. *J. Funct. Foods*3(4):255–266.

Patel, S. 2012. Therapeutic importance of sulfated polysaccharides from seaweeds: Updating the recent findings. *Biotech*2(3):171–185.

Pereira, H., Barreira, L., Figueiredo, F., Custódio, L., Vizetto-Duarte, C., Polo, C., Rešek, E., Engelen, A. and Varela, J. 2012. Polyunsaturated fatty acids of marine macroalgae: Potential for nutritional and pharmaceutical applications. *Mar. Drugs*, 10(9):1920–1935.

Pereira, L. 2018. Seaweeds as source of bioactive substances and skin care therapycosmeceuticals, algotheraphy, and thalassotherapy. *Cosmetics*5(4):68.

Pérez-Lloréns, J.L., Hernández, I., Vergara, J.J., Brun, F.G. and León, Á. 2018. *Those Curious and Delicious Seaweeds: A Fascinating Voyage from Biology to Gastronomy.* Cádiz: Servicio de Publicaciones de la Universidad de Cádiz.

Porse, H. and Rudolph, B. 2017. The seaweed hydrocolloid industry: 2016 updates, requirements, and outlook. *J. Appl. Phycol.*29(5):2187–2200.

Potts, T., Du, J., Paul, M., May, P., Beitle, R. and Hestekin, J. 2012. The production of butanol from Jamaica Bay macro algae. *Environ. Prog. Sustain. Energy*31(1):29–36.

Qi, H., Zhang, Q., Zhao, T., Chen, R., Zhang, H., Niu, X. and Li, Z. 2005. Antioxidant activity of different sulfate content derivatives of polysaccharide extracted from *Ulva pertusa* (Chlorophyta) *in vitro. Int. J. Biol. Macromol.*37(4):195–199.

Radulovich, R., Neori, A., Valderrama, D., Reddy, C.R.K., Cronin, H. and Forster, J. 2015. Farming of seaweeds. In Tiwari, B. K. and Troy, D. J. (Eds.), *Seaweed Sustainability.* San Diego, CA: Academic Press, pp. 27–59.

Rebours, C., Marinho-Soriano, E., Zertuche-González, J.A., Hayashi, L., Vásquez, J.A., Kradolfer, P., Soriano, G., Ugarte, R., Abreu, M.H., Bay-Larsen, I. and Hovelsrud, G. 2014. Seaweeds:an opportunity for wealth and sustainable livelihood for coastal communities. *J. Appl. Phycol.* 26:1939–1951.

Reddy, C.R.K., Dipakkore, S., Kumar, G.R., Jha, B., Cheney, D.P. and Fujita, Y. 2006. An improved enzyme preparation for rapid mass production of protoplasts as seed stock for aquaculture of macrophytic marine green algae. *Aquaculture*260(1–4):290–297.

Reddy, C.R.K., Gupta, M.K., Mantri, V.A. and Jha, B. 2008. Seaweed protoplasts: Status, biotechnological perspectives and needs. *J. Appl. Phycol.*20(5):619–632.

Rey, F., Cartaxana, P., Melo, T., et al. 2020. Domesticated populations of *Codium tomentosum* display lipid extracts with lower seasonal shifts than conspecifics from the wild-relevance for biotechnological applications of this green seaweed. *Mar. Drugs*18(4):188.

Roberts, D.A. and de Nys, R. 2016. The effects of feedstock pre-treatment and pyrolysis temperature on the production of biochar from the green seaweed *Ulva. J. Environ. Manag.*169:253–260.

Sanjeewa, K.K.A., Lee, W. and Jeon, Y.J. 2018. Nutrients and bioactive potentials of edible green and red seaweed in Korea. *Fish. Aquat. Sci.*21(1):19.

Santelices, B. 1999. A conceptual framework for marine agronomy. *Hydrobiologia*398:15–23.

Schmid, M., Kraft, L.G.K., van der Loos, L., Kraft, G.T., Virtue, P., Nichols, P.D. and Hurd, C.L. 2018. Southern Australian seaweeds: A promising resource for omega-3-fattyacids. *Food Chem.*265:70–77.

Schuenhoff, A., Shpigel, M., Lupatsch, I., Ashkenazi, A., Msuya, F.E. and Neori, A. 2003. A semi-recirculating, integrated system for the culture of fish and seaweed. *Aquaculture*221(1–4):167–181.
Seghetta, M. and Goglio, P. 2020. Life cycle assessment of seaweed cultivation systems. In Spilling, K. (Ed.), *Biofuels from Algae: Methods in Molecular Biology, vol 1980*. New York: Humana, pp. 103–119.
Segi, T. and Kida, W. 1968. Synopsis of biological data on *Monostromalatissimum* Wittrock in Japanese cultivation. *FAO Fish. Synop.*39:1–14.
Shpigel, M., Guttman, L., Shauli, L., Odintsov, V., Ben-Ezra, D. and Harpaz, S. 2017. *Ulva lactuca* from an integrated multi-trophic aquaculture (IMTA) biofilter system as a protein supplement in gilthead seabream (*Sparus aurata*) diet. *Aquaculture*481:112–118.
Siddhanta, A.K., Shanmugam, M., Mody, K.H., Goswami, A.M. and Ramavat, B.K. 1999. Sulphated polysaccharides of *Codiumdwarkense* Boergs. from the west coast of India: Chemical composition and blood anticoagulant activity. *Int. J. Biol. Macromol.*26(2–3):151–154.
Skrzypczyk, V., Hermon, K., Norumbuena, F., Turchini, G., Keast, R. and Bellgrove, A. 2019. Is Australian seaweed worth eating? Nutritional and sensorial properties of wild-harvested Australian versus commercially-produced Japanese seaweeds. *J. Appl. Phycol.*31(1):709–724.
Stabili, L., Acquaviva, M.I., Angilè, F., Cavallo, R.A., Cecere, E., Del Coco, L., Fanizzi, F.P., Gerardi, C., Narracci, M. and Petrocelli, A. 2019. Screening of Chaetomorpha linum lipidic extract as a new potential source of bioactive compounds. *Mar. Drugs*, 17(6):313.
Stabili, L., Acquaviva, M.I., Biandolino, F., Cavallo, R.A., De Pascali, S.A., Fanizzi, F.P., Narracci, M., Cecere, E. and Petrocelli, A. 2014. Biotechnological potential of the seaweed Cladophora rupestris (Chlorophyta, Cladophorales) lipidic extract. *N Biotechnol.* 31(5):436–444.
Suárez, Y., González, L., Cuadrado, A., Berciano, M., Lafarga, M. and Muñoz, A. 2003. Kahalalide F, A new marine-derived compound, induces oncosis in human prostate and breast cancer cells. *Mol. Cancer Ther.*2(9):863–872.
Sucaldito, M.R. and Camacho, D.H. 2017. Characteristics of unique HBr-hydrolyzed cellulose nanocrystals from freshwater green algae (*Cladophora rupestris*) and its reinforcement in starch-based film. *Carbohydr. Polym.*169:315–323.
Sugawara, T., Ganesan, P., Li, Z., Manabe, Y. and Hirata, T. 2014. Siphonaxanthin, a green algal carotenoid, as a novel functional compound. *Mar. Drugs*12(6):3660–3668.
Sun, S.M., Yang, S.H., Golokhvast, K.S., Le, B. and Chung, G. 2016. Reconstructing the phylogeny of *Capsosiphonfulvescens* (Ulotrichales, Chlorophyta) from Korea based on rbcL and 18S rDNA sequences. *BioMed Res. Int.*2016:1462916.
Taelman, S.E., Champenois, J., Edwards, M.D., De Meester, S. and Dewulf, J. 2015. Comparative environmental life cycle assessment of two seaweed cultivation systems in North West Europe with a focus on quantifying sea surface occupation. *Algal Res.*11:173–183.
Tanaka, K., Ohno, M. and Largo, D.B. 2020. An update on the seaweed resources of Japan. *Bot Mar.*63(1):105–117.
Tanduyan, S.N., Gonzaga, R.B. and Bensig, V.D. 2013. Off bottom culture of *Caulerpa lentillifera* in three different water levels in the marine waters of San Francisco, Cebu, Philippines. *Galaxea*15:123–132.
Tanna, B., Choudhary, B. and Mishra, A. 2018. Metabolite profiling, antioxidant, scavenging and anti-proliferative activities of selected tropical green seaweeds reveal the nutraceutical potential of Caulerpa spp. *Algal Res.*36:96–105.
Thurstan, R.H., Brittain, Z., Jones, D.S., Cameron, E., Dearnaley, J. and Bellgrove, A. 2018. Aboriginal uses of seaweeds in temperate Australia: An archival assessment. *J. Appl. Phycol.*30(3):1821–1832.
Titlyanov, E.A. and Titlyanova, T.V. 2010. Seaweed cultivation: Methods and problems. *Russ. J. Mar. Biol.*36(4):227–242.
Toskas, G., Heinemann, S., Heinemann, C., Cherif, C., Hund, R.D., Roussis, V. and Hanke, T. 2012. Ulvan and ulvan/chitosan polyelectrolyte nanofibrous membranes as a potential substrate material for the cultivation of osteoblasts. *Carbohydr. Polym.*89(3):997–1002.
Toskas, G., Hund, R.D., Laourine, E., Cherif, C., Smyrniotopoulos, V. and Roussis, V. 2011. Nanofibers based on polysaccharides from the green seaweed *Ulva rigida*. *Carbohydr. Polym.*84(3):1093–1102.
Tsubaki, S., Oono, K., Hiraoka, M., Onda, A. and Mitani, T. 2016. Microwave-assisted hydrothermal extraction of sulfated polysaccharides from *Ulva* spp. and *Monostromalatissimum*. *Food Chem.*210:311–316.
Tsutsui, I., Miyoshi, T., Aue-umneoy, D., Songphatkaew, J., Meeanan, C., Klomkling, S., Sukchai, H., Pinphoo, P., Yamaguchi, I., Ganmanee, M. and Maeno, Y. 2015. High tolerance of Chaetomorpha sp. to salinity and water temperature enables survival and growth in stagnant waters of central Thailand. *Int. Aquat. Res.* 7:47–62.
Tziveleka, L.A., Ioannou, E. and Roussis, V. 2019. Ulvan, a bioactive marine sulphated polysaccharide as a key constituent of hybrid biomaterials: A review. *Carbohydr. Polym.*218:355–370.

Tziveleka, L.A., Pippa, N., Georgantea, P., Ioannou, E., Demetzos, C. and Roussis, V. 2018. Marine sulfated polysaccharides as versatile polyelectrolytes for the development of drug delivery nanoplatforms: Complexation of ulvan with lysozyme. *Int. J. Biol. Macromol.*118(A):69–75.

Valero, M., Guillemin, M.L., Destombe, C., Jacquemin, B., Gachon, C.M., Badis, Y., Buschmann, A.H., Camus, C. and Faugeron, S. 2017. Perspectives on domestication research for sustainable seaweed aquaculture. *Perspect. Psychol. Sci.* 4(1):33–46.

van den Burg, S.W.K., Dagevos, H. and Helmes, R.J.K. 2019. Towards sustainable European seaweed value chains: A triple P perspective. *ICES J. Mar. Sci.*78(1):443–450.

Van der Wal, H., Sperber, B.L.H.M., Houweling-Tan, B., Bakker, R.R.C., Brandenburg, W. and López-Contreras, A.M. 2013. Production of acetone, butanol, and ethanol from biomass of the green seaweed *Ulva lactuca. Bioresour. Technol.*128:431–437.

van Oirschot, R., Thomas, J.B.E., Gröndahl, F., Fortuin, K.P.J., Brandenburg, W. and Potting, J. 2017. Explorative environmental life cycle assessment for system design of seaweed cultivation and drying. *Algal Res.*27:43–54.

Verbruggen, H. and Costa, J.F. 2015. Molecular survey of *Codium* species diversity in Southern Madagascar. *Cryptogam. Algol.*36(2):171–187.

Vroom, P.S., Smith, C.M. and Keeley, S.C. 1998. Cladistics of the Bryopsidales: A preliminary analysis. *J. Phycol.*34(2):351–360.

Wang, H.M.D., Chen, C.C., Huynh, P. and Chang, J.S. 2015. Exploring the potential of using algae in cosmetics. *Bioresour. Technol.*184:355–362.

Wells, M.L., Potin, P., Craigie, J.S., Raven, J.A., Merchant, S.S., Helliwell, K.E., Smith, A.G., Camire, M.E. and Brawley, S.H. 2017. Algae as nutritional and functional food sources: Revisiting our understanding. *J. Appl. Phycol.* 29:949–982.

Winberg, P.C., Fitton, H.J., Stringer, D., Karpiniec, S.S. and Gardiner, V.A. 2014. Controlling seaweed biology, physiology and metabolic traits in production for commercially relevant bioactives in glycobiology. In Bourgougnon, N. (Ed.), *Advances in Botanical Research*, Vol. 71. Sea Plants. London: Academic Press, pp. 221–252.

Wu, C., Jiang, P., Zhao, J. and Fu, H. 2018. High efficiency of protoplast preparation for artificially cultured *Ulva prolifera* (Ulvophyceae, Chlorophyta). *J. OceanolLimnol*36(5):1806–1811.

Xu, Y. and Lin, J. 2008. Effect of temperature, salinity, and light intensity on the growth of the green macroalga, *Chaetomorpha* linum. *J. World Aquac Soc.*39(6):847–851.

Yang, L.E., Lu, Q.Q. and Brodie, J. 2017. A review of the bladed Bangiales (Rhodophyta) in China: History, culture and taxonomy. *Eur. J. Phycol.*52(3):251–263.

Yang, P., Liu, D.Q., Liang, T.J., et al. Bioactive constituents from the green alga *Caulerpa racemosa. Bioorg. Med. Chem.*23(1):38–45.

Yu, H., Zhang, H., Dong, M., Wu, Z., Shen, Z., Xie, Y., Kong, Z., Dai, X. and Xu, B. 2017. Metabolic reprogramming and AMPKα1 pathway activation by caulerpin in colorectal cancer cells. *Int. J. Oncol.* 50(1):161–172.

Zamagni, A., Pesonen, H.L. andSwarr, T. 2013. From LCA to life cycle sustainability assessment: Concept, practice and future directions. *Int. J. Life Cycle Assess.*18(9):1637–1641.

Zan, J., Li, Z., Tianero, M.D., Davis, J., Hill, R.T. and Donia, M.S. 2019. A microbial factory for defensive kahalalides in a tripartite marine symbiosis. *Science*364(6445):eaaw6732.

Zemke-White, L.W. and Ohno, M. 1999. World seaweed utilisation: An end of century summary. *J. Appl. Phycol.*11(4):369–376.

Zhou, X. 2018. Notes from the aquaculture statistician. *FAO AquaculNewslet*58:6–8.

Zollmann, M., Robin, A., Prabhu, M., Polikovsky, M., Gillis, A., Greiserman, S. and Golberg, A. 2019. Green technology in green macroalgal biorefineries. *Phycologia*58(5):516–534.

Zubia, M., Draisma, S.G.A., Morrissey, K.L., Varela-Álvarez, E. and De Clerck, O. 2020. Concise review of the genus *Caulerpa* J.V. Lamouroux. *J. Appl. Phycol.*32(1):23–39.

Zulkifly, S.B., Graham, J.M., Young, E.B., Mayer, R.J., Piotrowski, M.J., Smith, I. and Graham, L.E. 2013. The genus *Cladophora* Kützing (Ulvophyceae) as a globally distributed ecological engineer. *J. Phycol.*49(1):1–17.

9 Marine Green Microalgae Biomass Production and Application

Isibor Patrick Omoregie, Kayode-Edwards Ifeoluwa Ihotu, and Agbontaen David Osagie

9.1 INTRODUCTION

Microalgae, also called microphytes or microscopic algae, are a diverse group of small unicellular, photosynthetic organisms typically found in aquatic environments (Yousuf, 2020; Russell *et al.*, 2022). These photosynthetic organisms produce a variety of biologically active compounds, including pigments, lipids, polysaccharides, and proteins (Wu *et al.*, 2021). Due to this ability as well as their rapid growth rate, microalgae are becoming increasingly important in a variety of applications, including food and feed production, biofuels, cosmetics, and pharmaceuticals (De Luca *et al.*, 2021; Fernández *et al.*, 2021; Siddiki *et al.*, 2022). Additionally, microalgae are being utilized for their potential in environmental applications such as bioremediation of wastewater and industrial effluents, carbon capture and utilization, and nutrient recycling and waste utilization (Alazaiza *et al.*, 2022; Goswami *et al.*, 2022).

Marine green microalgae biomass can be converted into biofuels such as biodiesel, bioethanol, and biogas. The lipids present in the biomass can be extracted and converted into biodiesel (Bhuyar *et al.*, 2021), while the carbohydrates can be fermented to produce bioethanol (Lakatos *et al.*, 2019). Additionally, the anaerobic digestion of microalgal biomass produces biogas (Díez-Montero *et al.*, 2020).

Microalgae biomass is a rich source of protein, vitamins, and minerals, and so serves as a food source for humans in the form of health supplements as well as a nutrient supplement in animal feed (Fernández *et al.*, 2021). The carbohydrates present in the microalgae biomass can be converted into bioplastics. Bioplastics are biodegradable and sustainable alternatives to traditional petroleum-based plastics. Biomass can also be used for bioremediation, as microalgae can absorb heavy metals, toxins, and other pollutants from water, making it safer for other marine organisms (Umamaheswari *et al.*, 2023).

With ongoing research and development, marine green microalgae have the potential to become essential resources in promoting sustainability and reducing dependence on fossil fuels. Hence, strategies have been employed to enhance biomass production, including genetic engineering, mixotrophic cultivation, nutrient recycling, and biofilm cultivation (Yadav *et al.*, 2021; Xu *et al.*, 2023).

Marine green microalgae play a critical role in marine ecosystems and in promoting environmental sustainability. The potential benefits of microalgae research are vast and have important implications for the well-being of the ecosphere.

9.2 FACTORS INFLUENCING BIOMASS PRODUCTION

Microalgae biomass production is influenced by various factors, including nutrient availability, light intensity and duration, temperature and pH, carbon dioxide (CO_2) supplementation, and harvesting

DOI: 10.1201/9781003369738-11

and biomass recovery methods. By carefully controlling these factors, optimizing microalgal biomass production in a cultivation system is possible.

9.3 HARVESTING AND BIOMASS RECOVERY METHODS

Harvesting and biomass recovery methods are techniques used to collect and process plant material for various purposes, such as bioenergy, food production, or industrial applications. The major methods available for harvesting and recovering microalgae include electrical, biological, chemical, mechanical, or a combination of two or more of these processes (Najjar and Abu-Shamleh, 2020). The effectiveness and efficiency of these methods can influence the productivity and overall yield of algal biomass. Selection of the most appropriate harvesting method can significantly impact the overall yield of algal biomass (Patel *et al.*, 2022). Some harvesting methods, such as centrifugation, can be very efficient but expensive and require significant energy input (Najjar and Abu-Shamleh, 2020). Other methods, such as sedimentation or filtration, may be less efficient but more cost-effective and easier to implement (Rakesh *et al.*, 2020).

The timing of harvesting can also impact algal biomass production (Xu *et al.*, 2005). Harvesting too early can result in lower biomass yield, while harvesting too late can lead to lower quality or reduced viability of the biomass.

The efficiency of biomass recovery methods can also impact biomass production. Some methods, such as solvent extraction or mechanical pressing, may be more efficient but more expensive or time-consuming (Karim *et al.*, 2020). Other methods, such as drying or freeze-drying, may be less efficient but more cost-effective and easier to implement (Chen *et al.*, 2015). The selection of appropriate harvesting and biomass recovery methods depends on various factors, including the algal species, cultivation system, growth conditions, and desired end-use application.

9.4 STRATEGIES TO ENHANCE MARINE GREEN MICROALGAE BIOMASS PRODUCTION

Several strategies can be employed to enhance marine green microalgae biomass production. These strategies include genetic engineering, mixotrophic cultivation, nutrient recycling, and biofilm cultivation.

9.4.1 Genetic Engineering

Genetic engineering is the manipulation of an organism's genetic material (DNA) using biotechnological techniques such as recombinant DNA technology and gene editing (Wu and Bazer, 2019). It involves adding to, deleting from, or modifying the genetic material of an organism to produce desired traits (Piotrowski-Daspit *et al.*, 2020).

Genetic engineering is a significant tool for enhancing the biomass production of marine green microalgae. Introducing specific genes or modifying existing genes can enhance the growth rate, nutrient utilization, and photosynthetic efficiency of these microorganisms.

One approach to genetic engineering in microalgae involves the manipulation of genes involved in the synthesis of key metabolic pathways. For example, engineering the genes involved in lipid biosynthesis can lead to increased lipid content in microalgae, which enhances biofuel production (Shahid *et al.*, 2020). Similarly, altering the genes involved in nitrogen or carbon fixation can enhance nutrient utilization and promote faster growth rates (Cheng *et al.*, 2019). Another approach to genetic engineering involves modifying genes that control photosynthesis. This can include increasing the expression of certain photosynthetic pigments (Simkin *et al.*, 2022) or altering the structure of the photosynthetic apparatus (Ho *et al.*, 2020) to improve light capture and energy conversion. Such modifications can result in increased biomass production and overall productivity.

However, the genetic engineering of microalgae for biomass production requires careful consideration of potential environmental impacts, such as the spread of genetically modified organisms into natural ecosystems (Abdullah *et al.*, 2019; Shokravi *et al.*, 2021). Therefore, it is critical to conduct thorough risk assessments and implement appropriate containment measures to minimize potential risks (Abdullah *et al.*, 2019). Overall, genetic engineering can be a powerful tool for enhancing the biomass production of marine green microalgae and has the potential to contribute to the development of sustainable biofuels and other bioproducts.

9.4.2 Mixotrophic Cultivation

Mixotrophic cultivation is a method of growing microorganisms that involves providing them with both organic carbon sources and light for photosynthesis (Marella *et al.*, 2021).

Mixotrophic cultivation is commonly used in the cultivation of microalgae, which are photosynthetic organisms that can also grow heterotrophically by consuming organic carbon sources (Abreu *et al.*, 2022). In this method, microalgae are cultivated in the presence of an inorganic carbon source, such as CO_2, in addition to an organic carbon source, such as glucose or glycerol (Javad *et al.*, 2019). This allows microalgae to grow in wide-ranging conditions and can result in higher biomass yields than autotrophic cultivation alone (Dragone, 2022).

The benefit of mixotrophic cultivation is that it allows a higher growth rate and biomass production compared with autotrophic cultivation (using only inorganic carbon sources) or heterotrophic cultivation (using only organic carbon sources) (Dragone, 2022). This is because mixotrophy provides alternative energy sources that allow the microalgae to grow and reproduce more rapidly and in various conditions (Dragone, 2022). In addition, mixotrophic cultivation can enhance the production of high-value compounds such as proteins, pigments, and lipids, which can be used for several applications, such as biofuels, feed, and food (Liaqat *et al.*, 2023).

Therefore, mixotrophic cultivation is a potentially lucrative strategy for enhancing the biomass production of marine green microalgae and improving their economic viability as sources of biomass and bioactive compounds.

9.4.3 Nutrient Recycling

Nutrient recycling is the process of reusing or recycling nutrients used in biological systems, such as agriculture, aquaculture, or wastewater treatment, to reduce waste and environmental pollution (Adegbeye *et al.*, 2020). Nutrient recycling can be used to enhance marine green microalgae biomass production by reusing the nutrients in the culture medium and reducing the need for expensive and environmentally harmful synthetic fertilizers (Hussain *et al.*, 2021). By recycling nutrients, a more sustainable and efficient system for biomass production of microalgae is created.

One approach to nutrient recycling is to use wastewater or effluent from aquaculture or other industries as a nutrient source for microalgal culture (Hussain *et al.*, 2021). Wastewater contains high levels of phosphorus and nitrogen, essential nutrients for microalgae growth (Alazaiza *et al.*, 2022). The microalgae can assimilate these nutrients and convert them into biomass, which can then be harvested and used for various applications (Hussain *et al.*, 2021). Using wastewater as a nutrient source can reduce the environmental repercussions of wastewater discharge and provide a cost-effective source of nutrients for microalgal culture (Alazaiza *et al.*, 2022). Another approach to recycling nutrients is to use the residual biomass from microalgal culture as a nutrient source for subsequent cultures (Liu *et al.*, 2019). The residual biomass contains high levels of nutrients and micronutrients, which can be reused as a fertilizer for successive cultures (Liu *et al.*, 2019). This can reduce the need for synthetic fertilizers, increase nutrient utilization efficiency, and reduce the environmental impact of microalgal biomass production.

Nutrient recycling is a valuable strategy for enhancing microalgal biomass production by providing a more efficient, sustainable, and cost-effective system for biomass production while reducing the environmental impact of the process.

9.4.4 Biofilm Cultivation

Biofilm cultivation is a method of growing microorganisms in a matrix of extracellular polymeric substances (EPS), which they produce themselves (Gross *et al.*, 2015). Biofilms are formed when microorganisms attach to a surface and secrete EPS to form a protective matrix around themselves. The ability of biofilms to create complex structures and protect microorganisms from harsh environments (Yin *et al.*, 2019) makes them useful for applications where the stability of microbial communities is important. Biofilm cultivation can be used to enhance marine green microalgae biomass production by providing a stable and protective environment for the microalgae to grow. Growing the microalgae in a biofilm matrix protects them from predation, grazing, and other environmental stresses that can limit their growth (Flemming *et al.*, 2023).

In addition, biofilm cultivation can provide a continuous nutrient source for the microalgae, as the EPS matrix can trap and concentrate nutrients from the surrounding water (Cheah and Chan, 2021). This can increase the nutrient availability for the microalgae, leading to higher biomass yields.

Biofilm cultivation can also promote the formation of diverse microbial communities, enhancing microalgal growth and remediation capabilities through symbiotic relationships (Zhao *et al.*, 2023). Overall, biofilm cultivation can be a useful strategy for enhancing marine green microalgae biomass production, particularly in situations where environmental stresses or nutrient limitations may limit their growth. By providing a protective and nutrient-rich environment for the microalgae to grow, biofilm cultivation can increase the efficiency and sustainability of microalgal biomass production.

9.5 APPLICATIONS OF MARINE GREEN MICROALGAE BIOMASS

9.5.1 Biofuel Production

In recent times, particularly in the developing nations of the world, energy shortages and environmental pollution have created rising concerns necessitating the development of renewable energy sources, such as solar, bioenergy, wind, and geothermal energy. Among the renewable energy sources, biofuels, particularly bioethanol, have superior properties that favor their use as a liquid fuel in all vehicle engines, distributed via the existing fossil fuel system, thereby encouraging the rural economy. However, the increasing demand for bioethanol has strained the availability of food material and arable land for production, resulting in the threat of rising food shortages, thereby restricting the growth of the bioethanol industry (Song *et al.*, 2015). A sustainable alternative to bioethanol production is the development of marine resource–based bioethanol production.

9.5.2 Application of Marine Microalgae in Bioethanol Production

Marine green microalgae are considered a potential source of bioenergy due to their high growth rate, high lipid content, and ability to fix carbon dioxide. One of the possible applications of marine green microalgae biomass is in the production of bioethanol, a renewable and sustainable fuel. Marine green microalgae biomass is a promising feedstock for bioethanol production due to its high sugar content and rapid growth rate. The use of microalgae as a feedstock for bioethanol production has gained significant attention in recent years due to its potential as a renewable and sustainable source of energy (Sanchez Rizza *et al.*, 2017).

The bioethanol production process from microalgae biomass involves several steps, such as cultivation, harvesting, cell disruption, and fermentation (Ghardiryanfar *et al.*, 2016). The process of bioethanol production from marine green microalgae biomass involves several steps. Firstly, the biomass is harvested and dried, and then subjected to pre-treatment to break down the cell wall and release the sugar content. This pre-treatment step can be done through physical, chemical, or enzymatic means. Next, the hydrolysate obtained is fermented using yeasts or bacteria to convert the sugar into ethanol. Finally, the ethanol is distilled and purified to obtain the final product (Meinita *et al.*, 2017).

The use of marine green microalgae biomass in bioethanol production offers several advantages. Firstly, microalgae can be cultivated in saltwater or freshwater, and they do not compete with food crops for land or resources, unlike other bioenergy crops such as corn or sugarcane. Secondly, microalgae can be grown year-round and have a higher growth rate compared with other crops, making them a more efficient source of biomass. Lastly, microalgae have a higher lipid content compared with other crops, making them useful for the production of biodiesel, another renewable fuel (Parab *et al.*, 2017).

However, there are also several challenges in the use of marine green microalgae biomass in bioethanol production. The high cost of cultivation and harvesting of microalgae remains a major obstacle to scaling up the technology for commercial production. Additionally, the low sugar content in microalgae biomass means that a large amount of biomass is required to produce a significant amount of ethanol, leading to higher processing costs. In conclusion, the application of marine green microalgae biomass in bioethanol production offers a promising solution to the growing energy demand and environmental concerns. However, further research and development are needed to address the challenges and optimize the process for commercial viability.

9.5.3 Application of Marine Microalgae in Biodiesel Production

Marine microalgae are a promising feedstock for the production of biodiesel due to their high oil content, fast growth rates, and ability to grow in a variety of environments. Compared with traditional oil crops, such as soybean or palm oil, marine microalgae have the potential to produce higher yields of oil per unit of land, while also reducing competition for land use (Ishika *et al.*, 2017). The process of marine microalgae biodiesel production involves several steps. Firstly, the microalgae must be cultured and harvested, which can be done in open ponds, closed photobioreactors, or other specialized systems. Next, the oil is extracted from the microalgae cells, typically through mechanical or chemical methods. Once the oil is extracted, it is converted into biodiesel through a process called transesterification, which involves reacting the oil with an alcohol and a catalyst to produce fatty acid methyl esters (FAME).

One of the advantages of using marine microalgae for biodiesel production is their high oil content. Some marine microalgae species can contain up to 60% oil by weight, which is much higher than most oilseed crops. Additionally, marine microalgae can be grown on non-arable land, such as deserts or wastewater, reducing competition with food crops for land use. Another advantage of marine microalgae is their ability to capture carbon dioxide during their growth. This makes them a potential carbon-neutral fuel source, as the carbon dioxide released during biodiesel combustion can be offset by the carbon dioxide captured during microalgae growth (Song *et al.*, 2015).

However, there are also challenges associated with biodiesel production from marine microalgae. One of the major challenges is the high cost of cultivation and processing. Additionally, the process of cultivating and harvesting marine microalgae requires significant energy inputs, which can offset some of the potential carbon benefits (Tredici *et al.*, 2015). Despite these challenges, marine microalgae show promise as a feedstock for biodiesel production, particularly in areas with high levels of sunlight and access to seawater (Ziolkowska and Simon, 2014). Further research and development are needed to improve the efficiency and economics of marine microalgae biodiesel production, but the potential benefits of this technology make it a promising area of study.

9.5.4 Application of Marine Microalgae in Biogas Production

Marine green microalgae can be used as a feedstock for the production of biogas. Biogas is a renewable energy source that is produced by the anaerobic digestion of organic matter. This process involves the breakdown of organic material by microorganisms in the absence of oxygen, which results in the production of a gas mixture primarily composed of methane and carbon dioxide.

Marine green microalgae can be used as a feedstock for biogas production because they are rich in carbohydrates, proteins, and lipids (Song *et al.,* 2015). These organic compounds can be easily broken down by microorganisms during the anaerobic digestion process, leading to the production of biogas. The use of marine green microalgae as a feedstock for biogas production has several benefits. Firstly, it reduces the dependence on fossil fuels for energy production, which helps to mitigate climate change. Secondly, it provides a solution for the disposal of excess microalgae biomass, which can be a problem in aquaculture and other industries. Finally, biogas production from marine green microalgae can be combined with other waste streams to increase the overall efficiency of the process.

However, there are also challenges associated with using marine green microalgae for biogas production. One of the main challenges is the high cost of microalgae cultivation and harvesting, which can make it less economically viable compared with other feedstocks for biogas production. Additionally, the composition of the microalgae biomass can affect the efficiency of the anaerobic digestion process and the quality of the resulting biogas. Overall, the application of marine green microalgae in biogas production has the potential to provide a renewable and sustainable source of energy, but further research and development are needed to optimize the process and make it more economically viable.

9.6 POTENTIAL HEALTH BENEFITS OF MICROALGAE-DERIVED PRODUCTS

Marine green microalgae are single-celled photosynthetic microorganisms that have gained attention as a source of high-quality nutrients and bioactive compounds; hence, they are used in food, cosmetics, and pharmaceutical products.

Some potential health benefits of microalgae-derived products include:

1. High nutritional value: Microalgae are a rich source of nutrients such as protein, fiber, vitamins, and minerals. For example, spirulina is a type of blue-green algae that is high in protein, beta-carotene, iron, and other nutrients.
2. Antioxidant activity: Microalgae contain a variety of antioxidants, which are compounds that protect cells from damage caused by free radicals. Some examples of microalgae with antioxidant properties include chlorella and astaxanthin.
3. Anti-inflammatory activity: Microalgae-derived products have been shown to have anti-inflammatory effects in animal and human studies. This may be due to the presence of compounds such as omega-3 fatty acids, phycocyanin, and polysaccharides.
4. Immune system support: Some microalgae-derived products have been shown to enhance the immune system in animal and human studies. For example, spirulina has been found to increase the production of antibodies and cytokines, which are important for immune function.
5. Potential anti-cancer activity: Some microalgae-derived compounds, such as fucoxanthin, have been shown to have anti-cancer properties in animal and cell studies. More research is needed to determine their effectiveness in humans.
6. Skin health: Microalgae-derived products have been used in cosmetics and skin care products due to their moisturizing and anti-aging effects. Some types of microalgae, such as dunaliella, contain high levels of carotenoids, which are believed to have protective effects against UV radiation.

Overall, microalgae-derived products have the potential to provide a wide range of health benefits. However, more research is needed to fully understand their effectiveness and safety in humans and to develop standardized methods for their production and use.

9.7 APPLICATION OF MARINE GREEN MICROALGAE IN BIOTECHNOLOGY

9.7.1 Application of Marine Green Microalgae in Bioplastics

Marine green microalgae can be used as a sustainable source of bioplastics. Bioplastics are biodegradable plastics made from renewable resources such as plants, animals, and microorganisms. Marine green microalgae, also known as seaweed, are a promising source of bioplastics because they grow quickly, have high productivity, and are rich in polysaccharides such as cellulose, starch, and agar.

The process of using marine green microalgae to produce bioplastics involves harvesting the seaweed, extracting the polysaccharides, and converting them into bioplastics using various chemical and physical processes. The resulting bioplastics have properties similar to traditional petroleum-based plastics, including flexibility, strength, and durability. However, they are biodegradable and do not contribute to the accumulation of plastic waste in the environment.

The use of marine green microalgae as a source of bioplastics has several benefits. It reduces the dependence on non-renewable resources, reduces greenhouse gas emissions, and provides a sustainable solution to the plastic waste problem. Additionally, the cultivation of marine green microalgae can have positive environmental impacts, such as improving water quality and providing habitat for marine life.

Overall, the use of marine green microalgae as a source of bioplastics has the potential to transform the plastic industry into a more sustainable and environmentally friendly industry. However, more research and development are needed to optimize the production processes and to scale up the production to meet the growing demand for bioplastics.

9.8 IMPLICATIONS FOR THE FUTURE OF MARINE GREEN MICROALGAE BIOMASS PRODUCTION AND APPLICATION

The future of marine green microalgae biomass production and application looks promising. As the demand for sustainable and environmentally friendly products continues to grow, microalgae biomass offers a viable alternative to traditional sources of food, feed, and fuel. Advancements in technology and cultivation methods have also improved the efficiency and scalability of microalgae production, making it more economically feasible. Additionally, research has uncovered the potential health benefits of microalgae consumption, which could further drive demand for its use in food and supplement products. However, further research is needed to fully explore the potential of microalgae and optimize their production and application. Overall, the future of marine green microalgae biomass production and application is likely to be characterized by continued growth and innovation.

REFERENCES

Abdullah, B., Muhammad, S.A.F.A.S., Shokravi, Z., Ismail, S., Kassim, K.A., Mahmood, A.N., and Aziz, M.M.A. (2019). Fourth generation biofuel: A review on risks and mitigation strategies. *Renewable and Sustainable Energy Reviews*, *107*, pp. 37–50.

Abreu, A.P., Morais, R.C., Teixeira, J.A., and Nunes, J. (2022). A comparison between microalgal autotrophic growth and metabolite accumulation with heterotrophic, mixotrophic and photoheterotrophic cultivation modes. *Renewable and Sustainable Energy Reviews*, *159*, p. 112247.

Adegbeye, M.J., Reddy, P.R.K., Obaisi, A.I., Elghandour, M.M.M.Y., Oyebamiji, K.J., Salem, A.Z.M., Morakinyo-Fasipe, O.T., Cipriano-Salazar, M., and Camacho-Díaz, L.M. (2020). Sustainable agriculture options for production, greenhouse gasses and pollution alleviation, and nutrient recycling in emerging and transitional nations–An overview. *Journal of Cleaner Production*, *242*, p. 118319.

Alazaiza, M.Y., Albahnasawi, A., Ahmad, Z., Bashir, M.J., Al-Wahaibi, T., Abujazar, M.S.S., Amr, S.S.A., and Nassani, D.E. (2022). Potential use of algae for the bioremediation of different types of wastewater and contaminants: Production of bioproducts and biofuel for green circular economy. *Journal of Environmental Management*, *324*, p. 116415.

Bhuyar, P., Yusoff, M.M., Rahim, M.H.A., Sundararaju, S., Maniam, G.P., and Govindan, N. (2021). Effect of plant hormones on the production of biomass and lipid extraction for biodiesel production from microalgae Chlorella sp. *Journal of Microbiology, Biotechnology and Food Sciences, 2021*, pp. 671–674.

Cheah, Y.T., and Chan, D.J.C. (2021). Physiology of microalgal biofilm: A review on prediction of adhesion on substrates. *Bioengineered, 12*(1), pp. 7577–7599.

Chen, C.L., Chang, J.S., and Lee, D.J. (2015). Dewatering and drying methods for microalgae. *Drying Technology, 33*(4), 443–454.

Cheng, J., Zhu, Y., Zhang, Z., and Yang, W. (2019). Modification and improvement of microalgae strains for strengthening CO_2 fixation from coal-fired flue gas in power plants. *Bioresource Technology, 291*, p. 121850.

De Luca, M., Pappalardo, I., Limongi, A.R., Viviano, E., Radice, R.P., Todisco, S., Martelli,G., Infantino, V., and Vassallo, A. (2021). Lipids from microalgae for cosmetic applications. *Cosmetics, 8*(2), p. 52.

Díez-Montero, R., Vassalle, L., Passos, F., Ortiz, A., García-Galán, M.J., García, J., and Ferrer, I. (2020). Scaling-up the anaerobic digestion of pretreated microalgal biomass within a water resource recovery facility. *Energies, 13*(20), p. 5484.

Dragone, G. (2022). Challenges and opportunities to increase economic feasibility and sustainability of mixotrophic cultivation of green microalgae of the genus *Chlorella*. *Renewable and Sustainable Energy Reviews, 160*, p. 112284.

Fernández, F.G.A., Reis, A., Wijffels, R.H., Barbosa, M., Verdelho, V., and Llamas, B. (2021). The role of microalgae in the bioeconomy. *New Biotechnology, 61*, pp. 99–107.

Flemming, H.C., van Hullebusch, E.D., Neu, T.R., Nielsen, P.H., Seviour, T., Stoodley, P., Wingender, J., and Wuertz, S. (2023). The biofilm matrix: Multitasking in a shared space. *Nature Reviews in Microbiology, 21*(2), pp. 70–86.

Goswami, R.K., Mehariya, S., Karthikeyan, O.P., Gupta, V.K., and Verma, P. (2022). Multifaceted application of microalgal biomass integrated with carbon dioxide reduction and wastewater remediation: A flexible concept for sustainable environment. *Journal of Cleaner Production, 13*, p. 130654.

Gross, M., Jarboe, D., and Wen, Z. (2015). Biofilm-based algal cultivation systems. *Applied Microbiology and Biotechnology, 99*(14), pp. 5781–5789.

Ho, M.Y., Niedzwiedzki, D.M., MacGregor-Chatwin, C., Gerstenecker, G., Hunter, C.N., Blankenship, R.E., and Bryant, D.A. (2020). Extensive remodeling of the photosynthetic apparatus alters energy transfer among photosynthetic complexes when cyanobacteria acclimate to far-red light. *Biochimica et Biophysica Acta (BBA) - Bioenergetics, 1861*(4), p. 148064.

Hussain, F., Shah, S.Z., Ahmad, H., Abubshait, S.A., Abubshait, H.A., Laref, A., Manikandan, A., Kusuma, H.S., and Iqbal, M. (2021). Microalgae an ecofriendly and sustainable wastewater treatment option: Biomass application in biofuel and bio-fertiliser production. A review. *Renewable and Sustainable Energy Reviews, 137*, p. 110603.

Ishika, T., Moheimani, N.R., and Bahri, P.A. (2017). Sustainable saline microalgae co-cultivation for biofuel production: A critical review. *Renewable and Sustainable Energy Reviews, 78*, pp. 356–368.

Karim, A., Islam, M.A., Khalid, Z.B., Faizal, C.K.M., Khan, M.M.R., and Yousuf, A. (2020). Microalgal cell disruption and lipid extraction techniques for potential biofuel production. In *Microalgae Cultivation for Biofuels Production* (pp. 129–147). Academic Press, London.

Lakatos, G.E., Ranglová, K., Manoel, J.C., Grivalský, T., Kopecký, J., and Masojídek, J. (2019). Bioethanol production from microalgae polysaccharides, *Folia Microbiologica, 64*(5), pp.627–644.

Liaqat, F., Khazi, M.I., Bahadar, A., He, L., Aslam, A., Liaquat, R., Agathos, S.N., and Li, J. (2023). Mixotrophic cultivation of microalgae for carotenoid production. *Reviews in Aquaculture, 15*(1), pp. 35–61.

Liu, F., Lane, P., Hewson, J.C., Stavila, V., Tran-Gyamfi, M.B., Hamel, M., Lane, T.W., and Davis, R.W. (2019). Development of a closed-loop process for fusel alcohol production and nutrient recycling from microalgae biomass. *Bioresource Technology, 283*, pp. 350–357.

Marella, T.K., Bhattacharjya, R., and Tiwari, A. (2021). Impact of organic carbon acquisition on growth and functional biomolecule production in diatoms. *Microbial Cell Factories, 20*(1), pp. 1–13.

Meinita, M.D.N., Marhaeni, B., Oktaviani, D.F., Jeong, G.T., and Hong, Y.K. (2017). Comparison of bioethanol production from cultivated versus wild Gracilaria verrucosa and Gracilaria gigas. *Journal of Applied Phycology*. https://dx.doi.org/10.1007/s10811-017-1297-x.

Najjar, Y.S., and Abu-Shamleh, A. (2020). Harvesting of microalgae by centrifugation for biodiesel production: A review. *Algal Research, 51*, p. 102046.

Patel, A.K., Vadrale, A.P., Tseng, Y.S., Chen, C.W., Dong, C.D., and Singhania, R.R. (2022). Bioprospecting of marine microalgae from Kaohsiung Seacoast for lutein and lipid production. *Bioresource Technology, 351*, p. 126928.

Parab, P., Khandeparker, R., Amberkar, U., and Khodse, V. (2017). Enzymatic saccharification of seaweeds into fermentable sugars by xylanase from marine Bacillus sp. strain BT21. *3 Biotech*, *7*(5), pp.1–7.
Piotrowski-Daspit, A.S., Kauffman, A.C., Bracaglia, L.G., and Saltzman, W.M. (2020). Polymeric vehicles for nucleic acid delivery. *Advanced Drug Delivery Reviews*, *156*, pp. 119–132.
Rakesh, S., TharunKumar, J., Sri, B., Jothibasu, K., and Karthikeyan, S. (2020). Sustainable cost-effective microalgae harvesting strategies for the production of biofuel and oleochemicals. *Highlights in Bioscience*, *3*(July), pp. 1–8.
Russell, C., Rodriguez, C., and Yaseen, M. (2022). High-value biochemical products & applications of freshwater eukaryotic microalgae. *Science of the Total Environment*, *809*, p. 151111.
Sanchez Rizza, L., Sanz Smachetti, M.E., Do Nascimento, M., Salerno, G.L., and Curatti, L. (2017). Bioprospecting for native microalgae as an alternative source of 363 sugars for the production of bioethanol. *Algal Research*, *364*, pp. 140–147. http://dx.doi.org/10.1016/j.algal. 2016.12.021.
Shahid, A., Rehman, A.U., Usman, M., Ashraf, M.U.F., Javed, M.R., Khan, A.Z., Gill, S.S., and Mehmood, M.A. (2020). Engineering the metabolic pathways of lipid biosynthesis to develop robust microalgal strains for biodiesel production. *Biotechnology and Applied Biochemistry*, *67*(1), pp. 41–51.
Shokravi, H., Shokravi, Z., Heidarrezaei, M., Ong, H.C., Koloor, S.S.R., Petrů, M., Lau, W.J., and Ismail, A.F. (2021). Fourth generation biofuel from genetically modified algal biomass: Challenges and future directions. *Chemosphere*, *285*, p. 131535.
Siddiki, S.Y.A., Mofijur, M., Kumar, P.S., Ahmed, S.F., Inayat, A., Kusumo, F., Badruddin, I.A., Khan, T.Y., Nghiem, L.D., Ong, H.C., and Mahlia, T.M.I. (2022). Microalgae biomass as a sustainable source for biofuel, biochemical and biobased value-added products: An integrated biorefinery concept. *Fuel*, *307*, p. 121782.
Simkin, A.J., Kapoor, L., Doss, C.G.P., Hofmann, T.A., Lawson, T., and Ramamoorthy, S. (2022). The role of photosynthesis-related pigments in light harvesting, photoprotection and enhancement of photosynthetic yield in planta. *Photosynthesis Research*, *152*(1), pp. 23–42.
Song, M., Duc Pham, H., Seon, J., and Chul Woo, H. (2015). Marine brown algae: A conundrum answer for sustainable biofuels production. *Renewable and Sustainable Energy Reviews*, *50*, pp. 782–792. http://dx.doi.org/10.1016/j.rser.2015.05.021.
Tredici, M., Bassi, N., Prussi, M., Biondi, N., Rodolfi, L., Zittelli, G.C., and Sampietro, G. (2015). Energy balance of algal biomass production in a 1-ha "Green Wall Panel" plant: How to produce algal biomass in a closed reactor achieving a high Net Energy Ratio. *Applied Energy*, *154*, pp. 1103–1111.
Umamaheswari, J., Rajamanickam, R., Vilvanathan, S., Shanthakumar, S., Venkateswarlu, K., Abinandan, S., and Megharaj, M. (2023). Removal of heavy metals and organic pollutants by marine microalgae. In Robert J. Cabin (Ed.) *Marine Organisms: A Solution to Environmental Pollution? Uses in Bioremediation and in Biorefinery* (pp. 29–64). Cham: Springer International Publishing.
Wu, G., and Bazer, F.W. (2019). Application of new biotechnologies for improvements in swine nutrition and pork production. *Journal of Animal Science and Biotechnology*, *10*(1), pp. 1–16.
Wu, J., Gu, X., Yang, D., Xu, S., Wang, S., Chen, X., and Wang, Z. (2021). Bioactive substances and potentiality of marine microalgae. *Food Science and Nutrition*, *9*(9), pp. 5279–5292.
Xu, C., Boyce, M.S., and Daley, D.J. (2005). Harvesting in seasonal environments. *Journal of Mathematical Biology*, *50*(6), pp. 663–682.
Xu, P., Li, J., Qian, J., Wang, B., Liu, J., Xu, R., Chen, P., and Zhou, W. (2023). Recent advances in CO_2 fixation by microalgae and its potential contribution to carbon neutrality. *Chemosphere*, *319*, p. 137987.
Yadav, G., Shanmugam, S., Sivaramakrishnan, R., Kumar, D., Mathimani, T., Brindhadevi, K., Pugazhendhi, A., and Rajendran, K. (2021). Mechanism and challenges behind algae as a wastewater treatment choice for bioenergy production and beyond. *Fuel*, *285*, p. 119093.
Yin, W., Wang, Y., Liu, L., and He, J. (2019). Biofilms: The microbial "protective clothing" in extreme environments. *International Journal of Molecular Sciences*, *20*(14), p. 3423.
Yousuf, A. (2020). Fundamentals of microalgae cultivation. In Abu Yousuf (Ed.) *Microalgae Cultivation for Biofuels Production* (pp. 1–9). Academic Press, London.
Zhao, D., Cheah, W.Y., Lai, S.H., Ng, E.P., Khoo, K.S., Show, P.L., and Ling, T.C. (2023). Symbiosis of microalgae and bacteria consortium for heavy metal remediation in wastewater. *Journal of Environmental Chemical Engineering*, *11*p. 109943.
Ziolkowska, J.R., and Simon, L. (2014). Recent developments and prospects for algae-based fuels in the US. *Renewable and Sustainable Energy Reviews*, *29*, pp. 847–853.

10 Edible Marine Greens
Uses and Limitations

Oyegbade Samuel Adeniyi, Abiodun Eniola Opeyemi, and Muritada Babatunde Oshinuga

10.1 INTRODUCTION

Seaweeds or algae that can be eaten by humans as food are known as edible marine greens. Usually taken from the ocean, they can be found in soups, salads, and sushi, among other foods. Edible seaweed, edible sea veggies, and edible marine algae are other names for edible marine greens. They are all referring to the same kind of marine plant, which can be eaten and is taken from the water. Multicellular marine species known as macroalgae, sometimes known as seaweeds, are regarded as non-vascular plants. Despite the fact that the name "seaweed" is frequently used, they are typically very different from "weeds," and yet they are the primary producers in the oceans and the foundation of the food chain. Seaweeds are utilized either directly or indirectly in food and household products (Babahan et al., 2020). Seaweeds, which are also known as benthic marine algae or sea vegetables, primarily inhabit the sea or brackish water as their preferred habitat, as stated by Yakhin et al. (2016). These simple, unique creatures are among the main products of Asian enterprises, and the primary goal of each industry is to make the most use of the output (Babahan et al., 2020). Although certain countries in the Far East consume seaweed as food, seaweed is more commonly employed in the West as a source of phycocolloids, thickeners, and gelling agents for a range of industrial uses, including food. The chemical composition of seaweeds varies based on their habitats, maturity, salinity, environmental habitat, and temperature (Pealver et al., 2020). This leads to different applications.

Edible marine greens can be utilized to prepare biodiesel, bioethanol, and hydrogen gases, as well as their use in food and several industries. In addition, seaweeds are used as virostatic, antibacterial, and antioxidant agents (Torregrosa-Crespo et al., 2018). The utilization of algae in the food industry, whether for animal or human consumption, has given rise to debates over the presence of toxic compounds such as cadmium. The extent of toxicity in algae is directly linked to the amount of bioactive substances and fiber present, which can affect the digestibility of such products and their use in the food industry. As noted by Torregrosa-Crespo et al. (2018), the key factor affecting digestibility is the consumption of nitrogen, both before and after digestion, and this process is influenced by the enzyme pepsin.

10.2 THE DISTRIBUTION OF EDIBLE MARINE GREENS

Since ancient times, people have consumed seaweed as food. In China and Japan, the sixth century saw the introduction of seaweeds as a food source. Seaweeds were particularly popular as a main or side dish or in soups among individuals who lived close to the seaside. Because of their stricter food laws and eating habits, Europeans often consume less seaweed than Asians. However, due to the significant health benefits of seaweeds that have been found over the past few decades, seaweed intake has steadily increased in European countries (Wells et al., 2017). Seaweeds are not only nutritious but also a rich source of structurally diverse bioactive compounds, such as pigments and

 DOI: 10.1201/9781003369738-12

sulfated polysaccharides. The bioactive properties of edible seaweeds and their potential health benefits are the subject of several scientific research studies that are published every year. Due to improved public knowledge, the demand for seaweeds and the commercial items made from them has marginally risen in international commerce (Sanjeewa et al., 2017).

Korea is among the top producers of edible marine greens for the global market. Natural collection and cultivation (maritime, agrarian, and agricultural practices) are the main methods of generating seaweed. Seaweeds are typically consumed by Koreans in their freshest forms, or they may be sun-dried to create various foods that are long-lasting. Seaweeds are frequently used to make soup (*mi-yeok-guk* and *mom-guk*), snacks (*kimbugak*), vegetables, pickles, salads, and *gim-bap* in traditional Korean cuisine. *Gim-bap* is a dish that combines steaming white rice, dried seaweed known as *gim*, and a variety of vegetables and meats. Koreans frequently eat *gim-bap* as a simple meal and as a side dish. As a side dish, most Koreans commonly eat fresh and dried *gim* (Wells et al., 2017; Debbarma et al., 2022).

10.3 CHEMICAL COMPONENTS OF EDIBLE SEAWEEDS

10.3.1 Polysaccharides

Polysaccharides make up the bulk of edible marine greens. Algae cell walls include a wide variety of polysaccharides that have both structural and storage functions (Ali et al., 2021). The physical and rheological properties of polysaccharides are directly impacted by the substitution of hydroxyl groups with sulfate esters, methyl groups, and pyruvic acid at different places (Pacheco-Quito et al., 2020). As one of the most crucial macromolecules, polysaccharides have notable impacts in immunomodulatory and anti-cancer processes. These impacts serve as a catalyst for extensive biochemical and medical study. Polysaccharides are prevalent in cell walls, and the season, age, species, and geographic location all have an impact on how they are composed. Their primary function in plants is food storage, but they can also offer strength and flexibility when dealing with wave motions and managing the ionic equilibrium within the cell. Alginates, fucoidans, and laminarans can all create various polysaccharides, depending on the seaweed (Gupta and Abu-Ghannam, 2011). Cellulose microfibrils combined with an amorphous polysaccharide make up the majority of the brown algal wall's constituents, which are connected by proteins. Sulfated fucans and alginic acid are the two types of acid polysaccharides found in the extracellular matrix of brown seaweeds.

10.3.2 Fucoidan

Fucoidan has the structure of a branched sulfate ester polysaccharide. L-fucose-4-sulfate or sulfate ester at C3 are the principal branches in this polysaccharide. According to Jia et al. (2011) and Pealver et al. (2010), the molecular weight of fucoidan ranges from 100 to 1600 kDa. Sulfated fucose, galactose, xylose, uronic acids, and fucose are the primary ingredients of fucoidan. The backbone of the sulfated fucans found in fucoidan is composed of several sugars, including fucose or uronic acid. Additionally, the backbone is branched to varying degrees. The species of algae have a significant impact on this structure.

Chordariales, Dictyotales, Dictyosiphonales, Ectocarpales, Scytosiphonales, Fucales, and Laminariales are the principal families of brown algae in which algal fucoidans, very prevalent sulfated polysaccharides, are found. Green, red, and freshwater algae, terrestrial plants, and even brown algae appear to lack algal material, despite the fact that brown algae does have it (Wang et al., 2019).

10.3.3 Carrageenans

Red edible seaweeds are used to produce carrageenans, linear sulfated polysaccharides. According to Necas and Bartosikova (2013), the word "carrageenan" comes from a kind of seaweed called

Chondrus crispus, also called "carrageen moss," "Irish moss," or "carraghin" in Ireland and England. With an average molecular weight of 100 kDa, this substantial and extremely flexible polysaccharide primarily consists of sulfate ester, which accounts for 15–40% of its composition. Anhydrogalactose (3,6-AG) and d-galactose appear alternately in the structural composition. These units are connected by 1,3- and 1,4-glycosidic linkages. Carrageenan comes in a variety of classes. Sulfate groups in all of these classes range from 22% to 35%. The main distinctions between various forms of carrageenans can be attributed to the quantity and location of sulfate esters. It should be noted that the chemical structures are not impacted by these nomenclatures. Sulfate ester content for the Kappa and Iota types is between 25% and 30%, as is the content of 3,6-AG, while the Lambda type has no 3,6-AG and a greater sulfate ester level of around 32–39%.

10.3.4 Alginic Acid

Alginic acid is also known as algin. Its building blocks are 1,4-linked β-d-mannuronic and α-l-guluronic acids, which are organized in an irregularly spaced manner. Alginic acid is produced from brown seaweed and is used mostly in the food and pharmaceutical industries. Due to their structural functions, they can attach to metal ions and when hydrated, produce highly viscous solutions. According to Wang et al. (2019), alginate's ability to absorb water makes it excellent for a variety of uses, particularly in biological investigations, where it may be used as an anticoagulant, anti-tumor, anti-viral, and antioxidant.

10.3.5 Laminarans

First discovered in Laminaria species, laminaran is the nutritional reserve of all brown algae. Laminaran is the primary sugar, structure, and component of the many Laminaria species, and it differs depending on the type of algae. According to Babahan et al. (2020), the majority of laminates, classified as dietary fiber because they are resistant to hydrolysis in the upper gastrointestinal tract (GIT), are stabilized by cross-chain hydrogen bonds. Water temperature, nutritional salt content, salinity, waves, sea movement, and plunge depth are a few environmental parameters that have an impact on the activity of laminarans. The antimicrobial and anti-cancer properties of laminaran are intriguing in addition to its function as a prebiotic and dietary fiber (Babahan et al., 2020).

10.3.6 Proteins

Given the growing difficulties in developing alternative and sustainable protein sources to increase food security, the protein content of seaweed has attracted a lot of interest (Harnedy and FitzGerald, 2011). Seaweeds are ingested in significantly lower amounts than beef while having protein and amino acid levels that are comparable on a gram-for-gram basis (Cherry et al., 2019). Furthermore, it is important to acknowledge that the Kjeldahl technique is typically utilized to determine the protein content of seaweed based on the total nitrogen present. The calculation involves a conversion factor of 6.25. So, according to Wells et al. (2017), seaweed has been assigned species-specific conversion coefficients that range from 3.57 to 5.72. The importance of proteins to the human diet, especially in terms of ensuring an appropriate consumption of essential amino acids, depends heavily on their amino acid makeup. However, the nutritional value of seaweed protein will be greatly impacted by how easily it can be absorbed by the body. This is because protein–polysaccharide interactions greatly hinder digestion.

10.3.7 Vitamins

The body cannot synthesize sufficient amounts of vitamins on its own; thus, it must acquire them from the diet. Vitamins are necessary organic micronutrients. Beriberi, which results from thiamine (vitamin B1) insufficiency, pernicious anemia, pellagra, and scurvy, which results from ascorbic acid (vitamin C) deficiency, are common human vitamin-deficiency disorders (Wells et al., 2017). These substances are required for vital metabolic processes and act as precursors for necessary cofactors for enzymes. Animals need to receive these cofactors from outside sources, as they are no longer able to generate them. Algal foods are nutrient-dense. While *Eisenia arborea*, a brown seaweed, has a vitamin C content that is similar to that of mandarin oranges, several sea vegetables, including sea spaghetti (*Himanthalia elongata*), laver (*Porphyra umbilicalis*), and *Gracilaria changii*, have vitamin C levels that are higher than those found in common vegetables like tomatoes and lettuce (Wells et al., 2017).

10.3.8 Lipids

All living organisms require lipids because they form membranes, serve as energy storage, and facilitate cellular communication. While some lipids can be synthesized by humans and other mammals, certain important lipids must be derived from oils or fats in the diet. Triacylglycerols (TAGs), which contain three fatty acid chains, are non-polar lipids that serve as a primary source of energy storage in cells. Unlike phospho- and glycolipids, which are necessary for membrane function, TAGs are not polar. Additionally, there are sterols such as fucosterol and beta-sitosterol that are abundant in lipid membranes and have been linked to beneficial health effects.

10.3.9 Dietary Fiber

According to Jiménez-Escrig et al. (2011), many communities don't consume enough dietary fiber on a daily basis. Dietary fiber's possible functional qualities are connected to its ability to bind water and become viscous in the gastrointestinal tract. Fiber has thus been hypothesized to support weight loss and satiety, delay gastric emptying to enhance glycemic control, enhance stool bulking to shorten gut transit time and boost frequency of bowel movements, and enhance bile acid excretion to lower levels of low-density lipoprotein cholesterol (LDL-C) in blood. Dietary fiber is hypothesized to be digested by the colonic microbiota, which may improve the composition of the gut flora and boost the production of volatile fatty acids connected to health, such as acetate, propionate, and butyrate. Gibson et al. (2017) found an increasing number of links between fiber-induced changes to the microbiota's makeup and the metabolites they produce and improvements in mental, gastrointestinal, cardiometabolic, and immunological health and bone development.

10.3.10 Alkaloids

Nitrogen atoms are found in the architecture of alkaloids, which are organic substances. In plants and other naturally occurring materials, there are numerous amines, cyclic nitrogen compounds, and organic molecules containing halogens. According to Babahan et al. (2020), the only places where cyclic nitrogen-containing alkaloids have been discovered are in marine creatures and seaweeds.

10.3.11 Phenylethylamine Alkaloids (PEAs)

2-Phenylethylamine, generally known as PEA, is a chemical molecule having a benzene ring and several ethylamine side chains. These significant alkaloids serve as natural and synthetic molecular building blocks. A range of pharmacological precursors can be created from substituted PEAs that are present in both plants and animals, such as simple phenylamine (tyramine, hordenine), and catecholamine (dopamine).

10.4 BIOACTIVE EFFECTS OF EDIBLE MARINE GREENS

10.4.1 Antioxidant Properties

According to Carocho and Ferreira (2013), extrinsic influences like smoking, radiation, medicines, and pesticides hasten the formation of free radicals. All kinds of molecules in biological materials, including proteins, amino acids, nucleic acids, and carbohydrates, can be harmed by an ongoing overproduction of free radicals in biological systems. In order to protect organisms from oxidative damage in biological systems, it is essential to avoid oxidative stress. Previous studies have indicated that the edible marine greens used in traditional Korean dishes have intriguing antioxidant capabilities.

A protein purified from *Capsosiphon fulvescens* was studied by Kim et al. (2015) to determine its antioxidant properties. The isolated protein has the capacity to reduce the generation of reactive oxygen species in HepG2 cells that are caused by H_2O_2 at concentrations of 0.01 to 10 g/mL, according to the authors' research. The antioxidant capabilities of the extracted and fractioned parts of *Enteromorpha prolifera* were examined by Kim et al. (2015).

10.4.2 Anti-Cancer Properties

In 2015, 8.8 million individuals worldwide were projected to have died from cancer. making it a significant global health issue. Previous research has shown that compounds in seaweed have the potential to prevent the growth of cancer cells and their spread (Sanjeewa et al., 2018). In order to lower the risk of cancer, encouraging the consumption of seaweed may be a viable strategy (Sanjeewa et al., 2018). However, the prevalence of the evidence comes from in vitro or animal research and is addressed elsewhere (Atashrazm et al., 2015). There is also a mass of evidence to indicate a role for fucoidan in the prevention of cancer. A polysaccharide extracted from *C. fulvescens* has also been reported by Kwon and Nam (2007) to have the ability to prevent gastric cancer cells from proliferating and progressing through the cell cycle by activating caspase-3 and decreasing the expression of Bcl-2 in cancer cells that have been exposed to polysaccharides.

10.4.3 Anti-Inflammation and Immunomodulatory Properties

An essential component of the host's defense mechanisms is inflammation, which is a defensive reaction. Inflammatory reactions are helpful in promoting tissue repair and eliminating infections. but unchecked and persistent inflammatory responses have been connected to cancer, obesity, and insulin resistance brought on by obesity (Sanjeewa et al., 2018).

10.4.4 Anti-Diabetic and Anti-Obesity Properties

According to Leung et al. (2017), diabetes, one of the most common and chronic metabolic illnesses, affects glucose metabolism, resulting in abnormally high plasma glucose levels. The metabolic disease known as obesity is characterized by a rise in body weight brought on by an excessive buildup of fat. According to Lee et al. (2016) and Leung et al. (2017), obesity puts people's health at risk by increasing their propensity for the pathogenesis of other chronic disease conditions like kidney diseases, diabetes, and cardiovascular disease. This raises the need for more frequent medical attention and reduces life expectancy. Recent studies on chemicals discovered in Korean edible seaweeds have revealed that they possess remarkable anti-diabetic and anti-obesity properties.

10.4.5 Antimicrobial Properties

Microbial infections are specifically avoided in the food processing sector and other associated businesses. As a result, the food processing industry and other allied sectors have paid close attention to food poisoning, food spoilage, and other illnesses related to food. There are various artificial food preservatives that have been created to prevent food contamination and increase shelf life. However, due to the undesirable side effects associated with those food preservatives, the market demand for goods containing synthetic food preservatives and antibacterial agents is progressively dropping. Numerous food preservatives have evolved from natural sources over the past few decades. *Pseudomonas aeruginosa*, *Bacillus subtilis*, *Escherichia coli*, *Staphylococcus aureus*, and *Salmonella* spp. may be susceptible to polysaccharides generated from *E. prolifera*, according to a recent study by Shao et al. (2017). As determined using 200 L of a 16 mg/mL drop, the findings revealed that the inhibition zone widths varied between 8.90 and 14 mm. The researchers also reported that the examined polysaccharides had minimum inhibitory concentrations against the aforementioned bacterial strains of 1 to 4 mg/mL. Patra et al. (2015) examined the efficacy of an essential oil derived from *Enteromorpha linza* against *Salmonella typhimurium* and *E. coli*. The results of this investigation (Patra et al., 2015) showed that the separated oils exhibited minimal inhibitory activity against *E. coli* and *S. typhimurium* of 12.5 and 25 mg/mL, respectively. In a 2016 study, Patra and Baek employed *E. linza* essential oil to look at how foodborne infections inhibit certain enzymes.

10.4.6 Anticoagulant Properties

The effective anticoagulant heparin is a sulfated polysaccharide. According to Sanjeewa, Lee, and Jeon (2018), thromboembolic illnesses have been treated and prevented with the help of the commercial anticoagulant medication heparin for more than 70 years. However, due to the fact that heparin has so many adverse side effects (alopecia, allergic reactions, thrombocytopenia, eosinophilia, heparin-induced thrombocytopenia, heparin-associated osteoporosis, and skin reactions), there is a growing need for some potent heparin alternatives (Sanjeewa et al., 2018).

10.5 LIMITATIONS OF EDIBLE MARINE GREENS

10.5.1 Heavy Metals

Consuming seaweed raises questions about the intake of heavy metals, such as tin, silicon, strontium, aluminum, arsenic, cadmium, lead, and rubidium (Desideri et al., 2016). Because seaweed heavy metal contamination varies depending on habitat or ecology, research findings have been conflicting. Seaweed grows in contaminated areas, which are commonly caused by industry or inadequate sewage systems, and accumulates heavy metals from the neighboring water and rocks (Wells et al., 2017). The health of people is not endangered by these concentrations. On the other hand, individuals who regularly consume perennial seaweeds may be at risk of heavy metal toxicity, as they are exposed to higher rates of pollution. About 426 seaweed samples were examined for levels of arsenic, lead, mercury, and cadmium in a Korean study. Maximum levels of arsenic were 17.4 mg/kg dry weight, and presuming an intake of 8.5 g of seaweed daily, the consumption was below the weekly amount advised by the WHO, the limit of which is 0.2–6.7%. The research verified the lowest likelihood of hazards to health from metals ingested through seaweed (Hwang et al., 2010). Due to species-specific changes in the bioabsorption of metals, Hwang et al. (2010) advocated for continuous monitoring of heavy metals in foods made from seaweed.

10.5.2 Iodine Content

Natural sources of iodine, which is required for thyroid function, can be found in seaweeds. However, excessive iodine consumption can cause thyroid illnesses, especially in persons who already have thyroid problems. Therefore, it's important to consume seaweed in moderation. It is still unclear from the epidemiological data if eating iodine-rich seaweeds has benefits or dangers. Consuming seaweed was linked to a higher incidence of thyroid papillary carcinoma among Japanese women who were postmenopausal, but not in those who were not. The iodine content should be disclosed, and cooking instructions should be included on product labels, in order to prevent the consumption of too much iodine from seaweed food products (Cherry et al., 2019).

10.5.3 Salt Content

Consuming *Laminaria digitata* and *Palmaria palmata* in small portions may be necessary to avoid excessive salt intake. A 5-g portion of *L digitata* can contain up to 0.35 g of salt and 0.26 g of sodium, while a 5-g portion of *P. palmata* can include up to 0.27 g of salt and 0.15 g of sodium. In comparison, an equivalent amount of the high-salt food bacon contains 0.144 g of salt and 0.0575 g of sodium, according to Scarborough et al. (2020). These figures are higher.

10.5.4 Digestibility

Complex carbohydrates and fibers found in seaweeds may be particularly difficult to digest for those who have digestive issues. It is essential to consume seaweed in moderation and to properly boil it in order to increase digestion. Dietary fiber is the component of food made from algae that human digestive enzymes are unable to completely break down. Dietary fibers can be categorized based on their solubility, viscosity, and fermentability, which affect how they are digested by the body. There are numerous chemical compositions for dietary fibers. Dietary fiber is made up of polysaccharides, oligosaccharides, lignin, cellulose, hemicellulose, and associated algal components, claim Demarco et al. (2022).

10.5.5 Availability and Affordability

Not all places of the world have access to edible marine greens, and they can be more expensive than other vegetables. For some communities, this may reduce their availability and consumption.

10.5.6 Allergic Reactions

Some persons, particularly those who have a history of seafood allergies, may be allergic to specific species of seaweeds. Seaweed should therefore be eaten with caution, especially if you have never done so previously. A number of substances have been identified that have been demonstrated to cause allergic reactions in people (James et al.,2023).

10.6 CONCLUSION

Regulations must be put in place to guarantee food product labeling and monitoring if edible marine greens are to contribute to future global food security. Even though they have the potential to be an excellent and sustainable supply of macronutrients (particularly dietary fiber) and micronutrients, it is imperative to prevent excessive intakes of salt, iodine, and heavy metals. An issue with arsenic bioaccumulation persists despite the fact that the levels of heavy metals in edible seaweeds are typically below the risk threshold, necessitating additional research on the toxicokinetics of heavy metals. The amount required may need to be balanced against the intake of iodine and/or heavy metals if large amounts of polyunsaturated fatty acids (PUFAs), protein, or dietary fiber are to be obtained through whole seaweed. It is acceptable to extract particular components from the intricate seaweed matrix in order to produce goods with added value, especially in light of the growing interest in novel bioactive seaweed components as potential therapeutics for noncommunicable diseases. It will be important to conduct future human intervention studies with well-defined health outcomes to ascertain the effects of chronic ingestion of whole seaweeds and the bioactive substances derived from them on human health. Clarifying the mechanisms of action is also required to enable applications in the food and nutraceutical industries and any anticipated health benefits linked to seaweed consumption.

REFERENCES

Ali, O., Ramsubhag, A. and Jayaraman, J. (2021) "Biostimulant properties of seaweed extracts in plants: Implications towards sustainable crop production," *Plants*, 10(3), p. 531. Available at: https://doi.org/10.3390/plants10030531.

Atashrazm, F., Lowenthal, R.M., Woods, G.M., Holloway, A.F. and Dickinson, J.L. (2015) "Fucoidan and cancer: A multifunctional molecule with anti-tumor potential," *Marine Drugs*, 13(4), pp. 2327–2346. Available at: https://doi.org/10.3390/md13042327.

Babahan, I., Kirim, B. and Mehr, H. (2020) "Major natural vegetation in coastal and marine wetlands: Edible seaweeds," in *Plant Communities and Their Environment*. IntechOpen. Available at: https://doi.org/10.5772/intechopen.88303.

Carocho, M. and Ferreira, I.C.F.R. (2013) "A review on antioxidants, prooxidants and related controversy: Natural and synthetic compounds, screening and analysis methodologies and future perspectives," *Food and Chemical Toxicology: An International Journal Published for the British Industrial Biological Research Association*, 51, pp. 15–25. Available at: https://doi.org/10.1016/j.fct.2012.09.021.

Cherry, P., O'Hara, C., Magee, P.J., McSorley, E.M. and Allsopp, P.J. (2019) "Risks and benefits of consuming edible seaweeds," *Nutrition Reviews*, 77(5), pp. 307–329. Available at: https://doi.org/10.1093/nutrit/nuy066.

Demarco, M., Oliveira de Moraes, J., Matos, Â.P., Derner, R.B., de Farias Neves, F. and Tribuzi, G. (2022) "Digestibility, bioaccessibility and bioactivity of compounds from algae," *Trends in Food Science and Technology*, 121, pp. 114–128. Available at: https://doi.org/10.1016/j.tifs.2022.02.004.

Desideri, D., Cantaluppi, C., Ceccotto, F., Meli, M.A., Roselli, C. and Feduzi, L. (2016) "Essential and toxic elements in seaweeds for human consumption," *Journal of Toxicology and Environmental Health: Part A*, 79(3), pp. 112–122. Available at: https://doi.org/10.1080/15287394.2015.1113598.

Gibson, G.R., Hutkins, R., Sanders, M.E., Prescott, S.L., Reimer, R.A., Salminen, S.J., Scott, K., Stanton, C., Swanson, K.S., Cani, P.D., Verbeke, K. and Reid, G. (2017) "Expert consensus document: The International Scientific Association for Probiotics and Prebiotics (ISAPP) consensus statement on the definition and scope of prebiotics," *Nature Reviews: Gastroenterology and Hepatology*, 14(8), pp. 491–502. Available at: https://doi.org/10.1038/nrgastro.2017.75.

Gupta, S. and Abu-Ghannam, N. (2011) "Bioactive potential and possible health effects of edible brown seaweeds," *Trends in Food Science and Technology*, 22(6), pp. 315–326. Available at: https://doi.org/10.1016/j.tifs.2011.03.011.

Harnedy, P.A. and FitzGerald, R.J. (2011) "Bioactive proteins, peptides, and amino acids from macroalgae(1): Macroalgae: Bioactive agent source," *Journal of Phycology*, 47(2), pp. 218–232. Available at: https://doi.org/10.1111/j.1529-8817.2011.00969.x.

Hwang, Y.O., Park, S.G., Park, G.Y., Choi, S.M. and Kim, M.Y. (2010) "Total arsenic, mercury, lead, and cadmium contents in edible dried seaweed in Korea," *Food Additives and Contaminants Part B – Surveillance*, 3(1), pp. 7–13. Available at: https://doi.org/10.1080/19440040903532079.

James, C.A., Welham, S. and Rose, P. (2023) "Edible algae allergenicity – A short report," *Journal of Applied Phycology*, 35(1), pp. 339–352. Available at: https://doi.org/10.1007/s10811-022-02880-2.

Jiao, G., Yu, G., Zhang, J. and Ewart, H.S. (2011) "Chemical structures and bioactivities of sulfated polysaccharides from marine algae," *Marine Drugs*, 9(2), pp. 196–223. Available at: https://doi.org/10.3390/md9020196.

Jiménez-Escrig, A., Gómez-Ordóñez, E. and Rupérez, P. (2011) "Seaweed as a source of novel nutraceuticals: Sulfated polysaccharides and peptides," in *Advances in Food and Nutrition Research*. Edited by S.-K. Kim, pp. 325–337. Available at: https://doi.org/10.1016/B978-0-12-387669-0.00026-0.

Kim, E.-Y., Choi, Y.H., Lee, J.I., Kim, I.-H. and Nam, T.-J. (2015) "Antioxidant activity of oxygen evolving enhancer protein 1 purified from Capsosiphon fulvescens: Antioxidant activity of OEE1 protein…," *Journal of Food Science*, 80(6), pp. H1412–H1417. Available at: https://doi.org/10.1111/1750-3841.12883.

Kwon, M.-J. and Nam, T.-J. (2007) "A polysaccharide of the marine alga Capsosiphon fulvescens induces apoptosis in AGS gastric cancer cells via an IGF-IR-mediated PI3K/Akt pathway," *Cell Biology International*, 31(8), pp. 768–775. Available at: https://doi.org/10.1016/j.cellbi.2007.01.010.

Lee, S.A., Moon, S.-M., Choi, Y.H., Han, S.H., Park, B.-R., Choi, M.S., Kim, J.-S., Kim, Y.H., Kim, D.K. and Kim, C.S. (2017) "Aqueous extract of Codium fragile suppressed inflammatory responses in lipopolysaccharide-stimulated RAW264.7 cells and carrageenan-induced rats," *Biomedecine & Pharmacotherapie [Biomedicine & Pharmacotherapy]*, 93, pp. 1055–1064. Available at: https://doi.org/10.1016/j.biopha.2017.07.026.

Leung, M.Y.M., Carlsson, N.P., Colditz, G.A. and Chang, S.-H. (2017) "The burden of obesity on diabetes in the United States: Medical expenditure panel survey, 2008 to 2012," *Value in Health: The Journal of the International Society for Pharmacoeconomics and Outcomes Research*, 20(1), pp. 77–84. Available at: https://doi.org/10.1016/j.jval.2016.08.735.

Pacheco-Quito, E.-M., Ruiz-Caro, R. and Veiga, M.-D. (2020) "Carrageenan: Drug delivery systems and other biomedical applications," *Marine Drugs*, 18(11), p. 583. Available at: https://doi.org/10.3390/md18110583.

Patra, J.K., Das, G. and Baek, K.-H. (2015) "Antibacterial mechanism of the action of Enteromorpha linza L. essential oil against Escherichia coli and Salmonella Typhimurium," *Botanical Studies (Taipei, Taiwan)*, 56(1), p. 13. Available at: https://doi.org/10.1186/s40529-015-0093-7.

Sanjeewa, K.K.A. and Jeon, Y.-J. (2018) "Edible brown seaweeds: A review," *Journal of Food Bioactives: An Official Scientific Publication of the International Society of Nutraceuticals and Functional Foods (ISNFF)*, 2, pp. 37–50. Available at: https://doi.org/10.31665/jfb.2018.2139.

Sanjeewa, K.K.A., Lee, J.-S., Kim, W.-S. and Jeon, Y.-J. (2017) "The potential of brown-algae polysaccharides for the development of anticancer agents: An update on anticancer effects reported for fucoidan and laminaran," *Carbohydrate Polymers*, 177, pp. 451–459. Available at: https://doi.org/10.1016/j.carbpol.2017.09.005.

Sanjeewa, K.K.A., Lee, W. and Jeon, Y.-J. (2018) "Nutrients and bioactive potentials of edible green and red seaweed in Korea," *Fisheries and Aquatic Sciences*, 21(1). Available at: https://doi.org/10.1186/s41240-018-0095-y.

Scarborough, P., Adhikari, V., Harrington, R.A., Elhussein, A., Briggs, A., Rayner, M., Adams, J., Cummins, S., Penney, T. and White, M. (2020) "Impact of the announcement and implementation of the UK Soft Drinks Industry Levy on sugar content, price, product size and number of available soft drinks in the UK, 2015–19: A controlled interrupted time series analysis," *PLOS Medicine*, 17(2), p. e1003025. Available at: https://doi.org/10.1371/journal.pmed.1003025.

Torregrosa-Crespo, J., Montero, Z., Fuentes, J., Reig García-Galbis, M., Garbayo, I., Vílchez, C. and Martínez-Espinosa, R. (2018) "Exploring the valuable carotenoids for the large-scale production by marine microorganisms," *Marine Drugs*, 16(6), p. 203. Available at: https://doi.org/10.3390/md16060203.

Wang, Y., Xing, M., Cao, Q., Ji, A., Liang, H. and Song, S. (2019) "Biological activities of fucoidan and the factors mediating its therapeutic effects: A review of recent studies," *Marine Drugs*, 17(3), p. 183. Available at: https://doi.org/10.3390/md17030183.

Wells, M.L., Potin, P., Craigie, J.S., Raven, J.A., Merchant, S.S., Helliwell, K.E., Smith, A.G., Camire, M.E. and Brawley, S.H. (2017) "Algae as nutritional and functional food sources: Revisiting our understanding," *Journal of Applied Phycology*, 29(2), pp. 949–982. Available at: https://doi.org/10.1007/s10811-016-0974-5.

Yakhin, O.I., Lubyanov, A.A., Yakhin, I.A. and Brown, P.H. (2016) "Biostimulants in plant science: A global perspective," *Frontiers in Plant Science*, 7, p. 2049. Available at: https://doi.org/10.3389/fpls.2016.02049.

11 Marine Greens as Nutritious Healthy Foods

Fadekemi O. Akinhanmi, Oluwadurotimi S. Aworunse, Opeyemi I. Ayanda, and Emmanuel Agiriga

11.1 INTRODUCTION

Marine greens (MGs) can be described as larger, macroscopic, complex, and multicellular marine algae, which are several meters in length. They are also referred to as marine macroalgae, sea greens, seaweeds (Peteiro, 2018), or sea vegetables (Mouritsen & Vinther, 2020; Pereira, 2021). Macroalgae have a body known as a thallus (Gutiérrez-Rodríguez *et al.*, 2018). The plant body lacks genuine roots, a vascular system, and a distinct stem and leaves (Pereira, 2021). Marine greens live on the bottom of continental shelves and washed-up intertidal zones, with some species drifting with plankton in the sunlit surface waters of the open sea. Marine greens are typically found on rocky sea fronts or free-floating along coastal areas (Pereira, 2021). They are classified into groups based on their pigmentation, the nature of their photosynthetic products, and other morphological features. Marine greens can be divided into the following three classification groups: red algae (Rhodophyta), green algae (Chlorophyta), and brown algae (Phaeophyta).

While marine greens form part of the main autotrophs at the bottom of the marine food web (Table 11.1), they can also be utilized extensively in various environmental, agricultural, industrial, and biomedical applications (Cerna, 2011); hence their ecological, commercial, and nutritional significance. Some seaweeds can occupy the sea surface, moistened by the sea foam, while some species can adhere to a substrate several meters deep (Fredericq & Schmidt, 2016). Marine greens and their debris can also serve as an essential food source for benthic organisms. The seaweed's fronds, which transport their leaves from the coast to deep ocean beds where they sink to the ocean floor without being re-mineralized by the organisms, can help to sequester carbon in the ocean as they absorb and release oxygen during photosynthesis (Moreira *et al.*, 2021).

Marine greens are highly versatile and nutritious products that can be directly consumed as food by humans. Despite being taxonomically considered algae, they belong to a food group that is rarely consumed in great quantities in the West in their unprocessed/raw state (Kaur *et al.*, 2023). Algae have been consumed by humans as food since the dawn of time, especially in the areas bordered by China, Japan, and the Korean peninsula (Ku *et al.*, 2013). Therefore, the usage of algae as a nutrient-rich food source is not surprising. After World War II, the use of this resource for commercial gain became a focus due to insufficient protein supply as a result of the rapid increase in the world population (Leandro *et al.*, 2019). Currently, the Asian countries are the nations that consume the most marine greens for food. However, the need for marine greens as food sources spread as citizens of these nations immigrated to other regions of the world; for example, in various regions of South and North America (Fredericq & Schmidt, 2016). Marine greens are now important coastal resources that are beneficial for human consumption in many nations, particularly in Asian regions, whether they are eaten fresh, dried, or as an ingredient in prepared dishes.

DOI: 10.1201/9781003369738-13

TABLE 11.1
Traditionally Consumed Marine Greens and Their Common Names

Marine greens	Classification	Common name
Porphyra sp.	Rhodophyta	Nori, laver, Folhuda, laverbread, sloak
Ulva lactuca, Ulva rigida	Chlorophyta	Sea lettuce, green laver, kwanpo, tahalib, luchi, sea grass, lettuce laver
Caulerpa lentillifera	Chlorophyta	Sea grape, latok, latoh, umibudo
Monostroma sp.	Chlorophyta	Hitoegusa, yaki-nori, hoshi-nori
Gelidium amansii	Rhodophyta	Umutgasari
Alaria esculenta	Phaeophyta	Dabberlocks, edible kelp, bladderlocks, henware, winged kelp
Laminaria digitata	Phaeophyta	Sea girdles, sea ware, kelp, oarweed, tangle
Saccharina japonica, Saccharina latissima	Phaeophyta	Kombu, sea belt, sugar kelp, hai tai, tasima, hirome, makombu
Undaria pinnatifida	Phaeophyta	Wakame, sea mustard, precious sea grass, miyeouk
Palmaria palmata	Rhodophyta	Dulse, dillesk, sheep dulse, sheep's weed, horse seaweed
Chondrus crispus	Rhodophyta	Irish moss, sea pearl moss, carrageen, sea moss, Dorset weed
Himanthalia elongata	Phaeophyta	Thong weed, sea spaghetti, sea thong

11.2 NUTRITIONAL COMPONENTS OF MARINE GREENS

Marine greens contain bioactive compounds that have been biologically assayed for interest in several vital remedial and medicinal areas. Good health outcomes have been associated with the activity of these bioactive chemicals, and it appears that the bioactive food ingredients can change the genetic expression of various cellular processes, impacting health outcomes or promoting advantageous enzyme activities (MacArtain *et al.*, 2007). As a result of growing commercial interest in health and wellness as well as public health initiatives, these substances are now common ingredients. The benefits of marine greens for health and nutrition have long been recognized in Asian cuisine, and the rest of the world is gradually catching up. In typical nutritional tests of marine greens, substantial concentrations of dietary fiber, proteins, phenols, vitamins, and trace elements were found (Gager *et al.*, 2021).

11.2.1 Polysaccharides

The primary component of marine greens is polysaccharides, which are often made up of repeated monosaccharide units connected together by glycosidic linkages or linear backbones made up of repetitive disaccharide units (Xu *et al.*, 2017; Gaignard *et al.*, 2018). Structural and storage polysaccharides are the two primary types found in the marine greens' cell walls. Similarly to terrestrial plants, structural polysaccharides mostly consist of cellulose, hemicellulose, and xylan. Storage polysaccharides, which are unique to certain species of macroalgae, are the components of marine greens that are most widely used for commercial purposes. They are extracted by the hydrocolloid industry and employed in food applications because they have textural and stabilizing qualities (Biancarosa *et al.*, 2018). Red algae are abundant in carrageenan and agar. Agars are composed of a mixture of agarose and the smaller molecule agaropectin, whereas carrageenans are linear polysaccharide chains with connected sulfate half-esters (Negreanu-Pirjol *et al.*, 2021). Ulvan, cellulose, amylose, amylopectin, inulin, mannan, pectin, and xylan are encountered in green algae. Ulvan is a hydro-soluble polysaccharide with a relative molecular mass that varies between 89 and 8200 kDa (Mo'o *et al.*, 2020). Brown algae are abundant in alginate, fucoidan, and laminaran. Alginates are comprised of α-l-guluronic and β-D-mannuronic acid, with a relative molecular mass varying from 500 to 1000 kDa (Moenne & Gonzalez, 2021).

11.2.2 Fatty Acids

Marine greens are a crucial source of very long-chain polyunsaturated fatty acids (PUFAs), which are important for animal and human health. Seaweeds from cold sea water always have a high content of PUFAs, with alpha-linolenic and linoleic acids as their basic precursors (Negreanu-Pirjol *et al.*, 2021). Omega-3 fatty acids (eicosapentaenoic and docosahexaenoic acid) and omega-6 fatty acids (γ-linolenic acid and arachidonic acid) are the major components of the macroalgae's lipid profiles (Zhang *et al.*, 2019). PUFAs have reportedly been shown to exhibit several beneficial health properties, including regulating blood pressure and blood coagulation as well as reducing the chance of developing chronic conditions including diabetes, obesity, and arthritis (Laye *et al.*, 2018).

11.2.3 Proteins

Asian-specific MG-rich diets have consistently been associated with a lower incidence of chronic diseases like cancer or cardiovascular disease. They are preferred over other plant sources of protein like rice or wheat because they meet the requirements for essential amino acids (Barbier *et al.*, 2019). The protein content of sea greens is particularly high in glutamic acid and aspartic acid (Figeuroa *et al.*, 2022). The primary ingredient of umami, the savory fifth basic flavor, is glutamate. *Spirulina* sp. and *Chlorella* sp. are the main cultivated species globally, due to their dry weight protein content of up to 70%. When compared with brown algae, red algae and green algae are especially rich in protein (Wells *et al.*, 2017). In vivo investigations have revealed that a number of bioactive amino acids and peptides, including taurine, carnosine, glutathione, and mycosporine-like, possess antioxidant capabilities and apoptosis-inhibition effects (Aydin *et al.*, 2016). Lectins are a valuable class of glycoproteins that manifest various beneficial effects (anti-viral, mitogenic, anti-inflammatory) because of their specificity in binding to glycan complexes (Silva *et al.*, 2010; Coelho *et al.*, 2017). It's significant to note that seasonal change limits the nutritional content, which influences the relative quantities of various amino acids during the summer (Wells *et al.*, 2017).

11.2.4 Amino Acids

Algal proteins are usually abundant in most amino acids, especially alanine, glycine, proline, glutamine, arginine, and aspartic acid (Cerna, 2011). In most marine greens, glutamic and aspartic acids are predominant, and they offer properties in the making of savory spices. Although the amino acid concentrations found in marine greens are higher than the levels in vegetables and cereals, their levels of lysine, tryptophan, threonine, and sulfur amino acids are limited (Holdt & Kraan, 2011). Among the free amino acids included in marine greens are taurine, alanine, amino butyric acid, ornithine, citrulline, and hydroxyproline. Diverse types of seaweed species also contain unusual amino acids, such as mycosporine-like amino acids (MAAs), which are well known for having antioxidant characteristics.

11.2.5 Phenols

Different phenolic compounds are derived from marine greens. Phenolic compounds are substances with different chemical structures of phenols: one aromatic ring (phenol) or multiple phenol units (polyphenols). The dominant polyphenols found in marine greens are phlorotannins and phloroglucoids. Phenolic acids, flavonoids, and bromophenols are predominant in red and green algae, while brown algae contain phlorotannins. Marine greens contain high concentrations of phenolic compounds, which have promising antioxidant, antiaging, anti-cancer, anti-microbial, and antitumoral effects when consumed in vegetable foods (Martinez-Valverde *et al.*, 2000; Ghosh *et al.*, 2019).

11.2.6 Dietary Fiber

Marine greens have been shown to contain slightly elevated levels of dietary fiber. Depending on the type of marine algae, the components are carrageenans, agar, and alginates. Although these fibers are not totally digested in the gut, some exhibit fermentative activity in the lower intestine, while naturally soluble fibers from marine greens pass through the gastrointestinal tract without digestion. Furthermore, fibers can enhance feelings of satisfaction and also aid digestive transit through their fecal bulking capacity (MacArtain *et al.*, 2007). Nori and Kombu are brown marine algae that contain slightly more fiber than some terrestrial food sources (MacArtain *et al.*, 2007). Based on the guidelines for daily amount of dietary fiber for humans (24 g), marine greens can provide up to 12.5% for an individual. This indicates that marine algae, in combination with their texturizing properties, may be an essential fiber source that is worthy of investigation. The potential nutritional function of alginates in brown algae has been described as beneficial to gut health and decreases the transit time of fecal bulk (Brownlee *et al.*, 2005). This aids in reduction of colon transit times, which is an advantageous factor in preventing colon cancer. The soluble but non-metabolized fiber content of marine greens enhances their potential applications in nutritionally balanced food sources.

11.2.7 Mineral Content

Marine greens absorb a wide diversity of minerals; hence, they are rich in minerals due to their marine habitat. Numerous nutritionally important minerals, including Na, K, Ca, Mg, Fe, Zn, Mn, and Cu, are found in them. For instance, *Gracilaria salicornia* and *Ulva lactuca* were discovered to have 948 and 2782 mg/100 g dry weight of calcium, respectively. These values exceed those of terrestrial plants including spinach (851 mg/100 dry weight), broccoli (503 mg/100 dry weight), and cabbage (369 mg/100 dry weight) by a significant margin (Tabarsa *et al.*, 2012). *Ulva lactuca* (sea lettuce) has relatively high levels of calcium, sodium, and potassium (MacArtain *et al.*, 2007). Iron and copper are present at high levels in *Palmaria palmata* (dulse). According to research on the bioavailability of iron from *Porphyra* sp., a type of red algae, animals fed with sea greens had reduced levels of iron absorption and retention, which may be due to the presence of polyphenolic chemicals or peptides from partially digested protein (Shaw & Liu, 2000). Brown algae, *Undaria pinnatifida* (wakame), and *Saccharina latissima* (kombu), bioaccumulate many elements and are a good source of copper, magnesium, iodine, iron, and other rare elements. While marine greens have been described as an ideal edible natural source of minerals, consuming them in large amounts could induce some undesirable effects (Teas *et al.*, 2004). Marine greens also contain some trace elements and heavy metals; whose levels are naturally below food safety limits.

11.2.8 Vitamin Content

Marine greens are rich in many forms of antioxidants, including vitamins and protective pigments, because their habitat varies from species to species. They contain vitamins that can be hydro- or fat-soluble, and many of them spend a lot of time in direct sunlight in an aquatic environment. These include vitamins A, B, C, and E (an important antioxidant). Hydro-soluble vitamins such as vitamin C are present in high levels in *Ulva lactuca* (sea lettuce), *Gracilaria* sp. (red ogo), and *Undaria pinnatifida* (wakame) (Norziah & Ching, 2000). This strengthens the evidence that sea greens can increase nutrient intake in a diet that includes a variety of foods. *Ulva lactuca* is also a source of vitamin B12 and can provide the guideline dietary allowance of this vitamin in excess. It provides an alternative source of vitamin B12 for vegans or vegetarians.

11.3 APPLICATIONS OF MARINE GREENS AS NUTRITIONAL SOURCES

The growing popularity of Asian cuisine and sushi in Western countries has encouraged the application of marine greens as nutritional sources of food. Migrant Asians across the world have stimulated the discovery of new ingredients from marine greens and have encouraged the creation of new and nutritious dishes (Turan & Cirik, 2018). Some of the species, such as *Porphyra* sp. (Nori) and *Ulva lactuca* (sea lettuce), can be eaten raw as fresh food, while others, such as *Undaria pinnatifida* (wakame) and *Saccharina* sp. (kombu), need cooking to overcome their tough and chewy composition. Optimizing and preserving the value of marine greens generally involves salting or drying processing treatments. The drying of marine greens is primarily a step to allow their storage and transportation. They can be air-dried, sun-dried, or dehydrated by salt addition.

Red macroalgae (*Gracilaria* sp.) are used as fresh food in Hawaii. *G. coronopifolia*, *G. tikvahiae*, *G. parvispora*, and *G. salicornia* are species that are commonly commercial, but they have a brief postharvest lifespan of about 4 days (Paul & Chen, 2008). *Ulva lactuca* is used in Scotland, where it is added to soups or used in salads. It can be prepared with rice, spicy meatballs, or fresh sardines, and in the preparation of mezze (a dish in the Middle East) (Figure 11.1). Sushi can also be made by combining *Ulva* sp. with *Porphyra* sp. Marine greens are rich in phytochemicals with anti-bacterial and antioxidant qualities. Mineral enrichment and salt reduction are made possible by the presence of fiber and minerals. The need for chemical preservatives can also be decreased by adding marine greens or their extracts to food products (Gupta & Abu-Ghannam, 2011).

The various bioactive compounds present in edible marine greens have potent health advantages, and their use as functional food substances channels new opportunities for meat product formulations and food processing. The food industry uses a few polysaccharides from marine greens as texture modifiers because of their high viscosity and gelling abilities. For centuries, marine macroalgae have been used in Asian cuisine in soups, salads, and in low-caloric diets. Their dietary fiber content is a major component and is primarily soluble fiber (Kilinc *et al.*, 2013). Particularly,

FIGURE 11.1 Use of *Ulva lactuca* in different foods. (A) Stuffed sea lettuce. (B) *Ulva* with rice. (C) In salad. (D) Sea lettuce mezze. (E) Spicy meatballs with sea lettuce. (F) Fresh pilchard wrapped in sea lettuce. (From Turan, G. & Cirik, S., 2018, http://doi.org//10.5772/intechopen.75014.)

Undaria pinnatifida (wakame) is often served in soups, salads, and side dishes. Wakame can be utilized as space food because it can be freeze-dried and gamma-irradiated without noticeable degeneration in the sensory quality, hence fulfilling the microbiological requirements for space food (Song *et al.*, 2012). Carrageenan was found to improve properties like sensory attributes, weight loss, lipid oxidation, and pH of low-fat fermented sausages, and had an advantageous effect on their microbiological and physicochemical properties (Koutsopoulos *et al.*, 2008).

11.4 NUTRITIONAL BENEFITS OF MARINE GREENS

As discussed earlier, marine greens are loaded with many bioactive components such as carbohydrates, proteins, fiber, vitamins, and minerals with essential nutritional benefits. This is a good reason to include them in the human diet. Vitamins A and E are important groups of vitamins for vision, strengthening the immune system, and every other physiological function. Many beneficial dietary fibers that can serve as prebiotics are also abundant in marine greens. These fiber-rich food products also aid in reducing glucose absorption into the bloodstream, thereby potentially helping in the prevention and management of diabetes for better control of blood sugar (Table 11.2) (Martinez-Hernandez *et al.*, 2018).

11.4.1 Gut Health

Prebiotics are compounds that help to maintain the gut bacteria, which are essential for overall health and well-being. Dietary fibers in marine greens are also beneficial in preventing and relieving constipation as well as improving bowel regularity. In addition, sea greens are potentially a great source of healthy fats, which are essential in preventing heart diseases and enhancing lung function. Some researchers have highlighted sea greens as sources of mitogens that are responsible for the delay or prevention of cell damage in the body (Kaur *et al.*, 2023).

11.5 LIMITATIONS TO THE USE OF MARINE GREENS AS FOOD MATERIALS

Since macroalgal cultivation technologies first emerged in Japan and China in the 20th century, global marine macroalgae aquaculture has rapidly evolved into a sector that generates an excess of US$13 billion annually, and is anticipated to further attract attention due to the increasing awareness of algae as sustainable and health-promoting foodstuffs, especially in the growing markets of Western societies (Circuncisão *et al.*, 2018; Moreira *et al.*, 2021). Whereas harvesting of naturally growing macroalgal stocks accounts for around 2.8%, macroalgae aquaculture presently constitutes about 51% of the world's marine aquaculture production figures, compared with 27% for mollusks, 13% for fish, and 9% for crustaceans (Moreira *et al.*, 2021; Lähteenmäki-Uutela *et al.*, 2021). Global production of macroalgae has in the last decade witnessed a two-fold increase to about 32 million tons of fresh weight per annum, with the exploitation of natural macroalgae remaining at an estimated 1 million tons annually (Lähteenmäki-Uutela *et al.*, 2021). Most of the world's macroalgae biomass production occurs in the Pacific and Southeast Asian countries, of which the Philippines, Indonesia, China, Japan, the Republic of Korea, and North Korea contribute approximately 98% (Moreira *et al.*, 2021). Chile and Norway are also bulk producers of macroalgae (Lähteenmäki-Uutela *et al.*, 2021). Humans' use of green macroalgae as food is an antiquated practice that dates back to 500 BC in China. It was not until 1000 years later that its use as food in Europe was documented (Pereira, 2021). As food, the green macroalgae are relished in varied forms, such as dried, fresh, powdered or flour form, in flakes, or included in other food products (Leandro *et al.*, 2019). Recently, green macroalgae have been in high demand for diverse industrial applications, including the production of premium-value food products, animal feed, flavorings, supplements, and food additives (García-Vaquero & Hayes, 2016; Moreira *et al.*, 2021; Ning *et al.*, 2022; Ummat *et al.*, 2021). This industrial exploitation of the green macroalgae is due to their nutritional qualities, as they are

TABLE 11.2
Nutritional Components and Benefits of Commonly Consumed Marine Greens

Marine greens	Nutritional components	Nutritional benefits	Reference
Ulva sp.	Proteins	Antioxidant	Steinbruch *et al.*, 2023
Monostroma nitidum, Monostroma undulatum	Fiber (polysaccharide), protein, lipids, minerals, and vitamin C	anticoagulant, antioxidant, anti-viral, anti-obesity, antithrombosis	Risso *et al.*, 2003, Kaur *et al.*, 2023
Arthrospira sp. (*Spirulina*)	Proteins, vitamins, and minerals	Weight loss, anti-cancer, anti-microbial	Gutiereez-Salmean *et al.*, 2015, Nuhu, 2013
Caulerpa lentillifera	Amino acids, fatty acids	Antioxidant, anti-cancer, anti-obesity	Nurkolis *et al.*, 2023
Porphyra sp.	Proteins/peptides, polysaccharides, fiber, polyunsaturated fatty acids,	Immunomodulating activity, anti-cancer, antioxidant, anti-hyperlipidemic	Cao *et al.*, 2016
Alaria esculenta, Saccharina latissima	Fatty acids, carotenoids, amino acids	Anti-inflammatory, antioxidant	Afonso *et al.*, 2021
Palmaria palmata	Proteins, fiber, amino acids, minerals, vitamins	Anti-inflammatory, antioxidant, anti-diabetic, improves gut health	Martinez-Hernandez *et al.*, 2018
Gelidium amansii	Polysaccharides, proteins, fatty acids, minerals	Anti-diabetic, anti-obesity, anti-inflammatory	Lee *et al.*, 2018
Laminaria digitata	Amino acids, proteins, vitamins, minerals, trace elements	Anti-cancer, weight loss, anti-hypertensive, anti-inflammatory, immunomodulation	Kolb *et al.*, 2004
Undaria pinnatifida	Proteins, amino acids, vitamins	Anti-diabetic, anti-cancer, cardiovascular disease treatment, anti-obesity	Kolb *et al.*, 2004
Himanthalia elongata	Fatty acids, amino acids, fiber, vitamins, macro and trace elements	Antioxidant, anti-bacterial, anti-hypoglycemic, neuroprotection	Ilyas *et al.*, 2023
Chlorella vulgaris	Fiber, proteins, vitamins and minerals, fatty acids, polysaccharides	Anti-hyperlipidemia, antioxidant, anti-cancer, anti-inflammatory	Panahi *et al.*, 2016
Aphanizomenon	Neuropeptides, amino acids, minerals, vitamins	Anti-tumoral, anti-allergic, anti-diabetic, anti-bacterial, anti-viral	Ku *et al.*, 2013

rich in polysaccharides, oligosaccharides, umami flavor, proteins, and essential amino acids, and abundant in trace elements and essential minerals required by the body to reach the recommended daily intake (Leandro *et al.*, 2019; Ummat *et al.*, 2021; Ning *et al.*, 2022). Although many green macroalgae are already available in the market and retailed under several brands (Leandro *et al.*, 2019), there are growing societal concerns about the safety of their consumption, considering their increased integration into diets around the world. Evidence is mounting to show that macroalgae bioaccumulate radioactive isotopes such as cesium-137 and radium-226 (Circuncisão *et al.*, 2018), as well as toxic heavy metals such as copper, lead, mercury, arsenic, nickel, antimony, strontium, aluminum, and cadmium to elevated levels that are up to 200–500 times those of terrestrial plants (Biancarosa *et al.*, 2018) (Table 11.3).

TABLE 11.3
Effects of Heavy Metals, Radio-Isotopes, Microplastics, and Pesticides Found in Green Macroalgae on Human Health

Harmful material	Effects on human health	References
Cadmium	Pulmonary damage, glomerular and tubular damage, proteinuria, loss of bone density and mineralization, Itai-Itai disease, renal cancer, testicular necrosis	Godt *et al.* (2006)
Arsenic	Dermatological abnormalities, pulmonary, reproductive, neurological, and cardiovascular complications; tumorigenesis in the liver, lungs, bladder, skin, and kidneys	Abdul *et al.* (2015)
Mercury	Infertility or sub-infertility, sperm DNA damage, abnormal sperm motility and morphology, hormonal and menstrual disruptions, and damaging reproductive abnormalities	Henriques *et al.* (2019)
	Neurodevelopmental disorder and increased risk of myocardial infarction	Khan *et al.* (2019)
Lead	Neuromuscular weakness, cognitive impairment, impaired auditory function, vascular and cardiac damage, liver and kidney damage	Assi *et al.* (2016)
Copper	Acute gastrointestinal symptoms and liver toxicity	Taylor *et al.* (2020)
Antimony	Cardiopulmonary dysfunctions, kidney and liver damage, chromosomal damage and reproductive disorders, carcinogenic and mutagenic effects	Wu *et al.* (2021)
Aluminum	Nervous system damage, neurofibril degeneration, osteomalacia, lymphopenia, and T lymphocyte dysfunction	Krupińska (2020)
	Dialysis dementia	Gupta *et al.* (2019)
Cesium-137	Birth defects, neurological disorders, infertility, and induction of various types of cancers	Hwang *et al.* (2020)
Radium-226	Cataracts, anemia, bone cancer, and death	Torres *et al.* (2018)
Microplastics	Endocrine disruption, coronary heart disease	Campanale *et al.* (2020)
	Neurotoxic effect	Sana *et al.* (2020); Prata *et al.* (2020)
	Immune function disruption, energy, and metabolism disruption	Prata *et al.* (2020)
	Gut microbiome disruption	Smith *et al.* (2018)
	Cellular damage and induction of inflammatory and immune system reactions	Vethaak and Legler (2021)
Pesticides	Neurodegenerative disease, developmental and reproductive toxicity	Hernández *et al.* (2013)

11.6 FUTURE PROSPECTS AND CONCLUSION

Large knowledge gaps remain to be filled, especially with respect to enhancing marine green productivity for food applications. Macroalgae are attractive species for food, as they do not compete with conventional agriculture for freshwater or land resources. Considering the challenge posed by climate change to agriculture, marine greens are gaining increasing global relevance (DeWeese & Osborne, 2021). Today, macroalgal mariculture depends on a proportionately small number of species from about seven genera, which together account for the majority of global production statistics (Moreira *et al.*, 2021). As the demand for marine greens continues to attract interest for various food and industrial applications around the world, the need to deploy innovative approaches to ensure diversification in mariculture and to scale up production for sustainable utilization becomes more compelling than ever. Whereas advances in modern biotechnologies have contributed to the development of new agricultural crops with improved traits, the macroalgae are relatively less studied globally and are rarely considered in strategies to elucidate gene function (Khan *et al.*, 2023). Commercial production of green macroalgae could benefit from biotechnological innovations such as tissue culture (Yong *et al.*, 2016) and genomic technologies (Wikandari *et al.*, 2021). To enhance productivity, tissue culture may be considered a viable option to overcome the challenges of traditional breeding methods such as the timely availability of propagules for planting and the destruction of seedlings by epiphytes, thereby facilitating the production of high-quality seaweeds (Yong *et al.*, 2016). Genomic technologies including RNA interference (RNAi), integrated multi-omics, and the clustered regularly interspaced short palindromic repeats-CRISPR-associated protein 9 (CRISPR-Cas9) gene editing could enable genetic modification to eliminate, for instance, anti-nutritional factors that limit the food utility of marine greens. Additionally, researchers could explore genome editing technologies to better understand the relationship between phenotype and genotype, and gene transformation strategies to identify promoter sequences. Furthermore, the introgression of modified morphological traits, improved characteristics, and enhanced yield represent a promising approach to significantly expand the production capacity of green macroalgae (Wikandari *et al.*, 2021). Whole genome sequencing of marine greens suited for aquaculture could allow researchers to gain insight into the network of genes and pathways that underpin yield for targeted breeding approaches. Akin to breeding efforts in mainstream agricultural crops, quantitative trait loci (QTL) mapping could be deployed to identify genomic loci or alleles that influence desirable phenotypes in marine greens. Similarly, genome-wide association studies (GWAS) may be used to detect single-nucleotide polymorphisms (SNPs) associated with traits of interest (DeWeese & Osborne, 2021). The development of molecular markers that target the plastid, nuclear, and mitochondrial DNA (mtDNA) in marine green germplasm could prove useful for varietal and species-specific identification, variety protection, and intergeneric and interspecific hybridization for economically important marine green species (Yong *et al.*, 2016). Harnessing macroalgae–microbiome symbiotic relationships through the introduction of beneficial microbial inoculants and ascertaining growth-promoting nutrients at specific stages of the macroalgal life cycle will be pivotal to future genomic approaches for the improvement and domestication of economically important marine-green species (DeWeese & Osborne, 2021). Although marine greens are excellent sources of minerals and high-quality lipids, bioavailability assessment of these nutrients in animal subjects is required.

Given that food and feed utilization of marine greens could be hampered due to the bioaccumulation of noxious heavy metals, microplastics, and iodine beyond the recommended concentrations, additional data on geographical and seasonal variability are required to evaluate the eligibility of marine macroalgae as next-generation food and feed (Biancarosa *et al.*, 2018).

REFERENCES

Abdul, K.S.M., Jayasinghe, S.S., Chandana, E.P., Jayasumana, C., & De Silva, P.M.C. (2015). Arsenic and human health effects: A review. *Environmental Toxicology and Pharmacology, 40*(3), 828–846.

Afonso, C., Matos, J., Guarda, I., Gomes-Bispo, A., Gomes, R., Cardoso, C., Gueifao, S., Delgado, I., Coelho, I., Castanheira, I., & Bandarra, N.M. (2021). Bioactive and nutritional potential of *Alaria esculenta* and *Saccharina latissima. Journal of Applied Phycology, 33*(1), 501–513.

Assi, M.A., Hezmee, M.N.M., Sabri, M.Y.M., & Rajion, M.A. (2016). The detrimental effects of lead on human and animal health. *Veterinary World, 9*(6), 660.

Aydin, A.F., Coban, J., Dogan-Ekici, I., Bet, L., & Uysal, M. (2016). Carnosine and taurine treatments diminished brain oxidative stress and apoptosis in D-galactose aging model. *Metabolic Brain Disease, 31*(2), 337–345.

Barbier, M., Charrier, B., Araujo, R., Holdt, S.L., Jacquemin, B., & Rebours, C. (2019). *C. pegasus* phycomorph European guidelines for a sustainable aquaculture of seaweeds, COST Action FA1406: Roscoff, France.

Biancarosa, I., Belghit, I., Bruckner, C.G., Liland, N.S., Waagbø, R., Amlund, H., Heesch, S., & Lock, E.J. (2018). Chemical characterization of 21 species of marine macroalgae common in Norwegian waters: Benefits of and limitations to their potential use in food and feed. *Journal of the Science of Food and Agriculture, 98*(5), 2035–2042.

Brownlee, I.A., Allen, A., Pearson, J.P., Dettmar, P.W., Haveler, M.E., Atherton, M.R., & Onsøyen, E. (2005). Alginate as a source of dietary fiber. *Critical Reviews in Food Science and Nutrition*, *45*(6), 497–410.

Campanale, C., Massarelli, C., Savino, I., Locaputo, V., & Uricchio, V.F. (2020). A detailed review study on potential effects of microplastics and additives of concern on human health. *International Journal of Environmental Research and Public Health*, *17*(4), 1212.

Cao, J., Wang, J., Wang, S., & Xu, X. (2016). *Porphyra* species: A mini-review of its pharmacological and nutritional properties. *Journal of Medicinal Food*, *19*(2), 111–119.

Cerna, M. (2011). Seaweed proteins and amino acids as nutraceuticals. *Advances in Food and Nutrition Research*, *64*, 297–312.

Circuncisão, A.R., Catarino, M.D., Cardoso, S.M., & Silva, A.M. (2018). Minerals from macroalgae origin: Health benefits and risks for consumers. *Marine Drugs*, *16*(11), 400.

Coelho, L.C., Silva, P.M., Lima, V.L., Pontual, E.V., Paiva, P.M., Napoleão, T.H., & Correia, M.T. (2017). Lectins, interconnecting proteins with biotechnological/pharmacological and therapeutic applications. *Evidence-Based Complementary and Alternative Medicine: eCAM*, 2017, 1594074, pp 22.

DeWeese, K.J., & Osborne, M.G. (2021). Understanding the metabolome and metagenome as extended phenotypes: The next frontier in macroalgae domestication and improvement. *Journal of the World Aquaculture Society*, *52*(5), 1009–1030.

Figueroa, V., Bunger, A., Ortiz, J., & Aguilera, J. M. (2022). Sensory descriptors for three edible Chilean seaweeds and their relations to umami components and instrumental texture. *Journal of Applied Phycology*, *34*(6), 3141–3156.

Fredericq, S., & Schmidt, W.E. (2016). Red algae. In: eLS. John Wiley & Sons, Ltd, Chichester, 1–7.

Gager, L., Lalegerie, F., Connan, S., & Stiger, V. (2021). Marine algal derived phenolic compounds and their biological activities for medicinal and cosmetic applications. In *Recent Advances in Micro and Macroalgal Processing*. https://doi.org/10.1002/9781119542650.ch11.

Gaignard, C., Macao, V., Gardarin, C., Rihouey, C., Picton, L., Michaud, P., & Laroche, C. (2018). The red microalga *Flintiella sanguinaria* as a new exopolysaccharide producer. *Journal of Applied Phycology*, *30*, 2803–2814.

García-Vaquero, M., & Hayes, M. (2016). Red and green macroalgae for fish and animal feed and human functional food development. *Food Reviews International*, *32*(1), 15–45.

Ghosh, C., Sarkar, P., Issa, R., & Haldar, J. (2019). Alternatives to conventional antibiotics in the era of antimicrobial resistance. *Trends in Microbiology*, *27*(4), 323–338.

Godt, J., Scheidig, F., Grosse-Siestrup, C., Esche, V., Brandenburg, P., Reich, A., & Groneberg, D.A. (2006). The toxicity of cadmium and resulting hazards for human health. *Journal of Occupational Medicine and Toxicology*, *1*(1), 1–6.

Gutiérrez-Rodríguez, A.G., Juárez-Portilla, C., Olivares-Bañuelos, T., & Zepeda, R.C. (2018). Anticancer activity of seaweeds. *Drug Discovery Today*, *23*(2), 434–447.

Gutiérrez-Salmeán, G., Fabila, C., Luis, H., & Chamorro, G. (2015). Nutritional and toxicological aspects of Spirulina (*Arthrospira*). *Nutricion Hospitalaria*, *32*(1), 34–40.

Gupta, S., Abu-ghannam, N., & Scannell, A.G.M. (2011). Growth and kinetics of *Lactobacillus plantarum* in the fermentation of edible Irish brown seaweeds. *Food and Bioproducts Processing*, *89*(4), 346–355.

Gupta, Y., Meenu, M., & Peshin, S. (2019). Aluminium utensils: Is it a concern? *The National Medical Journal of India*, *32*(1), 38–38.

Henriques, M.C., Loureiro, S., Fardilha, M., & Herdeiro, M.T. (2019). Exposure to mercury and human reproductive health: A systematic review. *Reproductive Toxicology*, *85*, 93–103.

Hernández, A.F., Parrón, T., Tsatsakis, A.M., Requena, M., Alarcón, R., & López-Guarnido, O. (2013). Toxic effects of pesticide mixtures at a molecular level: Their relevance to human health. *Toxicology*, *307*, 136–145.

Holdt, S.L., & Kraan, S. (2011). Bioactive compounds in seaweed: Functional food applications and legislation. *Journal of Applied Phycology*, *23*(3), 543–597.

Hwang, S.K., Kang, S.M., Rethinasabapathy, M., Roh, C., & Huh, Y.S. (2020). MXene: An emerging two-dimensional layered material for removal of radioactive pollutants. *Chemical Engineering Journal*, *397*, 125428.

Ilyas, Z., Redha, A.A., Wu, Y.S., Ozeer, F.Z., & Aluko, R.E. (2023). Nutritional and health benefits of the brown seaweed *Himanthalia elongata*. https://doi.org/10.1007/s11130-023-01056-8.

Kaur, M., Kala, S., Parida, A., & Bast, F. (2023). Concise review of green algal genus *Monostroma Thuret*. *Journal of Applied Phycology*, *35*(1), 1–10.

Khan, F., Momtaz, S., & Abdollahi, M. (2019). The relationship between mercury exposure and epigenetic alterations regarding human health, risk assessment and diagnostic strategies. *Journal of Trace Elements in Medicine and Biology: Organ of the Society for Minerals and Trace Elements*, *52*, 37–47.

Khan, N., Sudhakar, K., & Mamat, R. (2023). Seaweed farming: A perspective of genetic engineering and nano-technology application. *Heliyon*, *9*, e15168.

Kılınç, B., Cirik, S., Turan, G., Tekogul, H., & Koru, E. (2013). *Seaweeds for Food and Industrial Applications*. InTech. http://doi.org/10.5772/53172.

Kolb, N., Vallorani, L., Milanovi, N., & Stocchi, V. (2004). Evaluation of marine algae Wakame (*Undaria pinnatifida*) and kombu (*laminaria digitata japonica*) as food supplements. *Food Technology and Biotechnology*, *42*(1), 57–61.

Koutsopoulos, D. A., Koutsimanis, G. E., & Bloukas, J. G. (2008). Effect of carrageenan level and packaging during ripening on processing and quality characteristics of low-fat fermented sausages produced with olive oil. *Meat Science*, *79*(1), 188–197.

Krupińska, I. (2020). Aluminium drinking water treatment residuals and their toxic impact on human health. *Molecules*, *25*(3), 641.

Ku, C.S., Yang, Y., Park, Y., & Lee, J. (2013). Health benefits of blue-green algae: Prevention of cardiovascular disease and nonalcoholic fatty liver disease. *Journal of Medicinal Food*, *16*(2), 103–111.

Lähteenmäki-Uutela, A., Rahikainen, M., Camarena-Gómez, M.T., Piiparinen, J., Spilling, K., & Yang, B. (2021). European Union legislation on macroalgae products. *Aquaculture International*, *29*(2), 487–509.

Laye, S., Nadjar, A., Joffre, C., & Bazinet, R.P. (2018). Anti-inflammatory effects of omega-3 fatty acids in the brain: Physiological mechanisms and relevance to pharmacology. *Pharmaceutical Reviews*, *70*(1), 12–38.

Leandro, A., Pereira, L., & Gonçalves, A.M. (2019). Diverse applications of marine macroalgae. *Marine Drugs*, *18*(1), 17.

Lee, Y., Oh, H., & Lee, M. (2018). Anti-inflammatory effects of Agar free-*Gelidium amansii* (GA) extracts in high-fat diet-induced obese mice. *Nutrition Research and Practice*, *12*(6), 479–485.

MacArtain, P., Gill, C. I., Brooks, M., Campbell, R., & Rowland, I. R. (2007). Nutritional value of edible seaweeds. *Nutrition Reviews*, *65*(12), 535–543.

Martínez-Hernández, G.B., Castillejo, N., Carrión-Monteagudo, M.D.M., Artés, F., & Artés-Hernández, F. (2018). Nutritional and bioactive compounds of commercialized algae powders used as food supplements. *Food Science and Technology International*, *24*(2), 172–182.

Martinez-Valverde, I., Periago, M.J., & Ros, G. (2000). Nutritional importance of phenolic compounds in the diet. *Archivos Latinoamericanos de Nutrición*, *50*(1), 5–18.

Moenne, A., & Gonzalez, A. (2021). Chitosan, alginate-carrageenan derived oligosaccharides stimulate defense against biotic and abiotic stresses, and growth in plants: A historical perspective. *Carbohydrate Research*, *503*, 108298.

Mo'o, F.R.C., Wilar, G., Devkota, H.P., & Wathoni, N. (2020). Ulvan, a polysaccharide from macroalga *Ulva* sp.: A review of chemistry, biological activities and potential for food and biomedical applications. *Applied Sciences-Basel*, *10*(16), 5488.

Moreira, A., Cruz, S., Marques, R., & Cartaxana, P. (2021). The underexplored potential of green macroalgae in aquaculture. *Reviews in Aquaculture*, *14*(1), 5–26.

Mouritsen, O.G., & Vinther, S.C. (2020). A role for macroalgae and cephalopods in sustainable eating. *Frontiers in Psychology*, *11*, 1402.

Negranu-Pirjol, B.S., Negreanu-Pirjol, T., Prelipcean, A.M., & Popoviciu, D.R. (2021). Chapter vi marine algae an important resource for bioeconomy and biomedical applications. A review. In *Agricultural Studies on Different Subjects*. Iksad, Turkey, pp. 145–222.

Ning, L., Yao, Z., & Zhu, B. (2022). Ulva (Enteromorpha) polysaccharides and oligosaccharides: A potential functional food source from green-tide-forming macroalgae. *Marine Drugs*, *20*(3), 202.

Norziah, M.H., & Ching, C.Y. (2000). Nutritional composition of edible seaweed *Gracilaria changgi*. *Food Chemistry*, *68*(1), 69–76.

Nuhu, A.A. (2013). *Spirulina (Arthrospira)*: An important source of nutritional and medicinal compounds. *Journal of Marine Sciences*, *1*(8), 325636.

Nurkolis, F., Taslim, N.A., Qhabibi, F.R., Kang, S., Moon, M., Choi, J., Choi, M., Park, M.N., Mayulu, N., & Kim, B. (2023). Ulvophyte Ggreen algae *Caulerpa lentillifera*: Metabolites profile and antioxidant, anticancer, anti-obesity, and in vitro cytotoxicity properties. *Molecules*, *28*(3), 1365.

Panahi, Y., Darvishi, B., Jowzi, N., Beiraghdar, F., & Sahebkar, A. (2016). *Chlorella vulgaris*: A multifunctional dietary supplement with diverse medicinal properties. *Current Pharmaceutical Design*, *22*(2), 164–173.

Paul, I.R., & Chen, E.N.J. (2008). Postharvest handling and storage of the edible red seaweed *Gracilaria*. *Postharvest Biology and Technology*, *48*(2), 302–308.
Pereira, L. (2021). Macroalgae. *Encyclopedia*, *1*(1), 177–188.
Peteiro, C. (2018). Alginate production from marine macroalgae, with emphasis on kelp farming. In Rehm, B., Moradali, M. (Eds.) *Alginates and their Biomedical Applications, Springer series in biomaterials science and engineering*, vol. 11, Springer, Singapore, pp. 27–66.
Prata, J.C., da Costa, J.P., Lopes, I., Duarte, A.C., & Rocha-Santos, T. (2020). Environmental exposure to microplastics: An overview on possible human health effects. *Science of the Total Environment*, *702*, 134455.
Risso, S., Escudero, C., de Portela, M. L., & Fajardo, M. A. (2003). Chemical composition and seasonal fluctuations of the edible green seaweed, Monostroma undulatum, Wittrock, from the Southern Argentina coast. *Archivos Latinoamericanos de Nutrición*, *53*(3), 306–311.
Sana, S.S., Dogiparthi, L.K., Gangadhar, L., Chakravorty, A., & Abhishek, N. (2020). Effects of microplastics and nanoplastics on marine environment and human health. *Environmental Science and Pollution Research International*, *27*(36), 44743–44756.
Shaw, N.S., & Liu, Y.H. (2000). Bioavailability of iron from purple laver (*Porphyra* spp.) estimated in a rat hemoglobin regeneration bioassay. *Journal of Agriculture and Food Chemistry*, *48*(5), 1734–1737.
Silva, L.M., Lima, V., Holanda, M.L., Pinheiro, P.G., Rodrigues, J.A., Lima, M.E., & Benevides, N.M. (2010). Antinociceptive and anti-inflammatory activities of lectin from marine red alga *Pterocladiella capillacea*. *Biological and Pharmaceutical Bulletin*, *33*(5), 830–835.
Smith, M., Love, D.C., Rochman, C.M., & Neff, R.A. (2018). Microplastics in seafood and the implications for human health. *Current Environmental Health Reports*, *5*(3), 375–386.
Song, B., Park, J.S., Kim, G.J., Choi, H.J., Ahn, I.D., Hao, H.C., & Lee, J.W. (2012). Development of freeze-dried miyeokguk, Korean seaweed soup, as spacefood sterilized by irradiation. *Radiation Physics and Chemistry*, *81*(8), 1111–1114.
Steinbruch, E., Wise, J., Levkov, K., Chemodanov, A., Israel, Á., Livney, Y.D., & Golberg, A. (2023). Enzymatic cell wall degradation combined with pulsed electric fields increases yields of water-soluble-protein extraction from the green marine macroalga *Ulva* sp. *Innovative Food Science and Emerging Technologies*, *84*, 103231.
Tabarsa, M., Rezaei, M., Ramezanpour, Z., & Waaland, J.R. (2012). Chemical compositions of the marine algae *Gracilaria salicornia* (Rhodophyta) and *Ulva lactuca* (Chlorophyta) as a potential food source. *Journal of the Science of Food and Agriculture*, 92, 2500–2506.
Taylor, A. A., Tsuji, J. S., Garry, M. R., McArdle, M. E., Goodfellow, W. L., Adams, W. J., & Menzie, C. A. (2020). Critical review of exposure and effects: implications for setting regulatory health criteria for ingested copper. *Environmental management*, *65*, 131–159.
Teas, J., Pino, S., Critchley, A.T., & Braverman, L.E. (2004). Variability of iodine content in common commercially available edible seaweeds. *Thyroid: Official Journal of the American Thyroid Association*, *14*(10), 836–841.
Torres, L., Yadav, O.P., & Khan, E. (2018). Risk assessment of human exposure to Ra-226 in oil produced water from the Bakken Shale. *Science of the Total Environment*, *626*, 867–874.
Turan, G., & Cırık, S. (2018). Sea vegetables. InTech. http://doi.org//10.5772/intechopen.75014.
Ummat, V., Garcia-Vaquero, M., Poojary, M.M., Lund, M.N., O'Donnell, C., Zhang, Z., & Tiwari, B.K. (2021). Green extraction of proteins, umami and other free amino acids from brown macroalgae *Ascophyllum nodosum* and *Fucus vesiculosus*. *Journal of Applied Phycology*, *33*(6), 4083–4091.
Vethaak, A.D., & Legler, J. (2021). Microplastics and human health. *Science*, *371*(6530), 672–674.
Wells, M.L., Potin, P., Craigie, J.S., Raven, J.A., Merchant, S.S., Helliwell, K.E., Smith, A.G., Camire, M.E., & Brawley, S.H. (2017). Algae as nutritional and functional food sources: Revisiting our understanding. *Applied Phycology*, *29*, 949–982.
Wikandari, R., Manikharda, Baldermann, S., Ningrum, A., & Taherzadeh, M.J. (2021). Application of cell culture technology and genetic engineering for production of future foods and crop improvement to strengthen food security. *Bioengineered*, *12*(2), 11305–11330.
Wu, S., Zhong, G., Wan, F., Jiang, X., Tang, Z., Hu, T., Rao, G., Lan, J., Hussain, R., Tang, L., Zhang, H., Huang, R., & Hu, L. (2021). Evaluation of toxic effects induced by arsenic trioxide or/and antimony on autophagy and apoptosis in testis of adult mice. *Environmental Science and Pollution Research International*, *28*(39), 54647–54660.
Xu, S.Y., Huang, X., & Cheong, K.L. (2017). Recent advances in marine algae polysaccharides: Isolation, structure, and activities. *Marine Drugs*, *15*(12), 388.

Yong, W.T.L., Chin, G.J.W.L., & Rodrigues, K.F. (2016). Genetic identification and mass propagation of economically important seaweeds. In N. Thajuddin & D. Dhanasekaran (Eds.), *Algae– Organisms for Imminent Biotechnology*. InTech, pp. 277–305.

Zhang, T.T., Xu, J., Wang, Y.-M., & Xue, C.-H. (2019). Health benefits of dietary marine DHA/EPA-enriched glycerophospholipids. *Progress in Lipid Research*, *75*, 100997.

Section III

Industrial Applications

12 Biotechnological Potential of Marine Bacteria in the Production of Enzymes

Idris Abdullahi Dabban, Oluwafemi Adebayo Oyewole, Habiba Iliyasu Atta, Adejoh Suleiman Ocholi, Musa Bishir, and Ibrahim Isiaka Hussein

12.1 INTRODUCTION

The mining of natural product sources for novel metabolites such as enzymes has been prosperous and has produced over 50% of all drugs in the current market essentially from natural products (Novakova and Farkasovsky, 2013; Aransiola *et al.*, 2023). Laboratory and large-scale industrial sourcing and utilization of natural microbial products, by-products, and metabolites rely mainly on the cultivable representatives of prospective microorganisms, as these serve as easily accessible sources of microbial natural products. Conversely, the incredible diversity of uncultured bacterial communities in extreme habitats constitutes a huge reservoir of new and undiscovered natural products.

In corroboration, several studies (e.g., Trindade *et al.*, 2015; Albarano *et al.*, 2020) have shown that the uncultured microorganisms in diverse environments are promising sources of novel natural products such as beneficial genes, enzymes, antibiotic synthesis enzymes, biomolecules, antimicrobial compounds, chemical bioactive compounds, protein synthesis inhibitors, and many others. Bacteria are among the most essential sources of natural bioactive products, mostly for new medicinal discoveries (Novakova and Farkasovsky, 2013). Examples of the variety of marine-derived microbial genes coding for enzymes with distinctive capabilities are cellulose, lipases, proteases, alkane hydroxylase genes, high salt-tolerance esterases, and high-temperature optima metalloproteases (Lee *et al.*, 2007; Hardeman and Sjoling, 2007; Xu *et al.*, 2008; Chu *et al.*, 2008), also offering different worthwhile industrial applications and products. In this vein, there is no doubt that still uncultured bacteria present a rich source of novel bioactive molecules with potent therapeutic activity (Trindade *et al.*, 2015).

Studies of diverse microbial habitats reported that approximately 99% of microbes cannot be cultivated by routine conventional techniques (Amann *et al.*, 1995). For reference, microorganisms inhabiting both mangrove mud and microbial mat ecosystems have been discovered via the cultivation approaches (Al-Amoudi *et al.*, 2016), though the metagenomic approach combined with phylogenetic analysis revealed that only 1% of the bacterial diversity has been cultured to date (Amann *et al.*, 1995). Also, Vartoukian *et al.* (2010) stated that fewer than 1% of bacteria have been cultured *in vitro*, and of the nearly 61 bacterial phyla recognized, 31 lack cultivable representatives. Besides, seawater bacteria have a 10-fold lower representation of cultured microbial isolates when compared with other microbial habitats (Amann *et al.*, 1995).

DOI: 10.1201/9781003369738-15

12.2 BIOPROSPECTION/SCREENING FOR ENZYMES AND NATURAL PRODUCTS OF MARINE BACTERIA: CULTURE-DEPENDENT METHODS, GENE-BASED/METAGENOMICS, BIOACTIVITY-BASED

12.2.1 Culture-Dependent Techniques and Bioactivity

Culture-based methods for studying the diversity of bacteria in the environment still have their advantages in the field of microbial ecology, even though the molecular methods are favored in most studies. One of the most important merits is having pure isolates for use in further studies and applications. It is pertinent to be able to preserve these pure isolates to use them topically in industrial, medical, and biotechnological applications.

According to Gest (2008), the term, "uncultured bacteria" may not necessarily mean that these organisms cannot be cultured; it may be that they have not been specifically targeted; hence, the expressions "not-yet-cultured" and "yet-to-be-cultured" are more appropriate terminologies. The study of marine bacteria is no exception; their unique environment affords them the foundation for the production of novel bioactive products. One of the major limitations in cultivating marine bacteria in the laboratory is the challenges in the reproduction in the culture media of the exact quantities of elements that are found in the natural marine ecosystem (Zobell, 1941). Some important culture-dependent techniques are discussed in the following.

New and innovative culture techniques have been developed and used to grow novel microorganisms in vitro, including co-culture with other bacteria, density-based separation, dilution-to-extinction, and the use of selective inhibitors and diffusion chambers (Kaeberlein *et al.*, 2002; Nichols *et al.*, 2008; Kearney *et al.*, 2021; Lewis *et al.*, 2021).

12.2.2 Bioactivity-Based Screening of Marine Bacteria

The marine environment potentially harbors a diverse array of bacteria capable of producing numerous compounds important in the pharmaceutical, agricultural, biomedical, and food industries. Many of these bacteria are also key players in the environment, especially in biogeochemical cycles, bioremediation, and other environmental-related applications (Table 12.1). However, due to the difficulty of cultivating marine bacteria in the laboratory, studies on these organisms have often utilized metagenomics and whole genome sequencing. The isolation of bioactively important bacteria in pure cultures allows studies of their interaction with other microorganisms (syntrophic, cell-to-cell), physiological analyses, and analysis of *in situ* activity. Research on natural products can be classified as one of the Nature Based Solutions (NBS), and has led to breakthroughs in drug discovery and the field of biomedical research. The metabolic diversity of marine fungi and bacteria makes them more favorable candidates than microorganisms present in the marine environment. Their potential for the production of secondary metabolites is enhanced by their adaptation to survival in the unique environment of the marine ecosystem. This increases their chances of producing more unique bioactive compounds compared with microorganisms dwelling in the terrestrial environment. Their activity can be on the following basis: antimicrobial, antitumoral, cytotoxic, industrial application, biotechnology, antioxidant, or antiviral.

Screening of marine bacteria is carried out through extraction and separation methods: assisted extraction technology and chromatography. The assisted extraction technologies include microwave, ultrasonic, and enzyme-assisted technology. Other extraction technologies are as follows: ionic liquid extraction, supercritical fluid extraction, subcritical water extraction, and accelerated solvent (Li *et al.*, 2021).

Traditionally, antimicrobial activity is usually determined in the laboratory by tube dilution or disc diffusion methods; however, in screening marine bacteria, novel methods had to be devised to overcome the challenges faced, such as non-uniformity in the standards. A summary of the current

TABLE 12.1
Some Marine Bacteria Targeted for Cultivation Due to Their Potential Beneficial Uses

Group	Phylum/ Superphylum	Role/Function	References
Candidatus Actinomarinidae	Actinobacteria (OM1)	Streamlined genomes, genomes encode genes for rhodopsins	Ghai *et al.*, 2013
Dimethylsulfoniopropionate (DMSP) synthesis bacteria	Assorted	DMSP generation in the aerobic ocean	Curson *et al.*, 2017
Candidatus Electrothrix and Candidatus Electronema	Deltaproteobacteria	Cable bacteria and important role in biogeochemistry	Kjeldsen *et al.*, 2019; Muller *et al.*, 2020
Candidatus Marinimicrobia	Candidate phylum marine group A	Biogeochemical cycles in the marine ecosystem (particularly sulfur and nitrogen), likely absorb nitrous oxide	Hawley *et al.*, 2017; Bertagnolli *et al.*, 2017
Methylphosphonic acid synthesis bacteria	Assorted	Methane generation in the aerobic ocean	Metcalf *et al.*, 2012
Candidatus Poribacteria	Candidate phylum Poribacteria	Associated with marine sponges	Hawley *et al.*, 2017
SAR202	Chloroflexi	Possibly major role players in sulfur cycle	Mehrshad *et al.*, 2018
SAR324	Deltaproteobacteria	Metabolically diverse	Sheik *et al.*, 2014
SAR86	Gammaproteobacteria	Widespread globally	Dupont *et al.*, 2012
Candidate Phyla Radiation	Candidate Phyla Radiation	Very limited number of cultured representatives	
Dehalogenating bacteria	Chloroflexi, Firmicutes, and others	Bioremediation	Major *et al.*, 2002

methods used in screening marine bacteria for bioactivity against phytopathogens is presented in Table 12.2.

The constituents of antioxidants have mostly been obtained from fruits, plants, and seaweeds; however, the detection of natural products produced by marine microbes has expanded the scope of natural antioxidant biomolecule discovery (Hamidi *et al.*, 2019). According to Li *et al.* (2021), the antimicrobial activities of natural products from marine bacteria against plant pathogens can include affecting the permeability of the cell membrane, synthesis of the cell wall, fatty acid metabolism, quorum sensing, etc. In order to apply these bioactive marine bacteria in the mass production of biopesticides, synthetic biology, and chemical synthesis are often employed (Li *et al.*, 2021). The growth conditions of carotenoid-producing marine bacteria determine the type of carotenoid

TABLE 12.2
Screening Methods of Natural Products from Marine Bacteria for Biological Activity Against Plant Pathogens

Method	Advantage	References
Solid culture diffusion	Simple, requires small quantities of the compound, relatively affordable	Zhao *et al.*, 2010
Liquid culture dilution (tube dilution)	Can be used for the determination of minimum inhibitory concentration (MIC) of the compound, only requires small amounts of the compound	Huang *et al.*, 2011
Solid culture dilution (plate dilution)	Determination of minimum inhibitory concentration (MIC) of the compound, only requires small amounts of the compound	Cao *et al.*, 2019

TABLE 12.3
Some Bioactive Compounds Produced by Marine Bacteria

Bioactive compound (type)	Bacteria/Group	Function	References
Tauramamide (lipopolypeptide)	*Brevibacillus laterosporus* PNG276	Antimicrobial	Desjardine *et al.*, 2007
novel 2-alkylidene-5- alkyl-4-oxazolidinones, lipoxazolidinone A, B, and C	*Marinispora* strain NPS008920	Antimicrobial	Macherla *et al.*, 2007
Ayamycin	*Nocardia* sp.	Antifungal	El-Gendy *et al.*, 2008
Labrenzin (Pederin-like)	*Labrenzia alexandrii* PHM005		Schleissner *et al.*, 2017
Bagremycins	*Streptomyces* sp. ZZ745	Antimicrobial	Zhang *et al.*, 2018
2-Alkyl-4-hydroxyquinolines	*Streptomyces* sp.	Antifungal	Kim *et al.*, 2019
Exopolysaccharides	*Microbacterium aurantiacum* FSW-25	Antioxidant	Sran *et al.*, 2019

produced, as elucidated in the review by Hamidi *et al.* (2019). Examples of some genera of marine bacteria that produce carotenoids include *Haloferax*, *Agrobacterium*, and *Brevibacterium* (Nelis *et al.*, 1991; Yokoyama *et al.*, 1994; Asker and Ohta, 2002). Some bioactive compounds produced by marine bacteria are highlighted in Table 12.3.

12.3 ENZYME PRODUCTION BY MARINE BACTERIA

In the hunt for novel enzymes, biodiversity has been studied in a variety of habitats, such as forests, savannas, tundras, deserts, and lastly, the sea. The ability of marine microorganisms and their enzymes to function under conditions with a high salt content, a wide range of temperatures, and a high incidence of light and pressure makes them a significant source of special biocatalysts (Birolli *et al.*, 2019). Marine bacteria's enzymatic machinery is the product of significant evolution designed to help cells survive in the harsh and constantly varying circumstances prevalent in the marine environment (Rodrigues *et al.*, 2022; Abioye *et al.*, 2015b). Cellulase, amylase, alginate lyase, chitinase, inulinase, keratinase, ligninase, glucosidase, xylanase, and other enzymes have been reported to be produced by microbial strains isolated from a variety of substrates, including sediments, seawater,

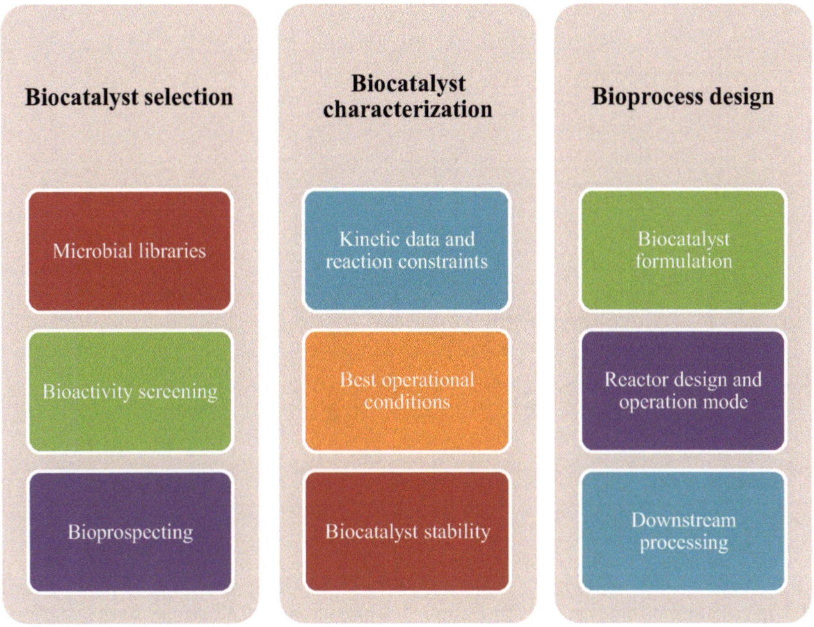

FIGURE 12.1 Biocatalytic process development in marine bacteria. (From Rodriguez, C.J.C. and de Carvalho, C.C.C.R., *Front. Bioeng.*, 10, 964589, 2022.)

and mangrove detritus (Bonugli-Santos *et al.*, 2015). The processes leading to the discovery and production of microbial enzymes from marine bacteria are outlined in Figure 12.1.

12.3.1 Techniques Used for Production of Enzymes from Marine Bacteria

12.3.1.1 Batch and Continuous Cultures

Batch growth is the cultivation of organisms in a vessel with an initial medium charge that is not changed by the addition or subtraction of nutrients. This method of growing is straightforward and popular in both the laboratory and industry. In continuous culture, new nutrient medium is continuously introduced to a well-stirred culture, and products and cells are simultaneously withdrawn. In discontinuous culture, growth, product production, and substrate consumption stop after a predetermined amount of time, while in continuous culture, growth and product creation can be sustained for extended periods of time. The system typically enters a steady state, where the concentrations of cells, products, and substrates are constant, after a specific amount of time. Continuous culture supplies products of a consistent quality and maintains growth and product creation conditions in the reactor.

12.3.1.2 Fed-Batch Methods

A production method in between batch and continuous operations is fed-batch cultivation. For a fed-batch operation, the optimum feed rate and composition are necessary. The feed of the necessary ingredients for growth and/or other substrates required for the production can never be depleted under controlled conditions, and the nutritional environment can be kept relatively constant throughout the batch with the necessary knowledge of the microorganisms involved in the cultivation. By limiting the amount of substrate to what is necessary only for the synthesis of the target chemical, it is possible to prevent the generation of by-products that are often associated with the presence of large substrate concentrations. Nutrients are continuously or semi-continuously provided in a fed-batch culture, while effluent is intermittently removed.

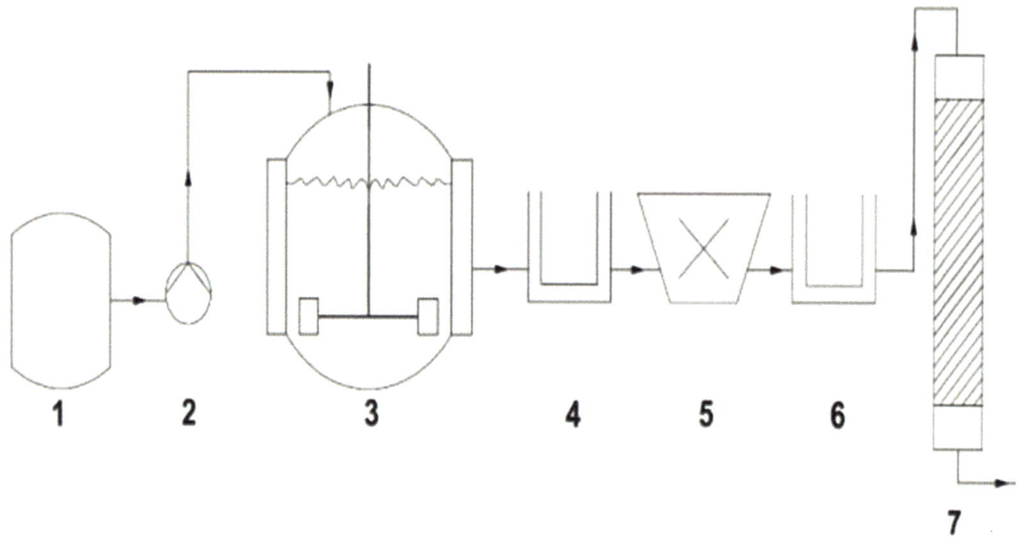

FIGURE 12.2 Fed-batch cultivation for production of sulfite oxidase from *Sulfitobacter pontiacus*. 1: Feed tank containing calcium sulfite solution. 2: Pump. 3: Bioreactor. 4: Centrifuge for separating *S. pontiacus* cells. 5: Sonicator for disintegrating cells for release of sulfite oxidase from *S. pontiacus*. 6: Centrifuge for removal of cell debris from cell-free extracts. 7: Sartobind S 100 membrane adsorber for purification of sulfite oxidase. (From Muffler, K. and Ulber, R., *Biotechnol. Bioeng.*, 99, 870, 2008.)

The substrate concentration can be kept at a specific level with fed-batch operation. The reactor volume grows as the substrate is fed into it. At some point, a portion of the reactor volume is removed, and the procedure is then carried out again. By bridging the gap between batch and continuous operations, this sort of operation lengthens batch cultivation and boosts reactor productivity (Figure 12.2) (Shluer and Kargi, 2003; Blanch and Clark, 1996).

12.3.1.3 Submerged Fermentation Processes

Microorganisms can be grown in liquid nutrient media using a technique called submerged fermentation. This entails growing the desired microorganism for industrial production in vessels called bioreactors that are sealed off and filled with nutrient broths. The system design for bioreactors allows the supply of oxygen as required by aerobic microorganisms as well as the monitoring and control of various parameters, including pH, temperature, viscosity, dissolved oxygen, foam formation, biomass formation, substrate utilization, and the formation of desired products (Figure 12.3) (Liu *et al.*, 2008).

Extracellular proteases have been made using the marine bacterium *Vibrio harveyi* by the submerged fermentation process (Estrada-Badillo *et al.*, 2003). Two Rushton disc turbines with two baffles were used to stir the bioreactor while it was in batch culture mode. Through a metallic tube with seven perforations, air was sparged. The overall oxygen mass transfer coefficient in seawater ranged from 0.254/s for reactors operating at 300 rpm and 0.2 L/L/min to 1.342/s for reactors operating at 700 rpm and 0.8 L/L/min. The lowest value was obtained when the reactor was operating at 300 rpm and 0.2 L/L/min; higher specific growth rates were the result of higher agitation rates. In this work, the authors unequivocally showed that choosing the right agitation and aeration rates in submerged processes produced high protease yields. An oxidative and sodium dodecyl sulfate (SDS)-stable alkaline protease produced by the marine haloalkalophilic *Bacillus clausii* (isolated from the tidal mud flats of the Korean Yellow Sea near Inchon City) was batch-cultivated under various process conditions (Kumar *et al.*, 2004; Abioye *et al.*, 2015a). The process time and agitation

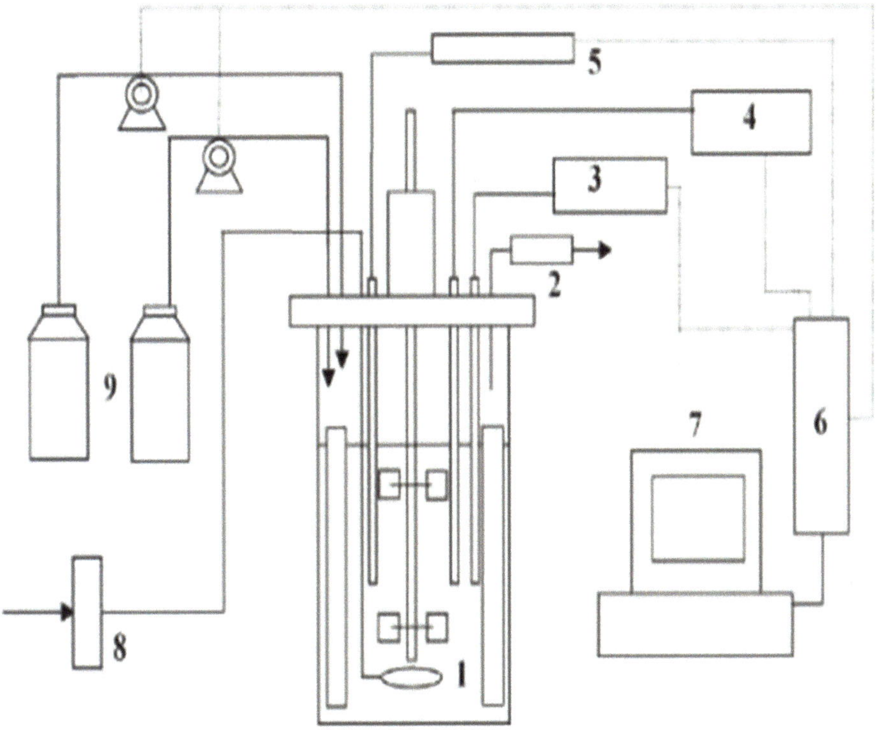

FIGURE 12.3 Experimental setup for a submerged fermentation process. (1) Bioreactor; (2) exhaust air condenser; (3) pH meter and controller; (4) DO meter and controller; (5) thermometer and controller; (6) interface box; (7) computer; (8) air filter; and (9) antifoam, acid, and base. (From Kao, P.M. *et al.*, *Biochem. Eng. J.*, 34, 172, 2007.)

rate increased along with the enzyme activity. Protease yields may have increased as a result of effective mass transfer coefficients, which depend on the right agitation and aeration rates.

A unique marine bacterium called *Hyphomonas jannaschiana*, which has exceptional temperature and/or pH stability, produces novel extracellular alkaline metalloproteases that were the subject of an invention published by Weiner *et al.* (1996). The submerged cultivation was carried out aerobically, using mechanical agitation or air spargers to mechanically shake, stir, aerate, or blow humidified air through the media. Any solid or liquid cultivation medium that was appropriate for the standard cultivation of microorganisms might be used. A successful cultivation typically lasted between 20 and 50 hours. Exoenzymes were best detected in the culture supernatant about 17 hours after the bioprocess began.

12.3.1.4 Immobilization Techniques

Increasing the synthesis of enzymes and growing the native bacterium could both be accomplished through cell immobilization (Elibol and Moreira, 2003). Enzymes are fixed to or contained by solid supports by the process of immobilization, resulting in a heterogeneous immobilized enzyme system. Enzymes that have been immobilized mimic the way that they normally function in living cells, where the majority of them are anchored to organelle, membrane, and cytoskeleton components (Kress *et al.*, 2002). Elibol and Moreira (2003) immobilized whole cells in calcium alginate beads that were utilized to make alkaline protease. This was the first report on the generation of protease by *Teredinobacter turnirae* cells trapped within calcium alginate beads, and the study's goal was to discover the cultivation conditions required to achieve this.

TABLE 12.4
Isolation Methods, Biochemical Reactions, and Physical Characteristics of Some Marine Bacterial Enzymes

Enzyme	Marine bacteria	Isolation technique	Activity	Optimal temperature and pH	References
Alkaline phosphatases	*Cobetia marina*	Liquid culture	Conversion of p-nitrophenylphosphate	50 °C; pH 10.3	Yu Plisova *et al.*, 2005
Alkaline protease	Marine bacterium SD11	Agar plate	Hydrolysis of carbohydrates	60 °C; pH 10	Cui *et al.*, 2015
α-Amylase	*Pseudoalteromonas* sp. *M175*	Flask culture	Cleavage of α-1,4-glycosidic linkages in starch molecules	25 °C; pH 8.0	Wang *et al.*, 2018
κ-Carrageenase	*Pseudoalteromonas porphyrae*	Medium containing carrageenan	κ-Carrageenan hydrolysis	55 °C; pH 8.0	Liu *et al.*, 2011
Chitinase	*Moritella marina*	Marine agar/broth containing colloidal chitin	p-Nitrophenyl-β-1,4-N,N0-diacetylchitobiose	28 °C; pH 5.0	Stefanidi, and Vorgias, 2008
Dye decolorizing peroxidase	Actinobacteria strain collection	Agar plate	Oxidization of several phenolic substrates	25 °C; pH 3–4	Cordas *et al.*, 2022
Esterase	*Thalassospira* sp. *GB04J01*	Flask culture	Hydrolysis of simple esters	45 °C; pH 8.5	De Santi *et al.*, 2016
Fibrinolytic enzymes	*Bacillus subtilis ICTF-1*	Agar plate	Conversion of N-succinyl-Ala-Ala-ProPhe-p-nitroanilide	50 °C; pH 9.0	Mahajan *et al.*, 2012
Inulinase (whole-cell)	*Bacillus subtilis*	Agar plate	Hydrolysis of inulin to reducing sugars	40 °C; pH 8.5	Rodriguez *et al.*, 2017
Lactase	*Pseudoalteromonas* sp. *KNOUC808*	Enrichments cultures	Hydrolysis of o-nitrophenyl β-D-galactopyranoside	20 °C; pH 7.8	Nam and Ahn, 2011
Lipase	*Pseudomonas* sp. *LSK25*	Growth in Luria-Bertani (LB) broth at 4 °C	Colorimetric assay with olive oil as substrate	10 °C; pH 7.0	Salwoom *et al.*, 2019
Sulfatase	*Pseudoalteromonas* sp.	Agar plate	Hydrolytic cleavage of sulfate ester groups on carbohydrates	68 °C; pH 6.5	Mikkelsen *et al.*, 2021
Xylose isomerase	*Fulvimarina pelagi*	Dilution-to-extinction	Interconversion of d-xylose and d-xylulose	35 °C; pH 7	Lajoie *et al.*, 2016

Beshay and Moreira (2003) conducted additional research. In this study, the authors determined the best conditions for immobilizing *T. turnirae* cells on various inorganic matrices and evaluated the immobilized biocatalysts in repeated batch cultivation for the production of the novel alkaline protease; 250-mL Erlenmeyer flasks were used for immobilization studies. Under the same growth circumstances, the alkaline protease activity produced by immobilized entire cells was approximately 2.3 times more than that produced by cells that were freely suspended. After four rounds, the alkaline protease activity created by immobilized cells was found to have reached a stable state.

The immobilization of microbes for the manufacture of enzymes has a number of benefits, including reducing pollutants in the product stream while maintaining biomass. Immobilized cells provide for higher productivity as well as operational stability and avoid cell wash-out in continuous operation. Marine bacteria are adapted to survive attached to submerged surfaces, which frequently leads to biofouling; however, this undesirable trait is also beneficial for processes involving immobilized cell structures (Sakar *et al.*, 2010).

12.3.1.5 Use of Organic Media

A strong antileukemic and flavor-improving compound utilized in the food business is known as L-glutamine amidohydrolase. Extracellular L-glutaminase has been produced by *Vibrio costicola*, an isolate from the marine environment, using a solid-state method (Nagendra and Chandrasekaran, 1996). Different organic substrates' applicability was assessed. When compared with sawdust, copra cake powder, and groundnut cake powder, wheat bran and rice husk produced the highest amounts of enzyme among the various solid substrates examined. Wheat bran and rice husk had the highest enzyme yield. The outcomes demonstrated the potential for solid-state marine *Vibrio costicola* to economically produce L-glutaminase from marine bacteria.

12.3.2 Purification of Enzymes from Marine Bacteria

Many marine microorganisms have produced enzymes such as alkaline proteases that have been tested and purified. Marine enzymes have been purified from marine microorganisms using techniques of ammonium sulfate precipitation (Subhashini *et al.*, 2012) or ultrafiltration (Baweja *et al.*, 2017). Enzymes released into the medium are first concentrated. Ammonium sulfate is most frequently used because it is inexpensive, is extremely soluble in water, and easily interacts with water in the precipitation of proteins. Moreover, ultrafiltration is a cheap, pressure-driven membrane separation technique that causes little enzyme activity loss (Rao *et al.*, 1998; Rani *et al.*, 2012).

Most often, non-specific chromatographic techniques like gel filtration, ion exchange chromatography, and hydrophobic interaction chromatography have been used in purification procedures. The two chromatographic methods that are most frequently employed for the purification of marine microbial enzymes are ion exchange and gel filtration (Annamalai *et al.*, 2014). Table 12.4 shows varieties of enzymes produced by marine bacteria under optimal conditions of pH and temperature as reported by different authors (Auta *et al.*, 2022).

12.4 APPLICATIONS OF ENZYMES FROM MARINE BACTERIA

12.4.1 Marine Bacteria as a Source of Novel Enzymes

The microbiological resources from marine habitats such as oceans, seas, and large lakes have continuously been exploited for a very long time. Recently, the focus has shifted to enzymes from marine bacteria with biotechnological, industrial, pharmaceutical, and biomedical applications (Imhoff *et al.*, 2011; Babaniyi *et al.*, 2023). A plethora of enzymes, mainly from the oxidoreductase, hydrolase, isomerase, lyase, and ligase classes (Birolli *et al.*, 2019), such as protease, mannase, pectinase, lipases, carbohydratases, phosphatases, carboxymethylcellulase, xylanase, γ-glutamyl transpeptidase (GGT), gelatinase, agarase, agaropectinase, β-N-acetylglucosaminidases, α-N-acetylgalactosaminidases, ketoreductase, etc., have been isolated successfully from marine bacteria such as *Bacillus subtilis*, *Bacillus pumilus*, *Bacillus horikoshii*, *Bacillus alveayuensis*, *Brevibacillus brevis*, *Exiguobacterium* sp., *Marinimicrobium* sp., *Aeromonas* sp., *Zunongwangia profunda*, *Agarivorans gilvus*, etc. (Lima and Porto, 2016). These enzymes demonstrate great industrial and biotechnological potential (Table 12.1).

The oxidoreductase, hydrolase, and lyase classes of marine bacterial enzymes are the most widely applied for the catalysis of major industrial and biochemical processes, kinetic resolution of various chemical compounds, production of biofuels and other important metabolites, hydroxylation of polyaromatic hydrocarbons (PAHs), and biotransformation reactions of environmentally hazardous contaminants to fine, high-volume compounds used in the chemical and biochemical industries. This ecologically friendly green chemistry approach is the current trend used in the reduction and elimination of environmental pollutants (Haritash and Kaushik, 2009; Birolli *et al.*, 2019; Aransiola *et al.*, 2022a) (Table 12.5).

12.4.2 Applications of Marine Bacterial Enzymes

The significance of marine bacterial enzymes cannot be overemphasized, as they are frequently and continuously being sought for, purified, characterized, and utilized in many biochemical, industrial, and biotechnological processes (Sana, 2013). Due to their many exceptional properties, enzymes from marine bacteria have several distinct merits over other conventional sources of enzymes. Marine bacterial enzymes have been applied in food processing, biofuel, pharmaceuticals, chemicals, leather, detergent, textile, paper, biotechnological research, and bioremediation (Figure 12.4).

12.4.2.1 Food Processing Industry

Bacterial enzymes from the marine ecosystem have displayed several relevant features in the food production and food processing industries (Figure 12.5). They have been used for food preservation, as part of food components, and for enhancement of the organoleptic properties of food, effective raw material exploitation, and an overall cutback in the cost of food processing (Synowiecki *et al.*, 2006; Babaniyi *et al.*, 2023). These enzymes are important in catalyzing reactions involved in several steps of food production processes at extremes of physical and chemical factors such as temperature, pH, or pressure (Sana, 2013).

The pH-stable and thermostable carbohydrases are the most frequently exploited enzymes in the food industry. The improvement of starch liquefaction is catalyzed by acid-stable enzymes, while the production of monosaccharides, disaccharides, and oligosaccharides (glucose, maltose, cyclodextrins, etc.) from the hydrolysis of starch is catalyzed by thermostable carbohydrases (Synowiecki *et al.*, 2006). Meanwhile, the use of proteases in the food processing industry, especially for baking, cheese production, hydrolysis of soy proteins, and meat tenderization has been recognized for a long time.

In addition, a number of proteases isolated from salt-tolerant and cold-adapted marine bacteria have also been employed as digestive enzymes with applications in cheese production, fermentation, and other food industries (Li *et al.*, 2010; Aransiola *et al.*, 2022a). The application of protease for baking and dairy processing usually requires inactivation of the enzyme at moderate temperatures. However, processing of meat, cheese, and seafood involves protease activity at a relatively high salinity of the fermentation medium (Li *et al.*, 2010). On account of their distinctive biocatalytic characteristics, such as hydrolysis, esterification, and transesterification reactions, lipases isolated from marine bacteria have several important applications in fat, oil, and other food processing industries (Sana, 2013). Lipases have been used in the manufacture of long-chain polyunsaturated fatty acids–enriched triglycerides (Barzkar *et al.*, 2021).

Enzymes that act on polysaccharides have been exploited in the manufacture of sweeteners and in brewing as well as baking industries (Lima and Porto, 2016). The conversion of glucose to fructose, a process known as isomerization, is catalyzed by glucose isomerase. Glucosidases, amylases, and glycosidases have been used in the breakdown of starch to produce maltose and glucose syrup. Also, the α-1,4- and α-1,6-glucosidic bonds of molecules of amylopectin are cleaved using

TABLE 12.5
Some Enzymes from Marine Bacteria, Their Activities and Applications

Enzyme	Bacterial source	Activities	Application	References
Proteases	*Bacillus alveayuensis*, *Exiguobacterium* sp., *Bacillus horikoshii*, *B. subtilis*	223.1–62745 U/mg	Synthesis of Bacillamide C, elimination of blood stains	Joo and Choi, 2012; Mahajan *et al.*, 2012; Annamalai *et al.*, 2013; Huang *et al.*, 2013, Raval *et al.*, 2014; Lei *et al.*, 2014
Mannase	*Brevibacillus brevis*		Hemicelluloses, hydrolysis of β-1,4-mannosidic bonds	Zhou *et al.*, 2013
Pectinase	*Bacillus subtilis*		Cleaning, softening, and brightening of cotton fibers, preservation of cotton structure	Joshi *et al.*, 2013
Carboxymethyl-cellulase	*Marinimicrobium* sp.	2.5 U/mL	Hydrolysis of endo-1,4-β-glucanase bonds in cellulose	Zhao *et al.*, 2012
Lipase	*Aeromonas* sp.	440 U/mL	Hydrolysis of long carbon acyl esters linkages	Charoenpanich *et al.*, 2011
Xylanase	*Bacillus pumilus*, *Zunongwangia profunda*	112.42 U/mg	Treatment of xylan rich agricultural wastes, production of ethanol, biobleaching	Menon *et al.*, 2010; Liu *et al.*, 2014a
Gelatinase	*Bacillus* spp.	4.1 U/mg	Degradation of tissue in tumor metastasis and arthritis	Balan *et al.*, 2012
β-agarase	*Agarivorans gilvus*	0.781 U/mg	Cleavage of agarose to produce neoagarotetraose	Liu *et al.*, 2014b

pullulanases (Lima and Porto, 2016). These enzymes have been isolated from marine microorganisms, especially bacteria (Li *et al.*, 2007), with varying applications in the beverage and food industry for hydrolysis of several cereal grains (Li *et al.*, 2007; Sana, 2013). Species of *Bacillus* isolated from marine environments have been reported to produce α-amylase with the ability to catalyze the hydrolysis of raw corn, rice, cassava, potato, and maize starch to produce several mono-, di-, and oligosaccharides, including glucose, maltose, malto-triose, malto-tetraose, malto-pentose, and malto-hexaose (Vidilaseris *et al.*, 2009; Puspasari *et al.*, 2011; Aransiola *et al.*, 2022b).

A number of chitinases with potential applications in the food industry have been isolated from marine bacteria as well as hyper-thermophilic archaea. This very important enzyme catalyzes the hydrolysis of chitin to produce oligomeric unit compounds, which are considered valuable in the food and pharmaceutical industries. Due to the formation of a more consistent product, marine bacterial chitinases have substituted the acid hydrolysis of chitin (Lee *et al.*, 2000).

FIGURE 12.4 Biotechnological applications of marine bacterial enzymes.

12.5 CONCLUSION: CHALLENGES OF THE TECHNOLOGY, AND FUTURE PERSPECTIVES

According to estimates, the present value of the worldwide industrial enzyme market, which is US$4.1 billion, is growing by 10% to 15% yearly (Guerrand, 2018). Although cold-adapted enzymes might help with energy-saving strategies, the unusual capacity of marine microbes to catalyze reactions at temperatures close to the freezing point of water has generated significant interest for both industrial usage and fundamental investigations (Morita *et al.*, 1997; Aransiola *et al.*, 2021). Moreover, marine enzyme biotechnology can provide novel biocatalysts that are easy to cultivate on a large scale and have traits like high salt tolerance, hyper-thermostability, barophilicity, and cold adaptivity (Debashish *et al.*, 2005).

Extracellular marine bacterial enzymes from a variety of species have been identified and described to date. They include vital enzymes used in pharmaceuticals, lignin degradation, and cellulolysis. The discovery of baro-tolerant protease from deep-sea bacteria is significant, since it has opened the door for the use of novel marine bacterial enzymes in more inventive applications. The identification of additional enzymes from bacteria that are specific to marine conditions may aid in

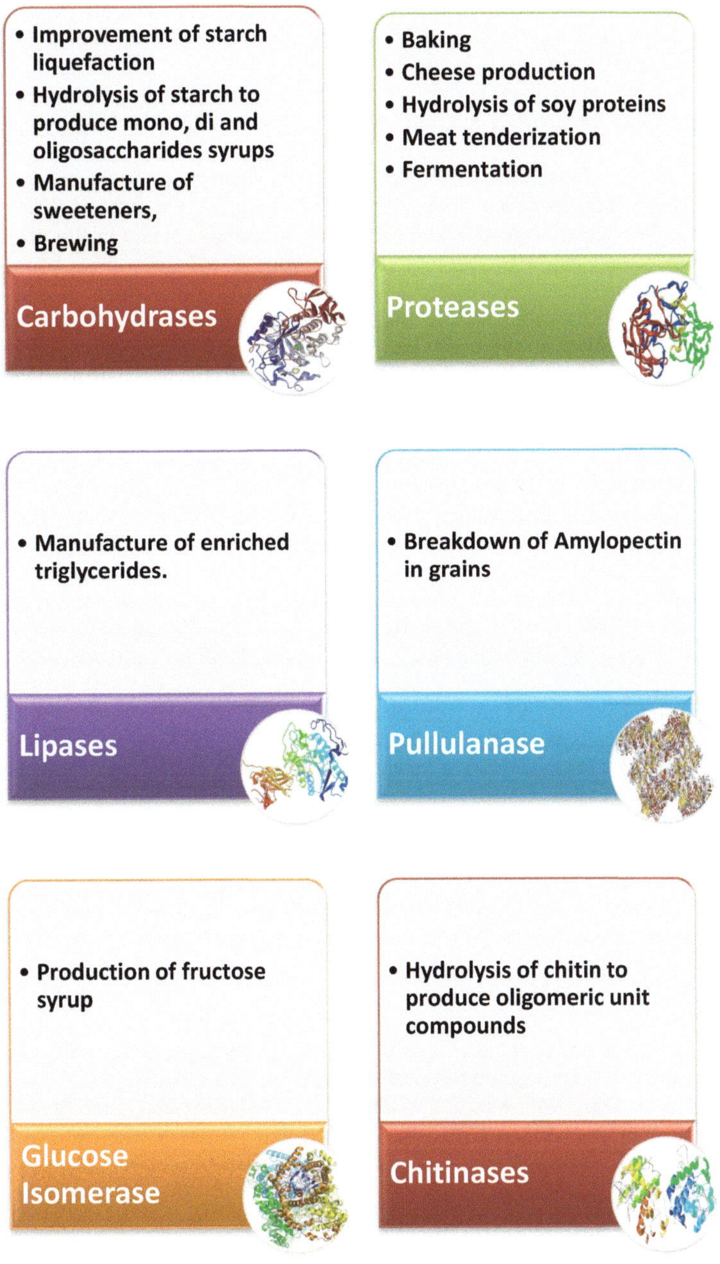

FIGURE 12.5 Food industry applications of some marine bacterial enzymes.

the creation of new industrial processes. To uncover novel and extremophilic enzymes with unique features, however, comprehensive molecular identification of marine bacterium enzymes is required.

REFERENCES

Abioye, O.P., Afolayan, E.O., Aransiola, S.A. (2015a). Treatment of pharmaceutical effluent by Saccharomyces cerevisiae and Torulaspora delbrueckii isolated from spoilt water melon. *Research Journal of Environmental Toxicology*, 1819–3420. https://doi.org/10.3923/rjet.2015.

Abioye, O.P., Ekundayo, P.O., Oyeleke, S.B., Aransiola, S.A. (2015b). Remediation of zinc contaminated soil using *Acalypha wilkesiana. Journal of Global Ecology and Environment*, 2(2), 82–89. http://www.ikpress.org/abstract/3906#.Vv4HqXok4dU.

Al-Amoudi, S., Razali, R., Magbubah, E., Mohammad, S.A., Salim, B., John, A.C.A., Feras, F.L., Vladimir, B.B. (2016). Metagenomics as a preliminary screen for antimicrobial bioprospecting. *Gene*, 594(2), 248–258.

Albarano, L., Esposito, R., Ruocco, N., Costantini, M. (2020). Genome mining as new challenge in natural products discovery. *Marine Drugs*, 18(4), 199. https://doi.org/10.3390/md18040199.

Amann, R., Ludwig, W., Schleifer, K. (1995). Phylogenetic identification and in situ detection of individual microbial cells without culturing. *Microbiology Revision*, 59, 143–169.

Annamalai, N., Rajeswari, M.V., Balasubramanian, T. (2013). Extraction, purification and application of thermostable and halostable alkaline protease from *Bacillus alveayuensis* CAS 5 using marine wastes. *Food and Bioproducts Processing*, 92(4), 335–342. http://doi.org/10.1016/j.fbp.2013.08.009.

Annamalai, N., Rajeswari, M.V., Balasubramanian, M.V. (2014). Extraction, purification and application of thermostable and halostable alkaline protease from Bacillus alveayuensis CAS 5 using marine wastes. *Food Bioproduction Process*, 92(4), 335–342.

Aransiola, S.A., Afolabi, F., Joseph, F., Maddela, N.R. (2023). Soil enzymes: Distribution, interactions and influencing factors. In P. Jeschke, E.B. Starikov (Eds.), *Agricultural Biocatalysis: Enzymes in Agriculture and Industry* (1st ed., pp. 303–333). Jenny Stanford Publishing. ISBN 9789814968478. https://www.routledge.com/Agricultural-Biocatalysis-Enzymes-in-Agriculture-and-Industry/Jeschke-Starikov/p/book/9789814968478.

Aransiola, S.A., Ijah, U.J.J., Abioye, O.P., Bala, J.D. (2022b). Vermicompost-assisted phytoremediation of toxic trace element-contaminated soil in Madaka, Nigeria using *Melissa officinalis* L and *Sida acuta. International Journal of Environmental Science and Technology.* http://doi.org/10.1007/s13762-022-04105-y.

Aransiola, S.A., Joseph, F., Oyedele, O.J., Maddela, N.R. (2022a). Editorial - Ecological Interplays in microbial enzymology. In N.R. Maddela, S.A. Aransiola, R. Prasad (Eds.), *Ecological Interplays in Microbial Enzymology* (1st ed.). Springer Nature Singapore Pte Ltd. ISSN: 2662-1681. https://www.springer.com/series/16324.

Aransiola, S.A., Victor-Ekwebelem, M.O., Ikhumetse, A.A., Abioye, O.P. (2021). Challenges and future prospects of biotechnology. In N.R. Maddela, L.C. García (Eds.), *Innovations in Biotechnology for a Sustainable Future.* Springer Nature Switzerland AG 2021. https://doi.org/10.1007/978-3-030-80108-3_20.

Asker, D., Ohta, Y. (2002). Production of canthaxanthin by Haloferax alexandrinus under non-aseptic conditions and a simple, rapid method for its extraction. *Applied Microbiology and Biotechnology*, 58(6), 743–750. https://doi.org/10.1007/s00253-002-0967-y.

Auta, H.S., Aboyeji, D.O., Bala, J.D., Abioye, O.P., Adabara, N.U., Aransiola, S.A., Hassana, A., Azize, A. (2022). Marine microbial enzymes - An overview. In N.R. Maddela, S.A. Aransiola, R. Prasad (Eds.), *Ecological Interplays in Microbial Enzymology* (1st ed.). Springer Nature Singapore Pte Ltd. ISSN: 2662-1681. https://www.springer.com/series/16324.

Babaniyi, G.G., Olagoke, O.E., Aransiola, S.A. (2023). Extracellular enzymatic activity of bacteria in aquatic ecosystems. In N.R. Maddela, L.K.W. Eller, R. Prasad (Eds.), *Microbiology for Cleaner Production and Environmental Sustainability.* CRC Press. ISBN: 9781032496061. https://www.routledge.com/Microbiology-for-Cleaner-Production-and-Environmental-Sustainability/Maddela-Eller-Prasad/p/book/9781032496061.

Balan, S.S., Nethaji, R., Sankar, S., Jayalakshmi, S. (2012). Production of gelatinase enzyme from *Bacillus* spp. isolated from the sediment sample of Porto Novo coastal sites. *Asian Pacific Journal of Tropical Biomedicine*, 2(3), S1811–S1816. http://doi.org/10.1016/S2221-1691(12)60500-0.

Barzkar, N., Sohail, M., Jahromi, S.T., Gozari, M., Poormozaffar, S., Nahavandi, R., Hafezieh., M. (2021). Marine bacterial esterases: Emerging biocatalysts for industrial applications. *Applied Biochemistry and Biotechnology*, 193(4), 1187–1214. https://doi.org/10.1007/s12010-020-03483-8.

Baweja, M., Singh, P.K., Sadaf, A., Tiwari, R., Nain, L., Khare, S.R., Shukla, P. (2017). Cost effective characterization process and molecular dynamic simulation of detergent compatible alkaline protease from *Bacillus pumilus* strain MP27. *Process Biochemistry*, 58, 199–203.

Bertagnolli, A.D., Padilla, C.C., Glass, J.B., Thamdrup, B., Stewart, F.J. (2017). Metabolic potential and in situ activity of marine Marinimicrobia bacteria in an anoxic water column. *Environmental Microbiology*, 19(11), 4392–4416.

Beshay, U., Moreira, A. (2003). Repeated batch production of alkaline protease using porous sintered glass as carriers. *Process Biochemistry*, 38(10), 1463–1469.

Birolli, W.G., Lima, R.N., Porto, A.L.M. (2019). Applications of marine-derived microorganisms and their enzymes in biocatalysis and biotransformation, the underexplored potentials. *Frontiers in Microbiology*, 10, 1453. https://doi.org/10.3389/fmicb.2019.01453.

Blanch, H.W., Clark, D.S. (1996). *Biochemical Engineering*. Marcel Dekker, New York, pp. 276–331.

Bonugli-Santos, R.C., Vasconcelos, M.R.D., Passarini, M.R.Z., Vieira, G.A.L., Lopes, V.C.P., Mainardi, P.H. (2015). Marine-derived fungi: Diversity of enzymes and biotechnological applications. *Froniers Microbiology*, 6, 269. https://doi.org/10.3389/fmicb.2015.00269.

Cao, F., Meng, Z.H., Mu, X., Yue, Y.F., Zhu, H.J. (2019). Absolute configuration of bioactive azaphilones from the marine-derived fungus *Pleosporales* sp. CF09-1. *Journal of Natural Products*, 82(2), 386–392.

Charoenpanich, J., Suktanarag, S., Toobbucha, N. (2011). Production of a thermostable lipase by *Aeromonas* sp. EBB-1 isolated from marine sludge in Angsila, Thailand. *Scienceasia*, 37(2), 105–114. http://doi.org /10.2306/scienceasia1513-1874.2011.37.105.

Cui, H., Wang, L., Yu, Y. (2015). Production and characterization of alkaline protease from a high yielding and moderately halophilic strain of SD11 marine bacteria. *Journal of Chemistry*.

Chu, X., He, H., Guo, C., Sun, B. (2008). Identification of two novel esterases from a marine metagenomic library derived from South China Sea. *Applied Microbiology and Biotechnology*, 80(4), 615–625.

Cordas, C.M., Nguyen, G.S., Valério, G.N., Jønsson, M., Söllner, K., Aune, I.H., Moura, J.J. (2022). Discovery and characterization of a novel Dyp-type peroxidase from a marine actinobacterium isolated from Trondheim fjord, Norway. *Journal of Inorganic Biochemistry*, 226, 111651.

Curson, A.R.J., Liu, J., Martinez, A.B., Green, R.T., Chan, Y.H., Carrion, O., Williams, B.T., Zhang, S.H., Yang, G.P., Page, P.C.B., Zhang, X.H., Todd, J.D. (2017). Dimethylsulfoniopropionate biosynthesis in marine bacteria and identification of the key gene in this process. *Nature Microbiology*, 2, 17009.

Debashish, G., Malay, S., Barindra, S., Joydeep, M. (2005). Marine enzymes. *Advances in Biochemistry and Engineering Biotechnology*, 96, 189–218.

De Santi, C., Leiros, H.K.S., Di Scala, A., de Pascale, D., Altermark, B., Willassen, N.P. (2016). Biochemical characterization and structural analysis of a new cold-active and salt-tolerant esterase from the marine bacterium Thalassospira sp. *Extremophiles*, 20, 323–336.

Desjardine, K., Pereira, A., Wright, H., Matainaho, T., Kelly, M., Andersen, R.J. (2007). Tauramamide, a lipopeptide antibiotic produced in culture by Brewibacillus laterosporus isolated from a marine habitat: Structure elucidation and synthesis. *Journal of Natural Products*, 70(12), 1850–1853.

Dupont, C.L., Rusch, D.B., Yooseph, S., Lombardo, M.J., Richter, R.A., Valas, R., Novotny, M., Yee-Greenbaum, J., Selengut, J.D., Haft, D.H., Halpern, A.L., Lasken, R.S., Nealson, K., Friedman, R., Venter, J.C. (2012). Genomic insights to SAR86, an abundant and uncultivated marine bacterial lineage. *ISME Journal*, 6(6), 1186–1199.

El-Gendy, M.M.A., Hawas, U.W., Jaspars, M. (2008). Novel bioactive metabolites from a marine derived bacterium *Nocardia* sp. ALAA 2000. *Journal of Antibiotics*, 61(6), 379–386.

Elibol, M., Moreira, A.R. (2003). Production of extracellular alkaline protease by immobilization of the marine bacterium *Teredinobacter turnirae*. *Process Biochemistry*, 38(10), 1445–1450.

Estrada-Badillo, C., Marquez-Rocha, F.J. (2003). Effect of agitation rate on biomass and protease production by a marine bacterium Vibrio harveyi cultured in a fermentor. *World Journal of Microbiology and Biotechnology*, 19(2), 129–133.

Gest, H. (2008). The modern myth of "unculturable" bacteria/scotoma of contemporary microbiology.

Ghai, R., Mizuno, C.M., Picazo, A., Camacho, A., Rodriguez-Valera, F. (2013). Metagenomics uncovers a new group of low GC and ultra-small marine Actinobacteria. *Science Reports*, 3(1), 2471.

Guerrand, D. (2018). Economics of food and feed enzymes: Status and prospectives. In *Enzymes in Human and Animal Nutrition* (pp. 487–514). Academic Press.

Hamidi, M., Kozani, P.S., Kozani, P.S., Pierre, G., Michaud, P., Delattre, C. (2019). Marine bacteria versus microalgae: Who is the best for biotechnological production of bioactive compounds with antioxidant properties and other biological applications? *Marine Drugs*, 18(1), 28.

Hardeman, F., Sjoling, S. (2007). Metagenomic approach for the isolation of a novel low-temperature active lipase from uncultured bacteria of marine sediment. *FEMS Microbial Ecology*, 59(2), 524–534.

Haritash, A.K., Kaushik, C.P. (2009). Biodegradation aspects of polycyclic aromatic hydrocarbons (PAHs): A review. *Journal of Hazardous Materials*, 169(1–3), 1–15.

Hawley, A.K., Nobu, M.K., Wright, J.J., Durno, W.E., Morgan-Lang, C., Sage, B. et al. (2017). Diverse Marinimicrobia bacteria may mediate coupled biogeochemical cycles along eco-thermodynamic gradients. *Nature Communications*, 8(1), 1–10.

Huang, H., Yao, Y., He, Z., Yang, T., Ma, J., Tian, X., et al. (2011). Antimalarial β-carboline and indolactam alkaloids from *Marinactinospora thermotolerans*, a deep sea isolate. *Journal of Natural Products*, 74(10), 2122–2127. https://doi.org/10.1021/np200399t.

Huang, S., Pan, S., Chen, G., Huang, S., Zhang, Z., Li, Y., Liang, Z. (2013). Biochemical characteristics of a fibrinolytic enzyme purified from a marine bacterium, *Bacillus subtilis* HQS-3. *International Journal of Biological Macromolecules*, 62, 124–130. http://doi.org/10.1016/j.ijbiomac.2013.08.048.

Imhoff, J.F., Labes, A., Wiese, J. (2011). Bio-mining the microbial treasures of the ocean: New natural products. *Biotechnology Advances*, 29(5), 468–482.

Joo, H.S., Choi, J.W. (2012). Purification and characterization of a novel alkaline protease from *Bacillus horikoshii. Journal of Microbiology and Biotechnology*, 22(1), 58–68. http://www.ncbi.nlm.nih.gov/pubmed/22297220.

Joshi, M., Nerurkar, M., Badhe, P., & Adivarekar, R. (2013). Scouring of cotton using marine pectinase. *Journal of Molecular Catalysis B: Enzymatic*, 98, 106–113. http://doi.org/10.1016/j.molcatb.2013.10.010.

Kao, P.M., Chen, C.I., Huang, S.C., Chang, Y.C., Tsai, P.J., Liu, Y.C. (2007). Effects of shear stress and mass transfer on chitinase production by *Paenibacillus* sp. CHE-N1. *Biochemical Engineering Journal*, 34(2), 172–178.

Kim, H., Hwang, J.Y., Chung, B., Cho, E., Bae, S., Shin, J., Oh, K.B. (2019). 2-Alkyl-4-hydroxyquinolines from a marine-derived Streptomyces sp. inhibit hyphal growth induction in Candida albicans. *Marine Drugs*, 17(2), 133.

Kjeldsen, K.U., Schreiber, L., Thorup, C.A., Boesen, T., Bjerg, J.T., Yang, T., Dueholm, M.S., Larsen, S., Risgaard-Petersen, N., Nierychlo, M., Schmid, M., Boggild, A., van de Vossenberg, J., Geelhoed, J.S., Meysman, F.J.R., Wagner, M., Nielsen, P.H., Nielsen, L.P., Schramm, A. (2019). On the evolution and physiology of cable bacteria. *Proceedings of the National Academy of Sciences of the United States of America*, 116(38), 19116–19125.

Kress, J., Zanaletti, R., Amour, A., Ladlow, M., Frey, J.G., Bradley, M. (2002). Enzyme accessibility and solid supports: Which molecular weight enzymes can be used on solid supports? An investigation using confocal Raman microscopy. *Chemistry – A European Journal*, 8(16), 3769–3772.

Kumar, C.G., Joo, H.S., Koo, Y.M., Paik, S.R., Chang, C.S. (2004). Thermostable alkaline protease from a novel marine haloalkalophilic *Bacillus clausii* isolate. *World Journal of Microbiology and Biotechnology*, 20(4), 351–357.

Lajoie, C.A., Kitner, J.B., Potochnik, S.J., Townsend, J.M., Beatty, C.C., Kelly, C.J. (2016). Cloning, expression and characterization of xylose isomerase from the marine bacterium Fulvimarina pelagi in Escherichia coli. *Biotechnology Progress*, 32(5), 1230–1237.

Lee, D.G., Jeon, J.H., Jang, M.K., Kim, N.Y., Lee, J.H., Lee, J.H., Kim, S.J., Kim, G.D., Lee, S.H. (2007). Screening and characterization of a novel fibrinolytic metalloprotease from a metagenomic library. *Biotechnology Letters*, 29(3), 465–472.

Lee, J.S., Joo, D.S., Cho, S.Y., Ha, J.H., Lee, E.H. (2000). Purification and characterization of extracellular chitinase produced by marine bacterium *Bacillus* sp. LJ-25. *Journal of Microbiology and Biotechnology*, 10, 307–311.

Li, H., Chi, Z., Wang, X., Duan, X., Ma, L., Gao, L. (2007). Purification and characterization of extracellular amylase from the marine yeast *Aureobasidium pullulans* N13d and its raw potato starch digestion. *Enzyme and Microbial Technology*, 40(5), 1006–1012.

Li, J., Peng, Y., Wang, X., Chi, Z. (2010). Optimum production and characterization of an acid protease from marine yeast *Metschnikowia reukaufii* W6b. *Journal of Ocean University of China*, 9(4), 359–364.

Li, X., Zhao, H., Chen, X. (2021). Screening of marine bioactive antimicrobial compounds for plant pathogens. *Marine Drugs*, 19(2), 69. https://doi.org/10.3390/md19020069.

Lima, R.N., Porto, A.L.M. (2016). Recent advances in marine enzymes for biotechnological processes. *Advances in Food and Nutrition Research*, 78, 153–192. https://doi.org/10.1016/bs.afnr.2016.06.005.

Liu, C., Sun, Z.T., Du, J.H., Wang, J. (2008). Response surface optimization of fermentation conditions for producing xylanase by *Aspergilllus niger* SL-05. *Journal of Indian Microbiology and Biotechnology*, 135(7), 703–711.

Liu, G.L., Li, Y., Chi, Z., Chi, Z.M. (2011). Purification and characterization of κ-carrageenase from the marine bacterium Pseudoalteromonas porphyrae for hydrolysis of κ-carrageenan. *Process Biochemistry*, 46(1), 265–271.

Liu, N., Mao, X., Du, Z., Mu, B., Wei, D. (2014b). Cloning and characterisation of a novel neoagarotetraose-forming-β-agarase, AgWH50A from *Agarivorans gilvus* WH0801. *Carbohydrate Research*, 388, 147–151. http://doi.org/10.1016/j.carres.2014.02.019.

Liu, X., Huang, Z., Zhang, X., Shao, Z., Liu, Z. (2014a). Cloning, expression and characterization of a novel cold-active and halophilic xylanase from *Zunongwangia profunda. Extremophiles: Life under Extreme Conditions*, 18(2), 441–450. http://doi.org/10.1007/s00792-014-0629-x.

Macherla, V.R., Liu, J., Sunga, M., White, D.J., Grodberg, J., Teisan, S., Lam, K.S., Potts, B.C. (2007). Lipoxazolidinones A, B, and C: Antibacterial 4-oxazolidinones from a marine actinomycete isolated from a Guam marine sediment. *Journal of Natural Products*, 70(9), 1454–1457.

Mahajan, P.M., Nayak, S., Lele, S.S. (2012). Fibrinolytic enzyme from newly isolated marine bacterium Bacillus subtilis ICTF-1: Media optimization, purification and characterization. *Journal of Bioscience and Bioengineering*, 113(3), 307–314.

Major, D.W., McMaster, M.L., Cox, E.E., Edwards, E.A., Dworatzek, S.M., Hendrickson, E.R., Starr, M.G., Payne, J.A., Buonamici, L.W. (2002). Field demonstration of successful bioaugmentation to achieve dechlorination of tetrachloroethene to ethane. *Environmental Science and Technology*, 36(23), 5106–5116.

Mehrshad, M., Rodriguez-Valera, F., Amoozegar, M.A., Lopez-Garcia, P., Ghai, R. (2018). The enigmatic SAR202 cluster up close: Shedding light on a globally distributed dark ocean lineage involved in sulfur cycling. *ISME Journal*, 12(3), 655–668.

Menon, G., Mody, K., Keshri, J., Jha, B. (2010). Isolation, purification, and characterization of haloalkaline xylanase from a marine *Bacillus pumilus* strain, GESF-1. *Biotechnology and Bioprocess Engineering*, 15(6), 998–1005. http://doi.org/10.1007/s12257-010-0116-x.

Metcalf, W.W., Griffin, B.M., Cicchillo, R.M., Gao, J., Janga, S.C., Cooke, H.A., Circello, B.T., Evans, B.S., Martens-Habbena, W., Stahl, D.A., van der Donk, W.A. (2012). Synthesis of methylphosphonic acid by marine microbes: A source for methane in the aerobic ocean. *Science*, 337(6098), 1104–1107.

Mikkelsen, M.D., Cao, H.T.T., Roret, T., Rhein-Knudsen, N., Holck, J., Tran, V.T.T., ... Meyer, A.S. (2021). A novel thermostable prokaryotic fucoidan active sulfatase PsFucS1 with an unusual quaternary hexameric structure. *Scientific Reports*, 11(1), 19523.

Morita, Y., Nakamura, T., Hasan, Q., Murakami, Y., Yokoyama, K., Tamiya, E. (1997). Cold-active enzymes from cold-adapted bacteria. *Journal of the American Oil Chemists' Society*, 74(4), 441–444.

Muffler, K., Ulber, R. (2008). Fed-batch cultivation of the marine bacterium *Sulfitobacter pontiacus* using immobilized substrate and purification of sulfite oxidase by application of membrane adsorber technology. *Biotechnology and Bioengineering*, 99(4), 870–875.

Muller, H., Marozava, S., Probst, A.J., Meckenstock, R.U. (2020). Groundwater cable bacteria conserve energy by sulfur disproportionation. *ISME Journal*, 14(2), 623–634.

Nam, E., Ahn, J. (2011). Antarctic marine bacterium Pseudoalteromonas sp. KNOUC808 as a source of cold-adapted lactose hydrolyzing enzyme. *Brazilian Journal of Microbiology*, 42, 927–936.

Nagendra Prabhu, G., Chandrasekaran, M. (1996). L-glutaminase production by marine *Vibrio costicola* under solid-state fermentation using different substrates. *Journal of Marine Biotechnology*, 4, 176–179.

Nelis, H.J., De Leenheer, A. (1991). Microbial sources of carotenoid pigments used in foods and feeds. *Journal of Applied Bacteriology*, 70(3), 181–191. https://doi.org/10.1111/j.1365-2672.1991.tb02922.x.

Nichols, D., Lewis, K., Orjala, J., Mo, S., Ortenberg, R., O'Connor, P., Zhao, C., Vouros, P., Kaeberlein, T., Epstein, S.S. (2008). Short peptide induces an "uncultivable" microorganism to grow in vitro. *Applied Environmental Microbiology*, 74, 4889–4897.

Nováková, J., Farkašovský, M. (2013). Bioprospecting microbial metagenome for natural products. *Biologia*, 6(68), 1079–1086. https://doi.org/10.2478/s11756-013-0246-7.

Puspasari, F., Nurachman, Z., Noer, A.S., Radjasa, O.K., Van derMaarel, M.J.E.C., Natalia, D. (2011). Characteristics of raw starch degrading a-amylase from *Bacillus aquimaris* MKSC 6.2 associated with soft coral Sinularia sp. *Starch – Stärke*, 63(8), 462–467.

Rani, K., Rana, R., Datt, S. (2012). Review on latest overview of proteases. *International Journal of Current Life Sciences*, 2(1), 12–18.

Rao, M.B., Tanksale, A.M., Ghatge, M.S., Deshpande, V.V. (1998). Molecular and biotechnological aspects of microbial proteases. *Microbiology and Molecular Biology Revolution*, 62(3), 597–635.

Raval, V.H., Pillai, S., Rawal, C.M., Singh, S.P. (2014). Biochemical and structural characterization of a detergent-stable serine alkaline protease from seawater haloalkaliphilic bacteria. *Process Biochemistry*, 49(6), 955–962. http://doi.org/10.1016/j.procbio.2014.03.014.

Rodrigues, C.J., Pereira, R.F., Fernandes, P., Cabral, J.M., de Carvalho, C.C. (2017). Cultivation-based strategies to find efficient marine biocatalysts. *Biotechnology journal*, 12(7), 1700036.

Rodrigues, C.J.C., de Carvalho, C.C.C.R. (2022). Cultivating marine bacteria under laboratory conditions: Overcoming the "unculturable" dogma. *Frontiers in Bioengineering and Biotechnology*, 10, 964589. https://doi.org/10.3389/fbioe.2022.964589.

Salwoom, L., Raja Abd Rahman, R.N.Z., Salleh, A.B., Mohd Shariff, F., Convey, P., Pearce, D., Mohamad Ali, M.S. (2019). Isolation, characterisation, and lipase production of a cold-adapted bacterial strain Pseudomonas sp. LSK25 isolated from Signy Island, Antarctica. *Molecules*, 24(4), 715.

Sana, B. (2013). Marine Microbial Enzymes: Biotechnological and Biomedical Aspects. In *Se-Kwon Kim (eds) Marine Microbiology: Bioactive Compounds and Biotechnological Applications* Wiley-VCH Verlag GmbH & Co. KGaA Publishers, 491–508.

Sarkar, S., Pramanik, A., Mitra, A., Mukherjee, J. (2010). Bioprocessing data for the production of marine enzymes. *Marine Drugs*, 8(4), 1323–1372.

Schleissner, C., Canedo, L.M., Rodríguez, P., Crespo, C., Zuniga, P., Peñalver, A., de la Calle, F., Cuevas, C. (2017). Bacterial production of a pederin analogue by a free-living marine alphaproteobacterium. *Journal of Natural Products*, 80, 2170–2173.

Sheik, C.S., Jain, S., Dick, G.J. (2014). Metabolic fexibility of enigmatic SAR324 revealed through metagenomics and metatranscriptomics. *Environmental Microbiology*, 16(1), 304–317.

Shluer, M.L., Kargi, F. (2003). *Bioprocess Engineering: Basic Concepts* (2nd ed.). Prentice Hall of India, New Delhi, pp. 245–301.

Sran, K.S., Bisht, B., Mayilraj, S., Choudhury, A.R. (2019). Structural characterization and antioxidant potential of a novel anionic exopolysaccharide produced by marine Microbacterium aurantiacum FSW-25. *International Journal of Biological Macromolecules*, 131, 343–352.

Stefanidi, E., Vorgias, C.E. (2008). Molecular analysis of the gene encoding a new chitinase from the marine psychrophilic bacterium Moritella marina and biochemical characterization of the recombinant enzyme. *Extremophiles*, 12, 541–552.

Subhashini, P., Annamalai, N., Saravanakumar, A., Balasubramanian, T. (2012). Thermostable alkaline protease from newly isolated *Vibrio sp.* extraction, purification and characterisation. *Biologia*, 67(4), 629–635.

Synowiecki, J., Grzybowska, B., Zdzieblo, A. (2006). Sources, properties and suitability of new thermostable enzymes in food processing. *Critical Reviews in Food Science and Nutrition*, 46(3), 197–205.

Trindade, M., van Zyl, L.J., Navarro-Fernández, J., Abd Elrazak, A. (2015). Targeted metagenomics as a tool to tap into marine natural product diversity for the discovery and production of drug candidates. *Frontiers in Microbiology*, 6, 890. https://doi.org/10.3389/fmicb.2015.00890.

Vartoukian, S., Palmer, R., Wade, W. (2010). Strategies for culture of 'unculturable' bacteria. *FEMS Microbiology Letters*, 309(1), 1–7. https://doi.org/10.1111/j.1574-6968.2010.02000.x.

Vidilaseris, K., Hidayat, K., Retnoningrum, D.S., Nurachman, Z., Noer, A.S., Natalia, D. (2009). Biochemical characterization of a raw starch degrading α-amylase from the Indonesian marine bacterium *Bacillus* sp. ALSHL3. *Biologia*, 64(6), 1047–1052.

Wang, X., Kan, G., Ren, X., Yu, G., Shi, C., Xie, Q., Betenbaugh, M. (2018). Molecular cloning and characterization of a novel α-amylase from Antarctic Sea ice bacterium Pseudoalteromonas sp. M175 and its primary application in detergent. *BioMed Research International.*

Weiner, R.M., Shi, J., Coyne, V.E. (1996). Thermostable alkaline metalloprotease produced by a *Hyphomonas* and preparation thereof. (5),589,373. US Patent.

Xu, M., Xiao, X., Wang, F. (2008). Isolation and characterization of alkane hydroxylases from a metagenomic library of Pacific deep-sea sediment. *Extremophiles: Life under Extreme Conditions*, 12(2), 255–262.

Yokoyama, A., Izumida, H., & Miki, W. (1994). Production of astaxanthin and 4-ketozeaxanthin by the marine bacterium, *Agrobacterium aurantiacum. Bioscience, biotechnology, and biochemistry*, 58(10), 1842–1844.

Yu Plisova, E., Balabanova, L.A., Ivanova, E.P., Kozhemyako, V.B., Mikhailov, V.V., Agafonova, E.V., Rasskazov, V.A. (2005). A highly active alkaline phosphatase from the marine bacterium Cobetia. *Marine Biotechnology*, 7, 173–178.

ZoBell, C.E. (1941). Studies on marine bacteria. I. The cultural requirements of heterotrophic aerobes. *Journal of Marine Research*, 4(1). https://elischolar.library.yale.edu/journal_of_marine_research/582

Zhang, D., Shu, C., Lian, X., Zhang, Z. (2018). New antibacterial bagremycins F and G from the marine-derived Streptomyces sp. ZZ745. *Marine Drugs*, 16(9), 330.

Zhao, K., Guo, L.Z., Lu, W.D. (2012). Extracellular production of novel halotolerant, thermostable, and alkali-stable carboxymethyl cellulase by marine bacterium *Marinimicrobium* sp. LS-A18. *Applied Biochemistry and Biotechnology*, 168(3), 550–567. http://doi.org/10.1007/s12010-012-9796-3.

Zhao, Z., Wang, Q., Wang, K., Brian, K., Liu, C., Gu, Y. (2010). Study of the antifungal activity of Bacillus vallismortis ZZ185 in vitro and identification of its antifungal components. *Bioresource Technology*, 101(1), 292–297.

Zhou, H., Yang, Y., Nie, X., Yang, W., Wu, Y. (2013). Comparison of expression systems for the extracellular production of mannanase Man23 originated from *Bacillus subtilis* B23. *Microbial Cell Factories*, 12(1), 78. http://doi.org/10.1186/1475-2859-12-78.

13 Marine Greens in Biorefinery

Comfort Okoji, S.I. Musa, Nathan Moses, Harrison Ogala, Ifeoma Sonia Onukwu, Ekene Chidozie, Okorie Maria Ezinne, and Sheila Ojie

13.1 INTRODUCTION

In recent years, environmental problems have challenged our ingenuity and ability to find lasting solutions to protect life on Earth as well as our own survival. The maintenance of fresh water supplies, farmland for crop production, the reduction of greenhouse gas emissions, and the use of non-fossil fuels to generate energy are some of the concerns (Demirbas, 2009; Hertel et al., 2010). With fossil fuels becoming scarcer and more expensive, biomass is becoming increasingly desirable as a source of liquid biofuels. Over the last few decades, efforts have been made to reduce the cost of biofuel, reduce greenhouse gas (GHG) emissions, improve vehicle compatibility, and reduce land and water resource use (Tian et al., 2010).

Marine microalgae, or seaweed, is emerging as a sustainable replacement for fossil fuels. It is believed that seaweeds will play a major role in the future bioeconomy, as they have the potential to serve as feedstock for biorefineries to produce fuels, chemicals, pharmaceuticals, food ingredients, and more (Tian et al., 2017). Algae absorb an enormous amount of CO_2, removing it from power plant emissions. In addition to this, algae are able to convert CO_2 into biomass, releasing more oxygen into the atmosphere via photosynthesis than forests. A further advantage of algae is that they filter the water by absorbing the urea expelled by aquatic animals while simultaneously increasing the conversion of CO_2 into biomass. A variety of biofuels can be produced from algae by liquefaction, pyrolysis, gasification, extraction, transesterification, fermentation, or anaerobic digestion (Ozkurt et al., 2009; Kita et al., 2010). This review evaluates the mass production of marine greens, potential yields, and production challenges, as well as the estimated production cost to compare it with the fossil fuel cost. The future prospects of marine greens as biofuel will also be discussed.

13.2 CURRENT POTENTIAL FOR USE OF MARINE GREENS AS BIOFUEL

Microalgae have the potential to produce biofuels in a way that is both economical and environmentally responsible. Marine microalgae capture carbon dioxide (CO_2) through photosynthesis and can be anaerobically digested to produce biogas, which may contain methane and hydrogen (Li et al., 2008; Mostafa, 2012; Hughes et al., 2012). It wasn't until the 1950s that algae were looked at as a potential source of energy and investigated for methane gas production by anaerobic digestion of their cell polysaccharides, even though some species of microalgae were known in the 1940s to produce large amounts of cellular lipids under selective growth pressures (Meier, 1955; Oswald and Golueke, 1960). When compared with traditional biodiesel feedstocks like soybean and maize, microalgae with a 30% oil content by weight showed land use efficiencies 130 and 338 times greater, respectively (Chisti, 2007).

Large-scale photoautotrophic production of these single-cell species for biofuel and byproducts is dependent on technological and economic feasibility, even if each species of microalga has its ideal growing conditions. Aquaculture feed additives or value-added products have now been used as cost-effective alternatives using photoautotrophic microalgal growth (Borowitzka,

DOI: 10.1201/9781003369738-16

1997; Slade and Bauen, 2013). In these operations, microalgae are grown in lit environments in objects like open raceway ponds. Microalgae are first de-watered before the lipids are extracted. The concentrated biomass is then processed to enhance solvent extraction through cell disruption, particle size reduction, and drying (U.S. Department of Energy, 2010). Defatted microalgal biomass, which is the microalgal skeleton left behind after lipid extraction, is utilized as animal feed. Without the envisioned use as feed, commercial microalgae processing and cultivation for the production of biofuels continue to be very inefficient (Chisti, 2013). Therefore, using the defatted biomass as animal feed would not only help lessen the current competition with the human food supply but would also aid in making the production of microalgae-based biofuels economically viable.

13.3 POTENTIAL YIELDS OF MARINE GREENS

Diverse organisms make up marine greens. Marine greens have historically been distinguished from other plants by their ability to carry out photosynthesis and survive in aquatic habitats, including lakes, seas, and oceans. (David et al., 2011). Most strains of marine greens require a light source for optimum growth, but some can still grow using dark reaction processes. Although some marine green species may store substantial amounts of oil within their cells, the majority of marine green cells are made up of lipids, proteins, and carbohydrates.

The cultivation and yield of marine greens are seen as the most important unit activity in marine green production. Based on the decisions taken regarding the cultivation system design, commercial algal biomass production activities may not be sustainable. The most important choices to be made while setting up an operation are the choice of strain, source of water and nutrients, reactor design, and location of the production facility. It has been documented that there are many types and forms of marine greens, and each marine green type has distinct properties and different growth requirements. For better marine green yield, it is advised that all stages of marine greens should be optimized using technology. Examples of factors to be considered when trying to increase the yield of marine greens include a high nutritional requirement, high and/or low carbon dioxide availability, temperature and salt tolerance, carbon dioxide sequestering ability, light requirement, biomass productivity (biomass weight/volume and or area), cell chemical composition (oil, protein, and carbohydrate content), and size, as well as potential metabolite production as a co-product. Depending on the use, the ideal marine green strain would generate the most oil, protein, or carbohydrates, grow the fastest, and thrive in harsh environments such as those with high salt, pH, or heat.

Nitrate, phosphate (in a ratio of about 6:1), and silicate for some strains, notably for diatoms, are macronutrients required for the growth of microalgae. Thiamin (B1), cyanocobalamin (B12), and biotin are three vitamins that are considered micronutrients. In large-scale systems, only the most essential nutrients are present in the growth medium. Fertilizers used in agriculture are frequently used as sources of nitrogen and potassium. Animal manure and municipal wastewater may also be effective sources of nutrients.

The following methods need to be considered if marine green yields are to be increased. These are:

1. The growth medium that the marine greens are using needs to be free of organisms that may compete with the marine greens. Organisms that may form a symbiotic relationship with marine greens, such as lichens, need to be discouraged from the production site to achieve the maximum yield of marine greens. The most prevalent types of species that compete with monospecific or axenic cultures of microalgae are phytoplankton, zooplankton, and bacteria.
2. It is necessary to treat the substrate on which marine greens are growing, since the starter culture, water, nutrients, air, and photobioreactor are thought to be the most common sources of contamination. By using chemical methods (chlorination and acidification),

filtration, sterilization, pasteurization, or UV irradiation, the growth medium's microbial load can be lowered.

13.4 USES OF MARINE GREENS IN BIOREFINERY

13.4.1 Bioenergy

Marine green extracts grown in wastewater can be turned into various biofuels, including biogas and biodiesel, by a series of conversion processes such as anaerobic digestion, transesterification, fermentation, gasification, pyrolysis, hydrothermal liquefaction (HTL), etc., as well as kerosene, bioethanol, and biohydrogen (Speranza et al., 2015; Tsavatopoulou et al., 2019). Proteins, lipids, carbohydrates, ash, and a variety of minor components including nucleic acids, pigments, etc. make up the components of algal biomass. Depending on the type of ocean green (and any interactions between distinct ocean greens) and growing conditions, lipids make up roughly 20–50% by weight of (dry) algae (Andersson et al., 2014). Ocean greens are therefore a good option for the generation of biofuel.

13.4.2 Liquid Biofuels

13.4.2.1 Supply of Biodiesel

Marine green biomass-derived bio-oils differ from fossil diesel fuels in several ways (Andersson et al., 2014; Speranza, 2015). These cannot be used directly as fuel in diesel engines; instead, they must first be extracted and then refined to produce high-quality biofuels. Lipids can be extracted from marine green biomass using a variety of techniques, including solvent extraction, osmotic shock, ultrasonic extraction, etc. (Singh et al., 2012; Chiang et al., 2015). Solvent extraction is a rapid and efficient process for directly extracting dry biomass. This entails extracting bio-oil from marine greens and obtaining an extraction of up to 98% of the fatty acids using an organic solvent (such as benzene, cyclohexane, hexane, or a combination of hexane and ethanol). While ultrasonic waves are used to induce cavitation bubbles in a solvent, osmotic shock is caused by shock waves and liquid jets breaching cell walls and releasing their contents into the solvent (Singh et al., 2012). The best method for refining bio-oil is transesterification because of its high conversion efficiency and low cost (Lin et al., 2011). There are numerous ways to refine bio-oil. Singh and Olsen (2011) claim that this process makes it possible to produce biodiesel that is remarkably similar to its fossil-based equivalent. The transesterification process can be completed in one or two phases using basic or acidic catalysts. The process and the type of marine greens used have a considerable impact on biodiesel production (Tsavatopoulou et al., 2019; Demirbas, 2007, 2009, 2010). The lipid content of marine greens increases when they are stressed, such as when nutrients are lacking (Ruiz et al., 2013; Zuliani, 2016; Xu et al., 2006). However, these culture conditions result in a lower biomass yield while also promoting lipid accumulation in the marine greens (Ruiz et al., 2013). Therefore, it is crucial to determine which marine green genera are best suited for growth on wastewater to produce lipids for biodiesel supply. HTL is another method for producing biofuels (Delrue et al., 2016; Hodaifa et al., 2013). This thermochemical process converts biomass into bio-oils (with yields ranging from 20% to 87%), gases, residual solids, and an aqueous phase. The obtained bio-oils can be upgraded into biofuels through hydrotreating or burned directly in a boiler (Delrue et al., 2016). The benefit of this method is that it converts the entire biomass, thereby avoiding processes like biomass drying and lipid extraction and selecting marine greens rich in lipids (Delrue et al., 2016; Ferro et al., 2018).

13.4.2.2 Bioethanol

Marine green biomass can also be used to make bioethanol. Bioethanol is produced as a result of this fermentation. Marine green biomass can be used to make biodiesel or to make bioethanol after the lipids have been extracted. The starch- and cellulose-rich algal remains can be used to produce bioethanol after the lipids have been extracted. An essential component in the production of bioethanol is the quantity of carbohydrates in the biomass (Singh and Olsen, 2011).

According to Maia et al. (2020), marine green can synthesize and accumulate a lot of carbohydrates that can be used to make bioethanol. The production of bioethanol can be facilitated by the accumulation of significant amounts of starch, cellulose, and glycogen in some genera, including *Chlorella*, *Chlamydomonas*, *Scenedesmus*, *Dunaliella*, and *Spirulina* (>50% dry weight) (Chen et al., 2009; Ungureanu et al., 2020). The bioethanol yield can be affected by the conditions for marine green culture. The biochemical composition of marine greens is established by changing the culture circumstances, favoring the accumulation of some components over others (Cabirol et al., 2014). The biochemical composition is principally regulated by the following environmental parameters: nutritional factors, salinity, temperature, light intensity, and pH (de Farias Silva and Bertucco, 2016). Generally, marine greens can be grown with strategies to reduce nutrient sources and enhance the proportion of carbohydrates (Dragone et al., 2011; Kim et al., 2014). When marine greens are produced in environments deficient in nitrogen and phosphate, the carbohydrate content of the biomass increases (Braga et al., 2018). According to Lam and Lee (2015), carbohydrates are stored in three different places: inside the cell (starch), in the inner layer of the cell wall (cellulose and hemicellulose), and in the outer layer of the cell wall (pectin, agar, and alginate). Pretreatment is necessary before extracting carbohydrates from marine green cells (Velazquez-Lucio, 2018). Starch and cellulose can be extracted from cells using water or organic solvents, which can then be fermented to produce bioethanol (John et al., 2011). Due to their ability to convert sugars into alcohol, yeasts dominate the variety of microorganisms used in fermentation. According to Walker and Stewart (2016), the yeast used most frequently in the industry for alcoholic fermentation is *Saccharomyces cerevisiae*. These microorganisms can be used to produce bioethanol in three ways: the conventional method, which uses bacteria or yeasts to ferment biomass by hydrolysis; the dark fermentation method, which uses modified cyanobacteria; or "photo-fermentation."

13.4.3 Gaseous Biofuels

The genetic, metabolic, and enzymatic features of cyanobacteria and marine green algae are necessary for the synthesis of biohydrogen gas. This is a sustainable and non-polluting energy source because the only byproduct of the combustion of H_2 is water (H_2O). H_2 is a fuel of the future because of its high conversion efficiency, between 122 and 142 kJ/g, which is 2.75 times higher than that of combustible hydrocarbons, according to Rashid et al. (2013). However, this resource does not occur naturally anywhere on Earth (Goswami et al., 2020; Preethi et al., 2019). In recent decades, it has been used to produce power using fuel cells or internal combustion engines. Conventional thermochemical and electrochemical methods, such as water electrolysis and gasification, use more energy and are more hazardous to the environment than biological hydrogen (bioH_2) production (Batista et al., 2015). Conventional techniques have a significantly bigger impact because of their high temperatures and high energy usage. Additionally, these procedures are not regarded as eco-friendly on a commercial scale due to the significant volumes of CO_2 they generate (Medisetty et al., 2020).

The techniques of direct bio-photolysis, indirect bio-photolysis, photo-fermentation, and dark fermentation can all be used to produce bioH_2 from marine greens (Goswami et al., 2020). Marine greens directly bio-photolyze water to produce oxygen (O_2) and protons (H^+) in the presence of light. H_2 is produced by the enzyme hydrogenase, which converts H^+ to H_2 (Show et al., 2011). Because H_2 and O_2 are produced simultaneously and instantly mix, this process produces little H_2 as a result because water is a byproduct. Additionally, hydrogenase production rates are slowed down by its sensitivity to oxygen (Show et al., 2011). This inhibiting effect can be eliminated by using indirect bio-photolysis. Indirect bio-photolysis involves two stages. Phase 1 begins with the evolution of oxygen as cells accumulate organic compounds through photosynthesis (primarily glucose). This phase is also known as the aerobic phase. Melis and Melnicki (2006) claim that during phase 2,

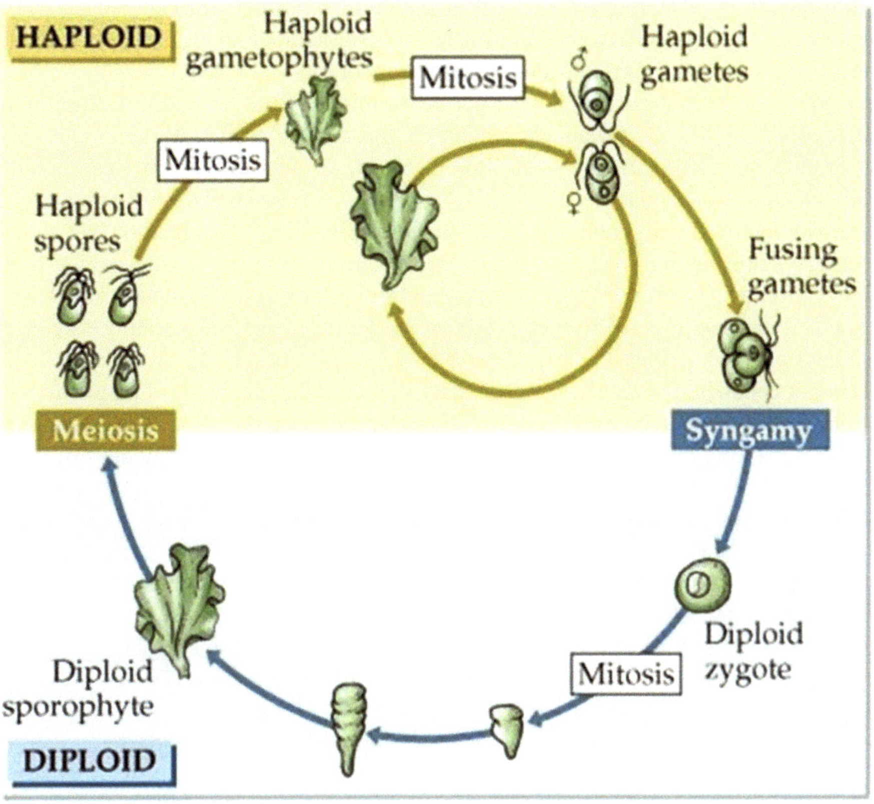

FIGURE 13.1 Life cycle of green algae (Ulva).

cells break down organic compounds that have been stored in anaerobic storage. The second stage is the anaerobic one. In the second step of the two-phase process, hydrogen and oxygen evolve independently. Hallenbeck (2005) asserts that the fermentation procedures are often faster and produce higher H_2 yields. One of the pioneering research projects on bioH_2 synthesis from marine green algae cultivated on wastewater (*Chlorella vulgaris*, *Scenedesmus obliquus*, *Ulva* sp and the Natural Algal Consortium) (Figure 13.1) was carried out by Batista et al. (2015).

13.5 BIONANOTECHNOLOGY IN IMPROVING MARINE GREEN PRODUCTION

Considering the challenges observed in employing marine greens in biorefinery, scientists are making several proposals on possible strategies to improve marine green production, especially to meet the commercial quantity (Singh et al., 2012). Nanomaterials having both magnetic and catalytic properties can be beneficial for marine green production through bionanotechnology (Spolaore et al., 2006). Bionanotechnology is the application of nanotechnology in solving biological problems (Ikhajiagbe and Musa, 2020). It involves the introduction of biological molecules into nano-artifacts (Rodionova et al., 2016). One such nanoparticle (NP) is FeO_4 silica, which could be used as a catalyst for the coupling process during algal biofuel production (Teresa et al., 2012). To this nanoparticle base, triazabicyclodecene was attached, which could add functionality to the NPs. Previous research showed that these NPs could attach to microalgae and therefore proved to be good micro-harvesters. This combination was then added to the chloroform: methanol solvent in which lipid extraction is performed. Since NP acts as a catalyst, the lipid will easily be converted to biodiesel at a rapid rate depending on the microalgae. From this, it is evident that NPs can enhance biofuel production by providing a greater surface area for the reaction to take place (Musa, 2020). Furthermore, magnetic NPs have been used when cultivating microalgae in a photobioreactor; such an arrangement distributes nutrients uniformly throughout.

Silver NPs have been used for higher light accessibility (Tran et al., 2010). Similar silver NPs have been used by Musa and Ikhajiagbe (2020) to improve the growth of rice under varying conditions through rhizoinoculation. During this research, an increased microalgae population was observed, which was likely due to the silver NP application. Magnetic NPs are a potential candidate to consider for harvesting algal biofuel (Ayesha and Malik, 2019). This method has gained interest in microalgae culturing due to its low-cost efficiency. Iron oxide nanoparticles have been used in this respect. Iron oxide NPs were used to cultivate two species of microalgae—*Scenedesmus ovalternus* and *Chlorella vulgaris*. This method showed a 95% increase in harvesting efficiency for both species (Converti et al., 2009).

In the work conducted by Cruz et al. (2018), they studied the efficiency of magnetic NPs (Fe_3O_4) as separating agents in the dissembling of cells during biofuel production. The magnetite was prepared by precipitating FeO and dissolving hydrated 3-4-Fe_3Cl_4 and FeCl under nitrogen pressure in water. After heating the solution, NH_4OH was added, while magnetite was precipitated further. Certain species of algae were grown in a modified medium. When the desired biomass was reached, magnetite was added. Under different stirring speeds, it was noted that harvested microalgae were greatly adsorbed over the magnetite. The recovery efficiency reached up to 99.9%. This suggests that nanoparticles boost microalgae recovery and can therefore produce more biomass for the generation of biofuel (Xu et al., 2011). For effective flocculation, Dautzenberg et al. (2011) synthesized magnetite with a polymer coating and polyamidoamine (PAMAM) (Wang et al., 2016).

One of the molecular techniques for employing bionanotechnology in marine green production is that the cells of marine greens must be the type that can allow free access to particles at nano levels (David et al., 2011). With the use of nanoparticles, marine green cell walls can be a restriction to easy disruption and extraction (Kumar et al., 2015). This has allowed the application of nanoparticles in biotechnology to carve out its niche in the renewable energy sector, especially as it improves the processes involved in algal biofuel production (Wang et al., 2016).

The ideas of bionanotechnology could be employed in the selection of the best strains of marine greens that could have better yield and efficiency in biorefinery. Silver nitrate-induced CRISPR-Cas9 has been used as a molecular tool in choosing species of organisms that are adapted to different ecotypes and ecoregions. With this technology, only marine greens that could produce a higher yield would be selected and cultivated. This same method could also be used to address the issue of the seasonality of marine green production. Furthermore, the use of genetically engineered marine greens that are capable of reproducing more rapidly is imperative. This would improve the scalability of the production of marine greens, thereby making them widely available in biorefinery. However, an efficient regeneration protocol must be in place before genetic transformation studies can be initiated. Improvement using nanoparticles and a continuous supply of nutrients in growing marine greens may also be considered; enabling environments and situations that would improve marine green multiplication would help in improving biofuel production. Since microalgae are very sensitive to environmental pollution (Wang et al., 2016), environments meant for growing marine greens can be monitored using biotechnological means prior to production. Among the most important products needed in biofuel production are the extracted carbohydrates and lipids; synthesized bionanoparticles should be used to improve their rapid extraction. Optimizing the culture conditions of the target plant cells or tissues and the transformation process are the two main elements that control the successful introduction of foreign genes into plant cells. Since researchers are only now exploring the field, there are only a few studies on transformation and transgenic development in algal production. However, transgenic nanoparticles have been described (Zhang et al., 2013). Further inputs are required for the establishment of efficient transgenic marine greens that would improve the supply of products needed in biofuel production. Judicious improvement of nanoparticles for the biotechnological production of biofuel from microalgae can bring more benefits to mankind throughout the world and reduce environmental pollution due to the use of fossil fuels.

13.6 CONCLUSION

The development of marine green biorefineries is entirely dependent on the conversion of residual biomass into a variety of sustainable products. The overall biorefinery approach is a broad strategy that starts with the selection of feedstock species and culminates in the global marketing of goods with added value, meeting consumer demand. The applications stated here in the establishment of the biorefinery plan represent zero waste management, which will benefit both the circular bioeconomy and the socioeconomic situation. The benefits of choosing green algae as a biorefinery feedstock depend on their rapid growth, high content of high-quality bio-components, and widespread use of macroalgal products in Western nations. However, the biorefinery method is a relatively young prospective commercialization approach, and it faces several challenges, which can be overcome by inventing novel technologies. Furthermore, advancements in strain selection, improved cultivation and harvesting strategies, yearly feedstock production, participation of seaweed farmers, waste recovery management, and bioenergy production as well as advancements in technology and procedures, will accelerate the growth of industrial seaweed commercialization.

REFERENCES

Andersson, V., Broberg Viklund, S., Hackl, R., Karlsson, M. and Berntsson, T. (2014). Algae-based biofuel production as part of an industrial cluster. *Biomass Bioenergy*, 71, 113–124.

Ayesha, M. and Malik, B. (2019). Magnetic nanoparticles: Eco-friendly application in biofuel production. In K. Abd-Elsalam, M. Mohammed, R. Prasad (Eds.), *Magnetic Nanostructures: Nanotechnology in the Life Sciences*. Springer, Cham, pp. 3–7.

Batista, A.P., Ambrosano, L., Graça, S., Sousa, C., Marques, P.A.S.S., Ribeiro, B., Botrel, E.P., Castro Neto, P. and Gouveia, L. (2015). Combining urban wastewater treatment with biohydrogen production - An integrated microalgae-based approach. *Bioresour. Technol.*, 184, 230–235.

Borowitzka, M. (1997). Microalgae for aquaculture: Opportunities and constraints. *J. Appl. Physiol.*, 9, 393–401.

Braga, V., Da, S., Mastrantonio, D.J., Costa, J.A.V. and Morais, M.G. (2018). Cultivation strategy to stimulate high carbohydrate content in Spirulina biomass. *Bioresour. Technol.*, 269, 221–226.

Cabirol, M., Ripa, M., Cocozza, A., Panfilo, C. and Ulgiati, S. (2014). Challenges and opportunities from circular wastewater management: The case of Campania region. Italy. *J. Environ. Manag.*, 2(2), 21–30.

Chen, W., Zhang, C., Song, L., Sommerfeld, M. and Hu, Q. (2009). A high throughput Nile red method for quantitative measurement of neutral lipids in microalgae. *J. Microbiol. Methods*, 77(1), 41–47.

Chiang, Y.D., Dutta, S., Chen, C.T., Huang, Y.T., Lin, K.S., Wu, J.C., Suzuki, N., Yamauchi, Y. and Wu, K.C. (2015). Functionalized Fe_3O_4 Silica core–shell nanoparticles as microalgae harvester and catalyst for biodiesel production. *ChemSusChem*, 8(5), 789–794

Chisti, Y. (2007). Biodiesel from microalgae. *Biotechnol. Adv.*, 25(3), 294–306.

Chisti, Y. (2013). Constraints to commercialization of algal fuels. *J. Biotechnol.*, 167(3), 201–214.

Converti, A., Casazza, A.A., Ortiz, E.Y., Perego, P. and Del Borghi, M. (2009). Effect of temperature and nitrogen concentration on the growth and lipid content of *Nannochloropsis oculata* and *Chlorella vulgaris* for biodiesel production. *Chem. Eng. Process. Process Intensif.*, 48(6), 1146–1151.

Cruz, Y.R., Aranda, D.G. and Seidl, P.R. (2018). Cultivation systems of microalgae for the production of biofuels. In Krzysztof Biernat (eds). *Biofuels – State of Development*, inTech.12(2), 12–24.

Dautzenberg, G., Gerhardt, M. and Kamm, B. (2011). Bio-based fuels and fuel additives from lignocellulose feedstock via the production of levulinic acid and furfural. *Holzforschung*, 65, 43–49.

David, S., Zelda, E., Penton, F. and Kitson, G. (2011). Hydrocarbons. In O.D. Sparkman, Z.E. Penton and F.G. Kitson (Eds.), *Gas Chromatography and Mass Spectrometry* (2nd Edition). Academic Press, pp. 331–339.

de Farias Silva, C.E. and Bertucco, A. (2016). Bioethanol from microalgae and cyanobacteria: A review and technological outlook. *Process Biochem.*, 51(11), 1833–1842.

Delrue, F., Álvarez-Díaz, P.D., Fon-Sing, S., Fleury, G. and Sassi, J.F. (2016). The environmental biorefinery: Using microalgae to remediate wastewater, a win-win paradigm. *Energies* 9-23.
Demirbas, A. (2007). Progress and recent trends in biofuels. *Prog. Energy Combus Sci.*, 33(1), 1–18.
Demirbas, A. (2009). Progress and recent trends in biodiesel fuels. *Energy Convers. Manag.*, 50(1), 14–34.
Demirbas, A. (2009). Political, economic and environmental impacts of biofuels: A review. *Appl. Energy*, 86, S108–117.
Demirbas, A. (2010). Social, economic, environmental and policy aspects of biofuels. *Energy Educ. Sci. Technol. B*, 2, 75–109.
Dragone, G., Fernandes, B.D., Abreu, A.P., Vicente, A.A. and Teixeira, J.A. (2011). Nutrient limitation as a strategy for increasing starch accumulation in microalgae. *Appl. Energy*, 88(10), 3331–3335.
Ferro, L., Gorzsás, A., Gentili, F.G. and Funk, C. (2018). Subarctic microalgal strains treat wastewater and produce biomass at low temperature and short photoperiod. *Algal Res.*, 35, 160–167.
Goswami, R.K., Mehariya, S., Verma, P., Lavecchia, R. and Zuorro, A. (2020). Microalgae-based biorefineries for sustainable resource recovery from wastewater. *J. Water Process Eng.*, 40, 101747.
Hallenbeck, P.C. (2005). Fundamentals of the fermentative production of hydrogen. *Water Sci. Technol.*, 52(1–2), 21–29.
Hertel, T.W., Tyner, W.E. and Birur, D.P. (2010). The global impacts of biofuel mandates. *Energy J.*, 31(1), 75–100.
Hodaifa, G., S´anchez, S., Martínez, M.E. and Órpez, R. (2013). Biomass production of *Scenedesmus obliquus* from mixtures of urban and olive-oil mill wastewaters used as culture medium. *Appl. Energy*, 104, 345–352.
Hughes, A.D., Kelly, M.S., Black, K.D. and Stanley, M.S. (2012). Biogas from microalgae: Is it time to revisit the idea? *Biotechnol. Biofuels*, 5(1), 86.
Ikhajiagbe, B. and Musa, S.I. (2020). Application of biosynthesized nanoparticles in the enhancement of growth and yield performances of rice (*Oryza sativa* var. Nerica) under salinity conditions in a ferruginous ultisol. *Fudma J. Sci.*, 4(1), 120–132.
John, R.P., Anisha, G.S., Nampoothiri, K.M. and Pandey, A. (2011). Micro and macroalgal biomass: A renewable source for bioethanol. *Bioresour. Technol.*, 102(1), 186–193.
Kim, K.H., Choi, I.S., Kim, H.M., Wi, S.G. and Bae, H.J. (2014). Bioethanol production from the nutrient stress-induced microalga *Chlorella vulgaris* by enzymatic hydrolysis and immobilized yeast fermentation. *Bioresour. Technol.*, 153, 47–54.
Kita, K., Okada, S., Sekino, H., Imou, K., Yokoyama, S. and Amano, T. (2010). Thermal pretreatment of wet microalgae harvest for efficient hydrocarbon recovery. *Appl. Energy*, 87(7), 2420–2423.
Kumar, R.R., Rao, P.H. and Arumugam, M. (2015). Lipid extraction methods from microalgae: A comprehensive review. *Front. Energy Res.*, 2, 1–9.
Lam, M.K. and Lee, K.T. (2015). Bioethanol production from microalgae. In *Handbook of Marine Microalgae Biotechnology Advances*. Elsevier Inc., pp. 197–208.
Li, Y., Horsman, M., Wu, N., Lan, C.Q. and Dubois-Calero, N. (2008). Biofuels from microalgae. *Biotechnol. Prog.*, 24(4), 815–820.
Lin, L., Cunshan, Z., Vittayapadung, S., Xiangqian, S. and Mingdong, D. (2011). Opportunities and challenges for biodiesel fuel. *Appl. Energy*, 88(4), 1020–1031.
Maia, J.L., da, Cardoso, J.S., Mastrantonio, D.J.S., Bierhals, C.K., Moreira, J.B., Costa, J.A.V. and Morais, M.G. (2020). Microalgae starch: A promising raw material for the bioethanol production. *Int. J. Biol. Macromol.*, 165(B), 2739–2749.
Medisetty, V.M., Kumar, R., Ahmadi, M.H., Vo, D.V.N., Ochoa, A.A.V. and Solanki, R. (2020). Overview on the current status of hydrogen energy research and development in India. *Chem. Eng. Technol.*, 43(4), 613–624.
Meier, R.L. (1955). Biological cycles in the transformation of solar energy into useful fuels. *J. Anim. Sci.*, 14, 179–183.
Melis, A. and Melnicki, M.R. (2006). Integrated biological hydrogen production. *Int. J. Hydrog. Energy*, 31(11), 1563–1573.
Mostafa, S. (2012). Microalgal biotechnology: Prospects and applications. In *Plant Science*. Edited by Nabin Kumar Dhal. INTech. http://www.intechopen.com/ books/plant-science/microalgal-biotechnology-prospects-and-applications.
Oswald, W.J. and Golueke, C.G. (1960). Biological transformation of solar energy. *Adv. Appl. Microbiol.*, 2, 223–262.
Ozkurt, I. (2009). Qualifying of safflower and algae for energy. *Energy Educ. Sci. Thecnol. Part A*, 23, 145–151.

Preethi, Usman, T.M.M., Rajesh Banu, J., Gunasekaran, M. and Kumar, G. (2019). Biohydrogen production from industrial wastewater: An overview. *Bioresour. Technol. Rep.*, 7, 100287.

Rashid, N., Rehman, M.S.U., Memon, S., Ur Rahman, Z., Lee, K. and Han, J.I. (2013). Current status, barriers and developments in biohydrogen production by microalgae. *Renew. Sustain. Energy Rev.*, 22, 571–579.

Rodionova, M., Poudyal, R., Tiwari, I., Voloshin, R., Zharmukhamedov, S., Nam, H., Zayadan, B., Bruce, B., Hou, H. and Allakhverdiev, S. (2016). Biofuel production: Challenges and opportunities. *Int. J. Hydrog. Energy*, 2, 1–12.

Ruiz, J., Álvarez-Díaz, P.D., Arbib, Z., Garrido-Pérez, C., Barragán, J. and Perales, J.A. (2013). Performance of a flat panel reactor in the continuous culture of microalgae in urban wastewater: Prediction from a batch experiment. *Bioresour. Technol.*, 127, 456–463.

Show, K.Y., Lee, D.J. and Chang, J.S. (2011). Bioreactor and process design for biohydrogen production. *Bioresour. Technol.*, 102(18), 8524–8533.

Singh, A. and Olsen, S.I. (2011). A critical review of biochemical conversion, sustainability and life cycle assessment of algal biofuels. *Appl. Energy*, 88(10), 3548–3555.

Singh, A., Pant, D., Olsen, S.I. and Nigam, P.S. (2012). Key issues to consider in microalgae based biodiesel production. *Energy Educ. Sci. Technol. A*, 29, 687–700.

Slade, R. and Bauen, A. (2013). Micro-algae cultivation for biofuels: Cost, energy balance, environmental impacts and future prospects. *Biomass Bioenerg.* 53, 29–38.

Speranza, L.G., Ingram, A. and Leeke, G.A. (2015). Assessment of algae biodiesel viability based on the area requirement in the European Union, United States and Brazil. *Renew. Energy*, 78, 406–417.

Spolaore, P., Joannis-Cassan, C., Duran, E. and Isambert, A. (2006). Commercial application of microalgae. *J. Biosci. Bioeng.*, 101(2), 87–96.

Teresa, M., Mata, A.A., Martins, B., Neto, M.L., Martins, R.R., Salcedo, C.A. and Costa, L. (2012). LCA tool for sustainability evaluations in the pharmaceutical Industry. *Chem. Eng. Trans.*, 23, 12–17.

Tian, Y., Zhao, L., Meng, H., Sun, L. and Yan, J. (2010). Estimation of un-used land potential for biofuels development in (the) People's Republic of China. *Appl. Energ.*, 86, S77–85.

Tran, N., Bartlett, J., Kannangara, G., Milev, A., Volk, H. and Wilson, M. (2010). Catalytic upgrading of biorefinery oil from micro-algae. *Fuel*, 189, 26–74.

Tsavatopoulou, V.D., Aravantinou, A.F. and Manariotis, I.D. (2019). Biofuel conversion of Chlorococcum sp. And Scenedesmus Sp. Biomass by one- and two-step transesterification. *Biomass Convers. Biorefin*, 2(1), 23–43.

U.S. Department of Energy (2010). *National Algal Biofuels Technology Roadmap: 9–10 December 2008.* Maryland Publication, pp 30–37.

Ungureanu, N., Vladut, V. and Biris, S.-S. (2020). Capitalization of wastewater-grown algae in bioethanol production Appl. Energ., 86, 66–91.

Velazquez-Lucio, J., Rodríguez-Jasso, R.M., Colla, L.M., Sáenz-Galindo, A., Cervantes-Cisneros, D.E., Aguilar, C.N., Fernandes, B.D. and Ruiz, H.A. (2018). Microalgal biomass pretreatment for bioethanol production: A review. *Biofuel Res. J.*, 5(1), 780–791.

Walker, G. and Stewart, G. (2016). *Saccharomyces cerevisiae* in the production of fermented beverages. *Beverages*, 2(4), 30.

Wang, T., Yang, W., Hong, Y. and Hou, Y. (2016). Magnetic nanoparticles grafted with amino-richeddendrimer as magnetic flocculant for efficient harvesting of oleaginous microalgae. *Chem. Eng. J.*, 297, 304–314.

Xu, H., Miao, X. and Wu, Q. (2006). High quality biodiesel production from a microalga Chlorella protothecoides by heterotrophic growth in fermenters. *J. Biotechnol.*, 126(4), 499–507.

Zhang, X.L., Yan, S., Tyagi, R.D. and Surampall, R. (2013). Biodiesel production from heterotrophic microalgae through transesterification and nanotechnology application in the production. *Ren. Sust. Energy Rev.*, 26, 216–223.

Zuliani, L., Frison, N., Jelic, A., Fatone, F., Bolzonella, D. and Ballottari, M. (2016). Microalgae cultivation on anaerobic digestate of municipal wastewater, sewage sludge and agro-waste. *Int. J. Mol. Sci.*, 17(10) 12–23.

14 Potential of Marine Greens for Biofuel Production

O.D. Ogundele, I.A. Amoo, A.O. Adesina, and V.A. Olagunju

14.1 INTRODUCTION

The global demand for energy is continuously increasing due to increase in population, industrialization, and modernization (International Energy Agency [IEA], 2020). The majority of the world's energy is generated from fossil fuels, leading to an increase in greenhouse gas (GHG) emissions, which contributes to climate change and environmental degradation (Ritchie & Roser, 2021). Recently, there has been rising concern over the development of sustainable and renewable energy sources, particularly biofuels, to ease dependency on fossil fuels and mitigate the negative impacts of GHG emissions (Demirbas, 2009).

Biofuels are derived from organic materials, such as plants and microorganisms, and can be used as alternatives to traditional fossil fuels (Balat, 2011). The potential of biofuels as a renewable energy source has been extensively studied, with particular emphasis on terrestrial plants, such as corn, sugarcane, and oilseed crops (Sims *et al.*, 2010). However, concerns have been raised regarding the use of agricultural land and resources for biofuel production, as it may compete with food production and contribute to deforestation (Tilman *et al.*, 2009). This has led to a search for alternative feedstocks that do not compete with food production, such as marine greens (Chisti, 2007).

Given the numerous advantages of marine greens for biofuel production, there has been a growing awareness of exploring their prospects for large-scale, sustainable, and economically viable biofuel production (Singh & Olsen, 2011). This chapter aims to provide a comprehensive overview of the potential of marine greens for biofuel production, highlighting their advantages and challenges, recent advances in research and technology, and real-world applications. Additionally, the review will examine the environmental and economic implications of using marine greens as feedstocks for biofuels, as well as future research directions and policy support needed for their successful implementation. This chapter will provide a comprehensive and critical analysis of the current state of knowledge on the potential of marine greens for biofuel production, offering valuable insights for researchers, policymakers, and industry stakeholders.

14.2 MARINE GREENS AS A POTENTIAL SOURCE FOR BIOFUEL PRODUCTION

Marine greens, comprising microalgae and macroalgae (seaweeds), have emerged as a promising feedstock for biofuel production due to their numerous advantages over traditional terrestrial sources, such as higher biomass productivity, non-competition with food crops, and low freshwater requirements (Wijffels & Barbosa, 2010). Biofuels derived from marine greens have the potential to play a substantial role in decreasing GHG emissions and enhancing energy security, thereby contributing to a more sustainable and renewable energy future (Chisti, 2007).

Microalgae are microscopic, photosynthetic organisms that can grow rapidly in a variety of environments, including marine, brackish, and freshwater systems (Mata *et al.*, 2010). These organisms have a high lipid content, making them ideal candidates for the production of biodiesel—a renewable and sustainable alternative to fossil-based diesel fuel (Chisti, 2007). Furthermore, some microalgae strains have a high carbohydrate content, which can be utilized to produce bioethanol

DOI: 10.1201/9781003369738-17

through fermentation processes (Harun *et al.*, 2010). Microalgae-based biofuels have the potential to deliver higher energy yields per unit area compared with terrestrial crops, such as corn, sugarcane, and oilseed rape, while avoiding the competition for arable land and freshwater resources (Dismukes *et al.*, 2008).

Macroalgae, commonly known as seaweeds, are large, multicellular marine organisms that grow predominantly in coastal areas and can be classified into three groups: green, red, and brown algae (Mann & Steinke, 2012). Seaweeds have a high carbohydrate content and can be used to produce bioethanol, biobutanol, and other biofuels through various fermentation and hydrolysis processes (Adams *et al.*, 2011). The advantages of using macroalgae for biofuel production include their rapid growth rates, ability to grow in a wide range of environmental conditions, and capacity to absorb nutrients and carbon dioxide from the surrounding water, thereby providing additional environmental benefits (Horn *et al.*, 2014).

One of the key challenges in using marine greens for biofuel production is the efficient and cost-effective extraction of lipids and carbohydrates from these organisms. Recent advancements in cell disruption, solvent extraction, and enzymatic hydrolysis techniques have shown promise in improving the extraction efficiency and reducing the costs related to the production of marine green–based biofuels (Lee *et al.*, 2012). Moreover, the development of genetically engineered strains of microalgae and macroalgae with enhanced lipid and carbohydrate content has the potential to further improve the productivity and feasibility of marine green–based biofuels (Radakovits *et al.*, 2010).

Another critical challenge in the large-scale production of marine green–based biofuels is the development of sustainable and efficient cultivation systems (Pires *et al.*, 2011). While open pond systems are relatively low-cost and easy to operate, they have limitations in terms of contamination risks, evaporation losses, and low biomass productivity. On the other hand, closed photobioreactors offer higher biomass productivity and better control over cultivation conditions, but at a higher capital and operational cost (Posten, 2009). Optimization of cultivation systems, including the integration of wastewater treatment and CO_2 capture technologies, can help improve the sustainability and cost-effectiveness of marine green–based biofuel production (Delrue *et al.*, 2016).

Marine greens offer significant potential as a renewable and sustainable feedstock for biofuel production. The utilization of marine greens for biofuel production can help address the global demand for energy while mitigating the negative environmental impacts associated with fossil fuel consumption. Further research and development efforts are needed to overcome the challenges related to the efficient extraction of lipids and carbohydrates from marine greens, as well as the design and optimization of sustainable cultivation systems (Singh *et al.*, 2011). By harnessing the potential of marine greens for biofuel production, we can contribute to a cleaner, greener, and more sustainable energy future.

14.3 BIOFUEL PRODUCTION PROCESSES

Biofuel production involves various processes to convert biomass into usable forms of energy, such as bioethanol, biodiesel, and biogas. These processes can be broadly categorized into biochemical, thermochemical, and physicochemical methods (Demirbas, 2009).

a) **Biochemical Processes:** Biochemical processes involve the use of enzymes, bacteria, or other microorganisms to break down biomass into simpler compounds that can be further processed into biofuels. Fermentation is a widely used biochemical process for the production of bioethanol from sugar- or starch-rich feedstocks, such as corn, sugarcane, and wheat (Soccol *et al.*, 2010). In this process, sugars are converted into ethanol by yeast or bacteria under anaerobic conditions. For the production of cellulosic ethanol, a pretreatment step is required to break down the complex lignocellulosic structure and release fermentable sugars, which are then fermented by microorganisms (Perlack & Stokes, 2011).

b) **Anaerobic Digestion:** Anaerobic digestion is another biochemical process used for the production of biogas, a mixture of methane and carbon dioxide. This process involves the decomposition of organic matter by various groups of microorganisms in the absence of oxygen, resulting in the production of biogas and nutrient-rich digestate (Angelidaki *et al.*, 2011). Anaerobic digestion can be applied to various feedstocks, including agricultural and animal wastes, sewage sludge, and organic fractions of municipal solid waste.
c) **Thermochemical Processes:** Thermochemical processes involve the application of heat to convert biomass into biofuels. Pyrolysis is a thermochemical process that involves the thermal decomposition of biomass in the absence of oxygen, resulting in the production of bio-oil, biochar, and non-condensable gases (Bridgwater, 2003). Bio-oil can be further upgraded and refined into advanced biofuels, such as drop-in biofuels and green chemicals, while biochar can be used as a soil amendment or for carbon sequestration (Lehmann & Joseph, 2009).
d) **Gasification:** Gasification is another thermochemical process, which involves the partial oxidation of biomass at high temperatures to produce a mixture of carbon monoxide, hydrogen, and methane, known as synthesis gas or syngas (McKendry, 2002). Syngas can be further processed into various biofuels and chemicals through catalytic processes, such as Fischer–Tropsch synthesis, methanol synthesis, and mixed alcohol synthesis (Huber *et al.*, 2006).
e) **Physicochemical Processes:** Physicochemical processes involve the application of physical and chemical methods to convert biomass into biofuels. Transesterification is a widely used physicochemical process for the production of biodiesel from vegetable oils, animal fats, or waste cooking oils (Ma & Hanna, 1999). In this process, triglycerides are reacted with an alcohol, usually methanol, in the presence of a catalyst, such as sodium or potassium hydroxide, to produce biodiesel and glycerol as a byproduct.
f) **Hydrothermal Liquefaction:** Hydrothermal liquefaction is another physicochemical process, which involves the conversion of biomass into bio-oil under high pressure and temperature in the presence of water (Toor *et al.*, 2011). This process is particularly suitable for the conversion of wet biomass, such as microalgae and sewage sludge, into bio-oil, which can be further upgraded into advanced biofuels and green chemicals (Elliott *et al.*, 2015).

Biofuel production processes can be broadly categorized into biochemical, thermochemical, and physicochemical methods, each with its advantages and limitations. The selection of the most appropriate process depends on the type and characteristics of the feedstock, as well as the desired biofuel product. Continuous research and development in these processes aim to improve the efficiency, sustainability, and economic viability of biofuel production (Sims *et al.*, 2010).

14.4 ADVANTAGES OF MARINE GREENS FOR BIOFUEL PRODUCTION

Marine greens, including macroalgae (seaweeds) and microalgae, have emerged as promising feedstocks for biofuel production due to their unique advantages over traditional terrestrial feedstocks, such as energy crops and agricultural residues (Wijffels & Barbosa, 2010). These advantages include high growth rates, high biomass productivity, reduced land and freshwater use, and the ability to capture and utilize CO_2 efficiently (Chisti, 2008).

One of the main advantages of marine greens is their high growth rate and biomass productivity, which can be several times higher than those of terrestrial plants (Mata *et al.*, 2010). This high productivity allows the production of larger amounts of biofuel per unit area compared with traditional feedstocks. Additionally, marine greens can be harvested year-round, providing a continuous and reliable supply of biomass for biofuel production (Potts *et al.*, 2012).

Another significant advantage of marine greens is their reduced land and freshwater use, which is particularly important in the context of global concerns over land and water resources. Unlike terrestrial feedstocks that require arable land and freshwater for irrigation, marine greens can be cultivated in saline water or wastewater, reducing the pressure on limited freshwater resources (Roesijadi *et al.*, 2010). Moreover, marine greens can be cultivated in open ponds, photobioreactors, or offshore systems, thereby avoiding competition for land with food production and other land-based activities (Xin *et al.*, 2017).

Marine greens are efficient at capturing and utilizing CO_2, which contributes to their high productivity and environmental benefits. Many species of microalgae and macroalgae are capable of fixing CO_2 at rates much higher than those of terrestrial plants (Beardall & Raven, 2004). This efficient CO_2 fixation can help mitigate GHG emissions and climate change, as well as potentially provide a sustainable solution for CO_2 capture and utilization from industrial sources, such as power plants and other emission-intensive industries (Markou & Nerantzis, 2013).

Marine greens also have the potential to produce a diverse range of biofuels, including bioethanol, biobutanol, and biogas, through various conversion processes, such as fermentation, anaerobic digestion, and thermochemical conversion (Harun *et al.*, 2010). The versatility of marine greens in terms of biofuel production provides an opportunity to develop integrated biorefinery concepts, where multiple products, including biofuels, chemicals, and materials, can be produced from a single feedstock, thereby enhancing the economic viability and sustainability of biofuel production (Enzing *et al.*, 2014).

14.5 RECENT ADVANCES AND INNOVATIONS IN MARINE GREENS FOR BIOFUEL PRODUCTION

In recent years, there has been growing interest in the development of marine greens, which include microalgae and macroalgae (seaweeds), as a sustainable and renewable feedstock for biofuel production (Wijffels & Barbosa, 2010). This is mainly due to their high photosynthetic efficiency, rapid growth rates, and ability to accumulate lipids and other energy-rich compounds that can be converted into various types of biofuels, such as biodiesel, bioethanol, and biogas (Hossain *et al.*, 2008; Chisti, 2007). Furthermore, marine greens can be cultivated in saltwater or brackish water, which reduces the demand for freshwater resources and mitigates potential conflicts with food production on arable land. This section will provide an overview of the recent advances and innovations in the field of marine greens for biofuel production, focusing on the areas of cultivation, harvesting, and processing technologies.

a) **Cultivation Technologies:** One of the key challenges in the large-scale production of marine greens for biofuel applications is the development of efficient and cost-effective cultivation systems that can optimize growth conditions, maximize biomass productivity, and minimize resource consumption (Singh & Olsen, 2011). Recent advances in this area include the design of innovative photobioreactors (PBRs) and open pond systems for microalgal cultivation, as well as novel cultivation techniques for macroalgae (Posten, 2009).

   For microalgae, PBRs offer several advantages over traditional open pond systems, such as improved light utilization, reduced contamination risks, and better control of cultivation parameters. Recent innovations in PBR design include the development of flat-panel, tubular, and column systems, which can optimize light distribution, gas exchange, and nutrient supply to enhance microalgal growth and lipid accumulation). Additionally, advances in materials science and engineering have led to the creation of PBRs with improved thermal and optical properties, which can further enhance the efficiency and sustainability of microalgal cultivation (Carvalho *et al.*, 2011; Grima *et al.*, 2003).

   Cultivation techniques have been a major area of innovation, with researchers exploring various methods to enhance the growth and biomass productivity of marine greens. One such approach is the development of photobioreactors, which are closed systems that provide a controlled environment for optimal growth conditions, including nutrients, light, and temperature (Chisti, 2007). These systems can be customized to improve the productivity of specific marine green species, while also minimizing water consumption and land requirements. Additionally, innovative photobioreactor designs, such as flat-panel, tubular, and raceway pond systems, have been developed to maximize light penetration and distribution, further enhancing algal growth and productivity (Grima *et al.*, 2003).

Another cultivation strategy involves the use of genetic engineering and selective breeding techniques to optimize marine green strains for biofuel production. By manipulating the genetic makeup of microalgae and macroalgae, researchers have been able to develop strains with higher lipid content, faster growth rates, and improved tolerance to environmental stressors, such as high salinity or nutrient limitation (Grima *et al.*, 2003). These advancements have not only improved the potential yield of biofuels from marine greens but also contributed to a better understanding of the underlying biological processes governing their growth and metabolism (Carvalho *et al.*, 2011).

b) **Harvesting Technologies:** Another critical aspect of marine green–based biofuel production is the development of efficient and cost-effective harvesting methods, which can separate the algal biomass from the cultivation medium and prepare it for further processing. Traditional harvesting techniques, such as sedimentation, centrifugation, and filtration, are often energy-intensive and expensive, which can significantly increase the overall cost of marine green–based biofuels (Milledge & Heaven, 2013).

Recent innovations in harvesting technologies include the development of novel methods, such as flotation, flocculation, and electrocoagulation, which can enhance the efficiency and sustainability of algal biomass recovery (Garg *et al.*, 2012). For example, the use of bio-flocculation agents, such as chitosan and natural polymers, can promote the aggregation and sedimentation of microalgal cells, reducing the energy requirements and environmental impacts of harvesting (Chen *et al.*, 2011). Similarly, advances in membrane technology, such as the development of forward osmosis and ultrafiltration systems, have enabled the high-throughput and low-energy separation of algal biomass from the cultivation medium (Bilad *et al.*, 2012; Lizzul *et al.*, 2014).

c) **Processing Technologies:** The conversion of marine green biomass into biofuels requires the development of efficient and cost-effective processing technologies, which can extract and convert the energy-rich compounds (such as lipids, carbohydrates, and proteins) into various types of fuels (Chisti, 2008; Mata *et al.*, 2010). Recent advances in this area include the optimization of traditional processing methods, such as transesterification for biodiesel production and fermentation for bioethanol production, as well as the development of novel techniques, such as hydrothermal liquefaction and gasification for biocrude and bio-syngas production, respectively (Dote *et al.*, 1994; Harun *et al.*, 2010; Jena *et al.*, 2011).

14.6 CASE STUDIES AND REAL-WORLD APPLICATIONS OF MARINE GREENS FOR BIOFUEL PRODUCTION

The potential of marine greens for biofuel production has prompted a variety of case studies and real-world applications, with both large-scale and small-scale projects being implemented to demonstrate the feasibility and benefits of this renewable energy source. This section will highlight some of these case studies, showcasing the diverse applications and successes of marine green-based biofuels in different regions and industries.

a) **Algenol Biofuels (United States):** Algenol, a Florida-based biotechnology company has developed an innovative, patented technology called the Direct to Ethanol® process, which uses photosynthetic microalgae to convert carbon dioxide directly into bioethanol

(Algenol Biofuels, 2016). This process is designed to be both energy-efficient and environmentally friendly, producing minimal waste and utilizing low-cost, non-arable land for algae cultivation. Algenol's integrated biorefinery in Fort Myers, Florida, has demonstrated the scalability and commercial viability of this technology, with the company aiming to expand its operations and further develop its algae-to-ethanol platform (Lane, 2014).

b) **Cellana (United States):** Cellana, a California-based company, is focused on the development of marine microalgae for the production of renewable fuels, feeds, and high-value products (Cellana, n.d.). The company operates a large-scale, outdoor algae cultivation facility in Hawaii, known as the Kona Demonstration Facility, which has produced over 10,000 gallons of algal oil since its inception (McNichol *et al.*, 2012). This facility uses proprietary, high-productivity algal strains to maximize biomass and lipid production, while utilizing saltwater and recycled CO_2 to minimize freshwater consumption and GHG emissions (Cellana, n.d.). Cellana has formed partnerships with industry leaders, such as Neste Oil and Living Ink Technologies, to develop and commercialize a range of algae-based products, including biofuels and bioproducts (Neste, 2013).

c) **Seaweed Energy Solutions (Norway):** SES, a Norwegian company, has developed an innovative seaweed cultivation system called the Seaweed Carrier, which is designed for the large-scale production of macroalgae for biofuels and other applications. The Seaweed Carrier is a flexible, modular system that can be deployed in a variety of ocean environments, allowing the efficient and cost-effective cultivation of different seaweed species (Buck *et al.*, 2016). SES has successfully demonstrated the scalability and productivity of this system through pilot projects in Norway and Portugal, with plans to expand its operations and commercialize its seaweed-based biofuel technology.

d) **AlgaePARC (Netherlands):** the Algae Production and Research Center (AlgaePARC) is a research facility in the Netherlands, which aims to develop and optimize microalgal cultivation and processing technologies for the production of biofuels and bioproducts. AlgaePARC utilizes a variety of innovative cultivation systems, including flat-panel photobioreactors, raceway ponds, and vertical bag systems, to study the growth and productivity of different microalgal strains under various environmental conditions (Wijffels & Barbosa, 2010). Through its research and demonstration activities, AlgaePARC seeks to overcome the technical and economic barriers to the large-scale production of microalgal biofuels, while promoting collaboration and knowledge exchange among researchers, industry partners, and policymakers.

e) **Bio Architecture Lab (Chile):** BAL, a biotechnology company based in Chile has developed a proprietary platform for the production of biofuels and bioproducts from macroalgae, specifically focusing on the abundant and fast-growing brown seaweed *Macrocystis pyrifera*. BAL's technology involves the enzymatic hydrolysis of the seaweed's complex carbohydrates, such as alginate, to produce fermentable sugars that can be converted into bioethanol and other value-added products (Enquist-Newman *et al.*, 2014). The company has successfully demonstrated the feasibility and scalability of this technology through pilot projects in Chile, with plans to commercialize its seaweed-based biofuel process and expand its operations to other regions with abundant macroalgae resources (Bio Architecture Lab, n.d.).

f) **Sapphire Energy (United States):** Sapphire Energy, a California-based company, has developed an innovative platform for the production of renewable crude oil, known as Green Crude, from photosynthetic microalgae (Chen *et al.*, 2011). The company operates a large-scale, integrated algal biorefinery in New Mexico, which utilizes open pond cultivation systems, proprietary algal strains, and advanced processing technologies to produce biofuels and bioproducts (Sapphire Energy, n.d.). Sapphire Energy's Green Crude has been successfully tested in various transportation applications, including on-road vehicles and commercial aircraft, demonstrating its potential as a drop-in replacement for conventional fossil fuels (Chen *et al.,* 2011).

g) **Solazyme (United States):** Solazyme, a California-based company, has developed a unique, heterotrophic algae-based platform for the production of renewable oils and other high-value products (Pyle *et al.*, 2008). Unlike most algal biofuel production methods, which rely on photosynthetic organisms, Solazyme's technology uses genetically engineered microalgae that can convert a wide range of sugar feedstocks into tailored lipids and other products through fermentation (Stephens *et al.*, 2010). The company has successfully demonstrated the commercial viability of its technology through partnerships with industry leaders, such as Chevron and United Airlines, and the construction of large-scale production facilities in the United States and Brazil.

There is a diverse range of technologies and applications being developed for marine green-based biofuel production, with varying degrees of success and commercialization. While some projects, such as Algenol Biofuels and Cellana, focus on the production of bioethanol or biodiesel from photosynthetic microalgae, others, like Seaweed Energy Solutions and Bio Architecture Lab, are exploring the potential of macroalgae as a feedstock for biofuels and bioproducts. Furthermore, companies like Solazyme have developed innovative, non-photosynthetic platforms for the production of tailored lipids and other products from engineered microalgae. Overall, these case studies highlight the significant progress and innovation occurring in the field of marine green-based biofuels, with a growing number of real-world applications and demonstration projects showcasing the feasibility, scalability, and benefits of this renewable energy source. As research and development in this area continue to advance, marine greens are poised to play an increasingly important role in the global transition towards more sustainable and renewable energy systems.

14.7 FUTURE PROSPECTS AND RESEARCH POTENTIAL OF MARINE GREENS FOR BIOFUEL PRODUCTION

Marine greens, including seaweed and microalgae, are a promising source for biofuel production due to their high productivity, abundant supply, and low environmental impact. These aquatic organisms are known for their ability to grow fast, have high lipid and carbohydrate content, and reduce GHG emissions by consuming carbon dioxide. In recent years, researchers have been exploring various ways to optimize the cultivation and processing of marine greens for biofuel production, as well as examining their potential to mitigate climate change and support sustainable development.

One area of research that has been gaining attention is the genetic engineering of marine greens to enhance their biofuel potential. Studies have shown that genetic modification can be used to increase lipid and carbohydrate production, improve stress tolerance and disease resistance, and create novel strains of marine greens for specific biofuel applications (Choi *et al.*, 2019). Researchers have also been investigating the use of gene editing technologies, such as CRISPR-Cas9, to precisely manipulate the genomes of marine greens and improve their traits for biofuel production (Kang *et al.*, 2021). However, the potential risks associated with the release of genetically modified marine greens into the environment still need to be carefully evaluated.

Another area of research is the development of sustainable and scalable cultivation systems for marine greens. Current cultivation methods, such as open ponds and photobioreactors, have limitations in terms of productivity, efficiency, and cost-effectiveness. Researchers have been exploring alternative cultivation systems, such as integrated multi-trophic aquaculture, which combines the cultivation of marine greens with the production of other aquatic organisms, such as fish and shellfish, to create a more sustainable and diversified food and energy system (Chopin *et al.*, 2020). Researchers have also been investigating the use of offshore cultivation systems, such as floating platforms and submerged structures, to expand the cultivation area and reduce the environmental impact of marine green cultivation (Gao *et al.*, 2021).

14.8 CONCLUSION

In conclusion, the potential of marine greens for biofuel production is vast and promising. These marine organisms offer several advantages over traditional terrestrial crops, including higher biomass production, faster growth rates, and the ability to grow in a wide range of environments. Furthermore, research has shown that marine greens are a rich source of lipids and carbohydrates, which can be converted into biofuels through various methods such as transesterification and hydrothermal liquefaction. However, significant challenges such as cultivation efficiency, nutrient availability, and processing costs need to be addressed to realize the full potential of marine greens for biofuel production.

REFERENCES

Adams, J. M. M., Gallagher, J. A., & Donnison, I. S. (2011). Fermentation study on Saccharina latissima for bioethanol production considering variable pre-treatments. *Journal of Applied Phycology*, 23(5), 877–886.

Algenol Biofuels. (2016). *Algenol Biofuels Fact Sheet*. Retrieved from http://www.algenol.com/sites/default/files/Algenol_Fact_Sheet_v1.pdf.

Angelidaki, I., Ellegaard, L., & Ahring, B. K. (2011). A comprehensive model of anaerobic bioconversion of complex substrates to biogas. *Biotechnology and Bioengineering*, 82(3), 291–300.

Balat, M. (2011). Potential alternatives to edible oils for biodiesel production – A review of current work. *Energy Conversion and Management*, 52(2), 1479–1492.

Beardall, J., & Raven, J. A. (2004). The potential effects of global climate change on microalgal photosynthesis, growth and ecology. *Phycologia*, 43(1), 26–40.

Bilad, M. R., Vandamme, D., Foubert, I., Muylaert, K., & Vankelecom, I. F. (2012). Harvesting microalgal biomass using submerged microfiltration membranes. *Bioresource Technology*, 111, 343–352.

Bridgwater, A. V. (2003). Renewable fuels and chemicals by thermal processing of biomass. *Chemical Engineering Journal*, 91(2–3), 87–102.

Buck, B. H., Troell, M., Krause, G., Angel, D., Grote, B., & Chopin, T. (2016). State of the art and challenges for offshore integrated multi-trophic aquaculture (IMTA). *Frontiers in Marine Science*, 3, 63.

Carvalho, A. P., Meireles, L. A., & Malcata, F. X. (2011). Microalgal reactors: A review of enclosed system designs and performances. *Biotechnology Progress*, 22(6), 1490–1506.

Cellana. (n.d.). *Algae: The Sustainable Source of Food, Feed, and Fuel*. Retrieved from https://www.cellana.com.

Chen, C. Y., Yeh, K. L., Aisyah, R., Lee, D. J., & Chang, J. S. (2011). Cultivation, photobioreactor design and harvesting of microalgae for biodiesel production: A critical review. *Bioresource Technology*, 102(1), 71–81.

Chisti, Y. (2007). Biodiesel from microalgae. *Biotechnology Advances*, 25(3), 294–306.

Chisti, Y. (2008). Biodiesel from microalgae beats bioethanol. *Trends in Biotechnology*, 26(3), 126–131.

Choi, S. A., Kim, S. H., Kwon, S. K., Lee, H. J., Kang, N. K., & Jeong, B. R. (2019). Enhancing biomass and lipid productivity of marine microalga Nannochloropsis oceanica IMET1 by metabolic engineering with the transcription factor Alg7. *Biotechnology and Bioengineering*, 116(1), 116–125.

Chopin, T., Robinson, S., Sawhney, M., Burridge, T., Ryan, J., Campbell, E., ... Vandermeulen, H. (2020). Multi-trophic integrated aquaculture and renewable energy systems for environmentally sustainable and socially acceptable food production. *Aquaculture*, 529, 735669.

Delrue, F., Setier, P. A., Sahut, C., Cournac, L., Roubaud, A., Pinton, A., & Isambert, A. (2016). An economic, sustainability, and energetic model of biodiesel production from microalgae. *Bioresource Technology*, 199, 710–719.

Demirbas, A. (2009). *Biofuels: Securing the Planet's Future Energy Needs*. Berlin: Springer.

Dismukes, G. C., Carrieri, D., Bennette, N., Ananyev, G. M., & Posewitz, M. C. (2008). Aquatic phototrophs: Efficient alternatives to land-based crops for biofuels. *Current Opinion in Biotechnology*, 19(3), 235–240.

Dote, Y., Sawayama, S., Inoue, S., Minowa, T., & Yokoyama, S. Y. (1994). Recovery of liquid fuel from hydrocarbon-rich microalgae by thermochemical liquefaction. *Fuel*, 73(12), 1855–1857.
Elliott, D. C., Biller, P., Ross, A. B., Schmidt, A. J., & Jones, S. B. (2015). Hydrothermal liquefaction of biomass: Developments from batch to continuous process. *Bioresource Technology*, 178, 147–156.
Enquist-Newman, M., Faust, A. M., Bravo, D. D., Santos, C. N., Raisner, R. M., Hanel, A., Sarvabhowman, P., Le, C., Regitsky, D. D., Cooper, S. R., Peereboom, L., Clark, A., Martinez, Y., Goldsmith, J., Cho, M. Y., Donohoue, P. D., Luo, L., Lamberson, B., Tamrakar, P., Kim, E. J., Villari, J. L., Gill, A., Tripathi, S. A., Karamchedu, P., Paredes, C. J., Rajgarhia, V., Kotlar, H. K., Bailey, R. B., Miller, D. J., Ohler, N. L., Swimmer, C., & Tremaine, M. (2014). Efficient ethanol production from brown macroalgae sugars by a synthetic yeast platform. *Nature*, 505(7482), 239–243.
Enzing, C., Ploeg, M., Barbosa, M., & Sijtsma, L. (2014). Microalgae-based products for the food and feed sector: An outlook for Europe. *JRC Scientific and Policy Reports*, EUR 26255 EN.
Gao, Y., Wang, L., Li, M., Zhang, X., Sun, X., & Xie, L. (2021). Offshore cultivation of macroalgae for bioenergy and bioremediation: Opportunities and challenges. *Renewable Energy*, 177, 1187–1198.
Garg, S., Li, Y., Wang, L., & Schenk, P. M. (2012). Flotation of marine microalgae: Effect of algal hydrophobicity. *Bioresource Technology*, 121, 471–474.
Grima, E. M., Belarbi, E. H., Fernández, F. G., Medina, A. R., & Chisti, Y. (2003). Recovery of microalgal biomass and metabolites: Process options and economics. *Biotechnology Advances*, 20(7–8), 491–515.
Harun, R., Singh, M., Forde, G. M., & Danquah, M. K. (2010). Bioprocess engineering of microalgae to produce a variety of consumer products. *Renewable and Sustainable Energy Reviews*, 14(3), 1037–1047.
Horn, S. J., Aasen, I. M., & Østgaard, K. (2014). Production of ethanol from mannitol by Zymobacter palmae. *Journal of Industrial Microbiology and Biotechnology*, 31(7), 377–384.
Hossain, A. B., Salleh, A., Boyce, A. N., Chowdhury, P., & Naqiuddin, M. (2008). Biodiesel fuel production from algae as renewable energy. *American Journal of Biochemistry and Biotechnology*, 4(3), 250–254.
Huber, G. W., Iborra, S., & Corma, A. (2006). Synthesis of transportation fuels from biomass: Chemistry, catalysts, and engineering. *Chemical Reviews*, 106(9), 4044–4098.
International Energy Agency [IEA]. (2020). Global energy review 2020. Retrieved from https://www.iea.org/reports/global-energy-review-2020.
Jena, U., Das, K. C., & Kastner, J. R. (2011). Effect of operating conditions of thermochemical liquefaction on biocrude production from Spirulina platensis. *Bioresource Technology*, 102(10), 6221–6229.
Kang, N. K., Jeong, B. R., Kwon, S. K., Kim, S. H., Lee, B., Kim, J. H., … Choi, H. I. (2021). Gene editing of microalgae: Scientific progress and future perspectives. *Algal Research*, 54, 102210.
Lane, J. (2014). Algenol biofuels: The digest's 2015 5-minute guide. *Biofuels Digest*. Retrieved from https://www.biofuelsdigest.com/bdigest/2014/12/28/algenol-biofuels-the-digests-2015-5-minute-guide/.
Lee, A. K., Lewis, D. M., & Ashman, P. J. (2012). Disruption of microalgal cells for the extraction of lipids for biofuels: Processes and specific energy requirements. *Biomass and Bioenergy*, 46, 89–101.
Lehmann, J., & Joseph, S. (2009). *Biochar for Environmental Management: Science and Technology*. London: Earthscan.
Lizzul, A. M., Hellier, P., Purton, S., Baganz, F., Ladommatos, N., & Campos, L. (2014). An integrated continuous microalgae recovery and fractionation process. *Algal Research*, 6, 271–277.
Ma, F., & Hanna, M. A. (1999). Biodiesel production: A review. *Bioresource Technology*, 70(1), 1–15.
Mann, K. H., & Steinke, M. (2012). Marine algae: A source of biomass for biotechnological applications. In Chisti, Y. (Ed.), *Biomass for Renewable Energy, Fuels, and Chemicals* (pp. 367–387). Academic Press, United States.
Markou, G., & Nerantzis, E. (2013). Microalgae for high-value compounds and biofuels production: A review with focus on cultivation under stress conditions. *Biotechnology Advances*, 31(8), 1532–1542.
Mata, T. M., Martins, A. A., & Caetano, N. S. (2010). Microalgae for biodiesel production and other applications: A review. *Renewable and Sustainable Energy Reviews*, 14(1), 217–232.
McKendry, P. (2002). Energy production from biomass (part 1): Overview of biomass. *Bioresource Technology*, 83(1), 37–46.
McNichol, J., MacDougall, K. M., Melanson, J. E., & McGinn, P. J. (2012). Suitability of Soxhlet extraction to quantify microalgal fatty acids as determined by comparison with in situ transesterification. *Lipids*, 47(2), 195–207.
Milledge, J. J., & Heaven, S. (2013). A review of the harvesting of micro-algae for biofuel production. *Reviews in Environmental Science and Bio/Technology*, 12(2), 165–178.
Neste. (2013). Neste Oil and Cellana to cooperate on developing algae oil for use as a feedstock for NExBTL renewable diesel. Retrieved from https://www.neste.com/en/neste-oil-and-cellana-cooperate-developing-algae-oil-use-feedstock-nexbtl-renewable-diesel.

Perlack, R. D., & Stokes, B. J. (2011). *U.S. Billion-Ton Update: Biomass Supply for a Bioenergy and Bioproducts Industry*. Oak Ridge: Oak Ridge National Laboratory.
Pires, J. C., Alvim-Ferraz, M. C., Martins, F. G., & Simões, M. (2011). Carbon dioxide capture from flue gases using microalgae: Engineering aspects and biorefinery concept. *Renewable and Sustainable Energy Reviews*, 16(5), 3043–3053.
Posten, C. (2009). Design principles of photo-bioreactors for cultivation of microalgae. *Engineering in Life Sciences*, 9(3), 165–177.
Potts, T., Du, J., Paul, M., May, P., Beitle, R., & Hestekin, J. (2012). The production of butanol from Jamaica Bay macro algae. *Environmental Progress and Sustainable Energy*, 31(1), 29–36.
Pyle, D. J., Garcia, R. A., & Wen, Z. (2008). Producing docosahexaenoic acid (DHA)-rich algae from biodiesel-derived crude glycerol: Effects of impurities on DHA production and algal biomass composition. *Journal of Agricultural and Food Chemistry*, 56(11), 3933–3939.
Radakovits, R., Jinkerson, R. E., Darzins, A., & Posewitz, M. C. (2010). Genetic engineering of algae for enhanced biofuel production. *Eukaryotic Cell*, 9(4), 486–501.
Ritchie, H., & Roser, M. (2021). CO2 and greenhouse gas emissions. Our world in data. Retrieved from https://ourworldindata.org/co2-and-other-greenhouse-gas-emissions.
Roesijadi, G., Jones, S. B., Snowden-Swan, L. J., & Zhu, Y. (2010). Macroalgae as a biomass feedstock: A preliminary analysis. *Pacific Northwest National Laboratory, PNNL*, 19944.
Sims, R. E. H., Mabee, W., Saddler, J. N., & Taylor, M. (2010). An overview of second-generation biofuel technologies. *Bioresource Technology*, 101(6), 1570–1580.
Singh, A., & Olsen, S. I. (2011). A critical review of biochemical conversion, sustainability, and life cycle assessment of algal biofuels. *Applied Energy*, 88(10), 3548–3555.
Soccol, C. R., Vandenberghe, L. P. S., Medeiros, A. B. P., Karp, S. G., Buckeridge, M., Ramos, L. P., Pitarelo, A. P., Ferreira-Leitão, V., Gottschalk, L. M., Ferrara, M. A., da Silva Bon, E. P., de Moraes, L. M., Araújo, J. A., & Torres, F. A. (2010). Bioethanol from lignocelluloses: Status and perspectives in Brazil. *Bioresource Technology*, 101(13), 4820–4825.
Stephens, E., Ross, I. L., King, Z., Mussgnug, J. H., Kruse, O., Posten, C., Borowitzka, M. A., & Hankamer, B. (2010). An economic and technical evaluation of microalgal biofuels. *Nature Biotechnology*, 28(2), 126–128.
Tilman, D., Socolow, R., Foley, J. A., Hill, J., Larson, E., Lynd, L., Pacala, S., Reilly, J., Searchinger, T., Somerville, C., & Williams, R. (2009). Beneficial biofuels—The food, energy, and environment trilemma. *Science*, 325(5938), 270–271.
Toor, S. S., Rosendahl, L., & Rudolf, A. (2011). Hydrothermal liquefaction of biomass: A review of subcritical water technologies. *Energy*, 36(5), 2328–2342.
Wijffels, R. H., & Barbosa, M. J. (2010). An outlook on microalgal biofuels. *Science*, 329(5993), 796–799.
Xin, L., Hong-ying, H., & Ke, G. (2017). Effects of different nitrogen and phosphorus concentrations on the growth, nutrient uptake, and lipid accumulation of a freshwater microalga Scenedesmus sp. *Bioresource Technology*, 98(8), 1507–1513.

Section IV

Biomedical Applications

15 Potential Therapeutic Role of Marine Greens for Liver Diseases

Thiyagarajan Gopal and Madan Kumar Arumugam

15.1 INTRODUCTION

Liver diseases are rapidly emerging as global health priorities (Mondal, Das, and Chowdhury 2022). The burden of liver disease in India and other developing countries is significant, because it alone contributed to 18.3% of the two million global liver disease–related deaths in recent years (Mondal, Das, and Chowdhury 2022). Major contributory causes include increasingly urbanized lifestyles and dietary changes involving high caloric contents. The overall prevalence of non-alcoholic fatty liver disease (NAFLD) and/or metabolic associated liver diseases has increased dramatically (Paschos and Paletas 2009; Qureshi and Abrams 2007); these metabolic associated liver diseases are most probably linked to the increased uptake of a high caloric diet and poor nutritional status, which leads to very high release of free fatty acids, abnormal secretion of factors produced by the adipose tissue, and low-grade chronic inflammation (Anty et al. 2007; Lomonaco et al. 2012; Milic, Lulic, and Stimac 2014). Non-alcoholic steatohepatitis (NASH) has become one of the leading causes of liver transplantation in the United States and is predicted to become the leading indication for liver transplantation in the near future (Wong, Cheung, and Ahmed 2014). NAFLD and NASH are associated with metabolic syndrome and the development of cardiovascular and kidney diseases (Yu et al. 2023; Pipitone et al. 2023; Ahmed et al. 2022). Alcoholic liver disease (ALD) is caused by heavy alcohol intake (Bruha, Dvorak, and Petrtyl 2012; Rehm and Shield 2013; Roerecke et al. 2019). As for liver injury, it has been postulated for many years that the type of alcoholic beverage makes little, if any, difference. Nevertheless, some authors have proposed that mortality from cirrhosis is associated more strongly with the consumption of spirits than with other alcoholic beverages (Wong, Cheung, and Ahmed 2014). Almost 50% of cirrhosis-related deaths are due to excessive alcohol consumption (Roerecke et al. 2019). Another common cause for chronic hepatitis worldwide is Hepatitis B virus (HBV) infection, which remains the primary cause of cirrhosis and hepatocellular carcinoma in most Asian nations (MacLachlan and Cowie 2015). Hepatitis C virus (HCV) is also related to the development of liver cirrhosis, and more than 71 million people are infected worldwide (Lazarus, Roel, and Elsharkawy 2020). HBV, HCV, NAFLD, and ALD are significant primary risk factors for the development of hepatocellular carcinoma (HCC), one of the most common malignant tumors worldwide (Gomaa et al. 2008). This chapter discusses the potential therapeutic applications of marine greens for studying liver diseases.

15.2 THERAPEUTIC BIOACTIVE MOLECULES FROM MARINE GREENS

Bioactive products from marine sources harbor a diversity of pharmacologically active metabolites, and marine greens are an attractive chief source for a novel drug candidate. Natural products such as vegetables, fruits, and traditional herbs have been developed as potential treatments for alcohol use disorders, based on the findings that they promote the elimination of alcohol and

DOI: 10.1201/9781003369738-19

alcohol metabolites, inhibit inflammatory mediators, and stimulate the expression of antioxidant enzymes (Wang et al. 2016). Many bioactive compounds isolated from marine greens, with antibacterial, antioxidant and anti-inflammatory, antidiabetic, anticancer, hepatoprotective, and neurologically protective properties, have been reported to be effective in alleviating chronic diseases (Nair, Weiskirchen, and Al-Musharafi 2015).

15.2.1 Antimicrobial Activity

Recently, secondary metabolites from marine-derived microbes have received more attention, as most of the metabolites exhibit multi-targeted pharmacological activities. Among these metabolites, a meleagrin alkaloid was isolated from deep sediment–derived *Penicillium chrysogenum* Strain S003 and has been reported to exert antibacterial action (Zheng et al. 2013). Hippolachnin A, a polyketide from the South China sea sponge *Hippospongia lachne*, was found to exhibit potent antifungal activity against three pathogenic fungi, *Trichophyton rubrum*, *Cryptococcus neoformans*, and *Microsporum gypseum* (Piao et al. 2013).

15.2.2 Anti-Inflammatory Activity

Oxidative stress is tightly connected with inflammation, which is evaluated through the nuclear factor erythroid 2-related factor 2 (Nrf2)/heme oxygenase-1 (OH-1) signaling pathway. Dysiarenone from *Dysidea arenaria* suppressed reactive oxygen species (ROS) production and inflammation through attenuation of 5-LOX/NF-κB/MAPKs and upregulation of Nrf-2/OH-1 expression in RAW 264.7 Macrophage cells. Unusual diterpenoids, Hipposponlachnins A and B, isolated from the marine sponge *Hippospongia lachne* showed anti-allergic properties by downregulating the release of β-hexosaminidase, a biomarker for degranulation and the stimulation of interleukin 4 (IL-4), a pro-inflammatory cytokine and lipid mediator of leukotrienes B_4 (LTB4), in dinitrophenyl (DNP)-IgE-induced RBL-2H3 mast cells. Jiao et al. demonstrated that inhibition of the release of β-hexosaminidase, IL-4, and LTB4 in mast cells was mediated by attenuating the phosphorylation of Syk and PLCγ1, involved in the IgE/FcεRI/Syk pathway (Jiao et al. 2017).

The anti-allergic effect of the polyketide phlorotannins 6,8′- and 8,8′-bieckol, isolated from the Japanese brown alga *Eisenia arborea*, was found to inhibit mouse ear swelling through the suppression of inflammatory mediators (histamine, leukotriene B4, and prostaglandin E2 [PGE2]) and mRNA-level expression of cyclooxygenase-2 (COX-2) in mice (Sugiura et al. 2018). In a similar study, Jiao et al. reported that the dimeric C21 meroterpenoid dysiarenone from the marine sponge *Dysidea arenaria* inhibited lipopolysaccharide (LPS)-stimulated macrophage production of PGE2 and COX-2 in the RAW264.7 cell line (Jiao et al. 2018). The shikimate 4-(hydroxymethyl) catechol, isolated from the marine sponge–derived fungus *Pestalotiopsis* sp., has been found to attenuate rheumatoid arthritis in tumor necrosis factor (TNF)-α-induced human synovial fibroblasts and the murine collagen-induced arthritis model by altering the PI3K/Akt/NF-κB signaling pathway (Lee et al. 2018). Isolated fatty acid monoolein from the marine brown alga *Ishige sinicola* inhibited LPS-induced cytokine production and nitric oxide release from both RAW264.7 macrophages and primary murine bone marrow–derived dendritic cells by targeting mitogen-activated protein kinase (MAPK) and NF-κB signaling (Ali et al. 2017). Further, Choi et al. isolated a novel bromophenol bis (3-bromo-4,5-dihydroxybenzyl) ether from the marine red alga *Polysiphonia morrowii*, which was found to attenuate LPS-induced inflammation in RAW 264.7 macrophages by inhibiting upregulation of the ERK signaling pathway (Choi et al. 2018). Also, scalarane-type sesterterpenes, namely, scalarinol, heteronemin, and 12-O-deacetyl-19-deoxyscalarin, isolated from the sponge *Hyrtios erectus* in the Saudi Red Sea demonstrated cytotoxic activities in cancer cell lines (Alarif et al. 2016). The 24-homoscalaranes derived from the marine sponge *Lendenfeldia* sp. were identified as potent anti-inflammatory agents (Peng et al. 2020). The novel bioactive molecule popolohuanone H, a dimeric sesquiterpene quinone from the Chinese marine sponge *Dactylospongia elegans*, was reported to exhibit anti-inflammatory activity, but the mechanism of action was not determined (Li et al. 2018).

15.2.3 Anticancer Activity

A bioactive extract from *Aspergillus terreus*, a symbiotic fungus derived from marine sponges and sediment collected off Obhur, Jeddah (Saudi Arabia), was found to inhibit the proliferation of human carcinoma cells, including colorectal carcinoma cells (HCT-116) and hepatocellular carcinoma cells (HepG2) (Asfour et al. 2019). Recently, De et al. demonstrated that the therapeutic potential of 2-chloro-6-phenyl-8H-quinazolino[4,3-b]quinazolin-8-one, a natural quinazoline derivative, isolated from the marine sponge *Hyrtios erectus* was mediated by the induction of apoptosis through ROS generation and activation of key markers involved in the extrinsic and intrinsic apoptosis pathways in breast cancer cells (Asfour et al. 2019). A meleagrin alkaloid was isolated from deep sediment–derived *Penicillium chrysogenum* Strain S003 and has been reported to show cytotoxic activities against different types of human cancer cell lines, such as the breast cancer cell line MCF-7, the lung cancer cell line A-549, the human leukemia cell line HL-60, the cervical cell line BEL-7402, and the T-lymphoblast cell line MOLT-4 (Mady et al. 2016). Bleomycin-induced pulmonary fibrosis was attenuated by meleagrin. Meleagrin also activated the Nrf2/HO-1 antioxidant signaling pathway and inhibited TLR4 and NF-κB gene expression, with a subsequent decreased release of pro-inflammatory cytokines (TNF-α, IL-6, and interferon [IFN]-γ). Additionally, meleagrin inhibited bleomycin-induced apoptosis by abating the activities of the pro-apoptotic proteins Bax and caspase-3 while elevating Bcl2 in mice against bleomycin-induced lung fibrosis (Elhady et al. 2022). Manoalide-type sesterterpenes from the sponge *Luffariella variabilis* in the South China Sea showed antiproliferative activities against different human cancer cell lines (Luo et al. 2021). Hippolides A, a cyclic manoalide derivative from the marine sponge *Hippospongia lachne*, showed cytotoxicity against A549, HeLa, and HCT-116 cell lines (Piao et al. 2011). BA6 (heteronemin), the marine sponge sesterterpene, displayed apoptosis in lung cancer cells through the modulation of mitochondrial reactive oxygen species and oxidative phosphorylation. Park et al. showed that the induction of mitochondrial apoptosis by (Z)-stellettic acid C, an acetylenic acid from the sponge *Stelletta* sp., was accompanied by inhibitory activity of telomerase in human leukemic cells (U937) (Park et al. 2007). Similarly, an unsaturated fatty acid, hydroxylated C18 (±)-(*E*)-12-hydroxyoctadec-10-enoic acid (HOEA), isolated from the red alga *Tricleocarpa jejuensis* exhibited cytotoxic effects as well as induction of apoptosis in human leukemia cells (33801204). Neopetrosidines A–D, pyridine alkaloids from the marine sponge *Neopetrosia chaliniformis*, have been shown to inhibit cell cycle progression of HeLa/fluorescent ubiquitination-based cell cycle indicator (Fucci)2 cells by reducing the mitochondrial membrane potential (34649068). Sesquiterpene quinone, neoisosmenospongine, nakijiquinone I, nakijiquinone N, and nakijiquinone Q isolated from *Dactylospongia metachromia* were observed to attenuate the cell cycle progression at S/G2/M phases in HeLa/Fucci2 cells (Hitora et al. 2021). Similarly, Metachromins X and Y isolated from *Spongia* sp. arrested cell cycle progression at S/G2/M phases in HeLa/Fucci2 cells (Hitora et al. 2020).

15.2.4 Antidiabetic Activity

Park et al. reported that a minor phlorotannin derivative, 8,8′-bieckol, isolated from the Korean brown marine alga *Ecklonia cava* competitively inhibited α-glucosidase, which is a molecular target for the treatment of type 2 diabetes (Park et al. 2018). Admassu et al. identified two bioactive peptides, Gly-Gly-Ser-Lys and Glu-Leu-Ser, from the red seaweed *Porphyra* spp., which exhibited α-amylase inhibitory activity (Admassu et al. 2018). Fernando et al. described the effect of diphlorethohydroxycarmalol, isolated from the South Korean edible brown alga *Ishige okamurae*, which suppressed high glucose–stimulated angiogenesis in human vascular endothelial cells and in zebrafish embryos via suppression of the vascular endothelial growth factor receptor 2 (VEGFR-2) signaling mechanism (Fernando et al. 2018). Protein tyrosine phosphatase-1B (PTP1B) is a negative regulator of insulin and leptin signaling and is a key target being extensively studied for its role in diabetes and obesity. Experimental studies have strongly implicated that identification of a PTP1B inhibitor is vital for modulation of the insulin and leptin signaling cascade (Fernando et al. 2018). A study by Han et al. highlighted the in vitro inhibition profile of new compounds isolated from the marine fungus *P. chrysogenum*, chrysopyrones A and B, on human cytoplasmic PTP1B (Han et al. 2020). Five new sesterterpenoids isolated from the sponge *Hippospongia lachne* off Yongxing island in the South China Sea displayed PTP1B inhibitory activities (Piao et al. 2014). Hippolides A, a cyclic manoalide derivative from the marine sponge *Hippospongia lachne*, showed moderate PTP1B inhibitory activity (Piao et al. 2011). These potential drug candidates from marine greens could be used as novel therapeutics for the management of diabetes mellitus.

15.2.5 Neurological Diseases

A marine edible polyphenol, triphlorethol A, isolated from the brown seaweed *Ecklonia cava* has been shown to reduce sleep latency and induce augmented sleep duration with increased non–rapid eye movement sleep in mice (Yoon and Cho 2018). A novel oxysterol isolated from the marine sponge *Inflatella* sp. was found to increase viability in a Parkinson's disease model of 6-hydroxydopamine-stimulated Neuro2a cells (Kolesnikova et al. 2018). To identify potential therapeutic applications for the early phase of Alzheimer's disease and age-related cognitive decline, an ethanolic extract of the bioactive compound arachidonic acid from the red alga *Gracilariopsis chorda* was investigated and found to induce synaptogenesis in a rat hippocampal neuron cell line; it protects against scopolamine-induced memory impairment in mice by increasing the expression of cell division control protein 42 (CDC42) and actin-related protein 2 (ARP2) (Mohibbullah et al. 2018). Astaxanthin, a xanthophyll carotenoid isolated from the marine alga *Haematococcus pluvialis*, was found to improve neuropathic pain in a chronic constriction injury rat model. A study using C6 cells treated with astaxanthin exhibited a reduction in the levels of nitrate, oxidative stress, and activation of astrocytes, suggesting the nociceptive and neuroprotective therapeutic potential of astaxanthin (Sharma et al. 2018). Fucosterol, a novel marine metabolite isolated by Oh et al. from the dietary brown alga *Ecklonia stolonifera*, has been shown to protect aged rats from amyloid beta–induced age-related cognitive impairment via upregulation of mature brain-derived neurotrophic factor (BDNF)-TrkB-ERK1/2 signaling and downregulation of GRP78 expression (Oh, Choi, and Nam 2018). Furthermore, the same group have identified a phycoerythrin-derived tryptic peptide from the red alga *Pyropia yezoensis* and found that it suppresses glutamate-induced endoplasmic reticulum stress and neuronal senescence in primary rat hippocampus through the TrkB receptor-associated ERK1/2 signaling pathway (Oh, Kim, and Nam 2018). Together, these data suggest that metabolites isolated from marine greens have therapeutic potential in neurodegenerative disorders.

15.3 HEPATOPROTECTIVE EFFECT OF MARINE GREENS

At present, for the management of chronic liver diseases, there are very limited targeted pharmacotherapeutics for NASH and no pharmacotherapeutics for AALD. Moreover, antioxidant supplementation is being used for the liver diseases to ameliorate the progression of liver function and reverse fibrosis. Here, we list the hepatoprotective drugs extracted and isolated from marine greens for the management of toxicity, including drug, alcoholic, and non-alcoholic liver diseases.

15.3.1 NAFLD and Hepatoprotective Activities

Recently, Zhang et al. identified three compounds, the indole-C-mannopyranoside alkaloids neopetrosin A, B, and D, isolated from the South China Sea marine sponge *Neopetrosia chaliniformis*, which exhibited hepatoprotective potential in a zebrafish model. Siphonaxanthin (SPX), a rare marine carotenoid from the green alga *Codium cylindricum* Holmes, was found to inhibit aggravated hepatic lipogenesis in HepG2 cells by suppressing activation of Liver X receptor alpha (LXRα) via downregulating SRBP1c, fatty acid translocase (CD36), and fatty acid-binding protein-1, making it a promising candidate in the prevention and alleviation of NAFLD (Zheng et al. 2018). An extract from the brown alga *Padina boergesenii* showed hepatoprotective activity against carbon tetrachloride (CCl_4)-induced oxidative damage and liver fibrosis in Wistar rats. The extract stabilized cellular membranes by establishing antiperoxidase activity. Similarly, a study by Makhmoor et al. showed the hepatoprotective roles of chemical constituents from the marine brown alga *Spatoglossum variabile* in CCl_4-induced hepatotoxicity in Wistar rats (Makhmoor et al. 2013). Additionally, administration of *Spirulina platensis* to CCl_4-challenged mice showed protection against hepatotoxicity and DNA damage with enhanced antioxidant potential. Hong et al. isolated fucoidan, a sulfated polysaccharide extracted from brown seaweeds, and investigated its effect against acetaminophen-induced liver damage. Fucoidan was found to suppress the expression of CYP2E1 and reduced inflammatory markers including TNFα, interleukin 1β, and inducible nitric oxide synthase (iNOS). In addition, fucoidan attenuated acetaminophen-induced hepatic apoptosis by modulating markers of cell death such as Bax, Bcl-2, and cleaved caspase-3 in rats (Hong et al. 2012). Similarly, acetaminophen-induced hepatotoxicity was prevented by hexane- and methanol-soluble fractions of the brown seaweed *Sargassum ilicifolium*, which showed enhanced hepatic antioxidant activity through increasing hepatic glutathione (GSH) and reversing lipid peroxidation. Sulfated polysaccharides isolated from *Sargassum swartzii* (Turn.) protected against acetaminophen-induced liver toxicity in rats by increasing GSH while reducing levels of lipid peroxidation. An extract of the marine-derived fungus *Trichurus spiralis* Hasselbr (TS) isolated from the sponge *Hippospongia communis* was shown to have hepatoprotective potential against liver toxicity induced by heavy metals in rats (Abdel-Monem et al. 2013). Similarly, investigation of a crude extract of *Hyrtios aff. Erectus* sponge has shown that it plays a hepatoprotective role against liver toxicity induced by a mixture of persistent organic pollutants (POPs) in BALB/c mouse liver by significantly increasing the levels of hepatic GSH and glutathione S-transferase (Abd El-Moneam et al. 2017).

Alcohol-induced oxidative stress is known to be an important component in the pathogenesis of AALD (Cederbaum, Lu, and Wu 2009). As noted earlier, alcohol metabolism produces high amounts of ROS, which in turn, affect the antioxidant systems and increases the level of oxidative stress in several organs, particularly in liver (Cederbaum, Lu, and Wu 2009; Huang et al. 2009; Tsermpini, Plemenitas Iljes, and Dolzan 2022). Marine greens have been reported to protect against liver injury (Zheng et al. 2019; Ge et al. 2018; Nair, Weiskirchen, and Al-Musharafi 2015). The administration of fucoxanthin, isolated from marine seaweed, was found to enhance antioxidant capacity by increasing glutathione peroxidase, superoxide dismutase, and catalase levels in the liver of mice fed on alcohol. Also, fucoxanthin suppressed alcohol-induced oxidative stress and alcoholic inflammatory responses by downregulating TLR4-activated NF-κ B expression and the Nrf2-mediated signaling pathway (Zheng et al. 2019). Ge et al. isolated Aplysin, a marine bromosesquiterpene, from the red alga *Laurencia tristicha* and found that it protected the liver against alcohol-induced oxidative injury by restoring the antioxidant system and attenuating hepatocellular DNA damage (Ge et al. 2018). Overexpression of cytoplasmic cytochrome c and caspase-3 activation were attenuated by Aplysin (Ge et al. 2013). Aplysin downregulated CYP2E1 expression and inhibited mitochondrial damage–associated apoptosis genes such as Bax and cytochrome c while upregulating Bcl-2, caspase-9, and caspase-3 in ethanol-induced rats (Ge et al. 2013, 2018). Also, Aplysin administration to chronic alcohol-exposed rats protected against alcohol-disturbed gut microbiota (Xue et al. 2017).

15.4 CONCLUSION

In the past year, intense research on marine greens as a resource has brought to light new prospects for developing marine-based drugs for various diseases. Liver diseases are among the life-threatening conditions that develop from poor lifestyle habits, poor health, low nutritional intake, high calorie intake, and alcohol consumption. Treatments for chronic liver diseases, such as protective treatment for liver injury, antifibrotic treatment, and treatment of decompensated complications of liver cirrhosis, are all effective and of great significance in preventing the progression of liver disease, maintaining liver function, reducing complications of portal hypertension, and preventing liver cancer. Potential pharmaceutical drugs and compounds from marine natural products have shown great effects on health-care sector products and medicinal values. This discovery of potential therapeutics for liver diseases from marine greens and in-depth discussions of their therapeutic mechanisms and applications will continue to expand and ultimately benefit more liver disease patients and other humans.

REFERENCES

Abd El-Moneam, N. M., M. A. Shreadah, S. A. El-Assar, and A. Nabil-Adam. 2017. "Protective role of antioxidants capacity of Hyrtios aff. Erectus sponge extract against mixture of persistent organic pollutants (POPs)-induced hepatic toxicity in mice liver: Biomarkers and ultrastructural study." *Environ Sci Pollut Res Int* 24(27):22061–72. doi: 10.1007/s11356-017-9805-8.

Abdel-Monem, N. M., A. M. Abdel-Azeem, S. H. el El-Ashry, D. A. Ghareeb, and A. Nabil-adam. 2013. "Pretreatment hepatoprotective effect of the marine fungus derived from sponge on hepatic toxicity induced by heavy metals in rats." *BioMed Res Int* 2013:510879. doi: 10.1155/2013/510879.

Admassu, H., M. A. A. Gasmalla, R. Yang, and W. Zhao. 2018. "Identification of bioactive peptides with alpha-amylase inhibitory potential from enzymatic protein hydrolysates of red seaweed (Porphyra spp)." *J Agric Food Chem* 66(19):4872–82. doi: 10.1021/acs.jafc.8b00960.

Ahmed, E. A., M. O. El-Derany, A. M. Anwar, E. M. Saied, and S. Magdeldin. 2022. "Metabolomics and lipidomics screening reveal reprogrammed signaling pathways toward cancer development in non-alcoholic steatohepatitis." *Int J Mol Sci* 24(1). doi: 10.3390/ijms24010210.

Alarif, W. M., S. S. Al-Lihaibi, M. A. Ghandourah, M. I. Orif, S. A. Basaif, and S. E. Ayyad. 2016. "Cytotoxic scalarane-type sesterterpenes from the Saudi Red Sea sponge Hyrtios erectus." *J Asian Nat Prod Res* 18(6):611–7. doi: 10.1080/10286020.2015.1115019.

Ali, I., Z. Manzoor, J. E. Koo, S. R. Moon, S. H. Byeon, E. S. Yoo, H. K. Kang, J. W. Hyun, N. H. Lee, and Y. S. Koh. 2017. "Monoolein, isolated from Ishige sinicola, inhibits lipopolysaccharide-induced inflammatory response by attenuating mitogen-activated protein kinase and NF-kappaB pathways." *Food Sci Biotechnol* 26(2):507–11. doi: 10.1007/s10068-017-0070-x.

Anty, R., P. Gual, P. M. Huet, Y. L. Marchand-Brustel, and A. Tran. 2007. "Metabolic fatty liver diseases: Hepatic consequences of the metabolic syndrome." *Gastroenterol Clin Biol* 31(12):1127–34. doi: 10.1016/s0399-8320(07)78350-1.

Asfour, H. Z., Z. A. Awan, A. A. Bagalagel, M. A. Elfaky, R. F. A. Abdelhameed, and S. S. Elhady. 2019. "Large-scale production of bioactive terrein by Aspergillus terreus Strain S020 isolated from the Saudi coast of the Red Sea." *Biomolecules* 9(9). doi: 10.3390/biom9090480.

Bruha, R., K. Dvorak, and J. Petrtyl. 2012. "Alcoholic liver disease." *World J Hepatol* 4(3):81–90. doi: 10.4254/wjh.v4.i3.81.

Cederbaum, A. I., Y. Lu, and D. Wu. 2009. "Role of oxidative stress in alcohol-induced liver injury." *Arch Toxicol* 83(6):519–48. doi: 10.1007/s00204-009-0432-0.

Choi, Y. K., B. R. Ye, E. A. Kim, J. Kim, M. S. Kim, W. W. Lee, G. N. Ahn, N. Kang, W. K. Jung, and S. J. Heo. 2018. "Bis (3-bromo-4,5-dihydroxybenzyl) ether, a novel bromophenol from the marine red alga Polysiphonia morrowii that suppresses LPS-induced inflammatory response by inhibiting ROS-mediated ERK signaling pathway in RAW 264.7 macrophages." *Biomed Pharmacother* 103:1170–7. doi: 10.1016/j.biopha.2018.04.121.

Elhady, S. S., M. S. Goda, E. T. Mehanna, M. A. Elfaky, A. E. Koshak, A. O. Noor, H. A. Bogari, R. T. Malatani, R. F. A. Abdelhameed, and A. S. Wahba. 2022. "Meleagrin isolated from the Red Sea fungus Penicillium chrysogenum protects against bleomycin-induced pulmonary fibrosis in mice." *Biomedicines* 10(5). doi: 10.3390/biomedicines10051164.

Fernando, K. H. N., H. W. Yang, Y. Jiang, Y. J. Jeon, and B. Ryu. 2018. "Diphlorethohydroxycarmalol isolated from Ishige okamurae represses high glucose-induced angiogenesis in vitro and in vivo." *Mar Drugs* 16(10). doi: 10.3390/md16100375.

Ge, N., H. Liang, Y. Liu, A. G. Ma, and L. Han. 2013. "Protective effect of Aplysin on hepatic injury in ethanol-treated rats." *Food Chem Toxicol* 62:361–72. doi: 10.1016/j.fct.2013.08.071.

Ge, N., H. Liang, Y. Y. Zhao, Y. Liu, A. J. Gong, and W. L. Zhang. 2018. "Aplysin protects against alcohol-induced liver injury via alleviating oxidative damage and modulating endogenous apoptosis-related genes expression in rats." *J Food Sci* 83(10):2612–21. doi: 10.1111/1750-3841.14320.

Gomaa, A. I., S. A. Khan, M. B. Toledano, I. Waked, and S. D. Taylor-Robinson. 2008. "Hepatocellular carcinoma: Epidemiology, risk factors and pathogenesis." *World J Gastroenterol* 14(27):4300–8. doi: 10.3748/wjg.14.4300.

Han, W., J. Cai, W. Zhong, G. Xu, F. Wang, X. Tian, X. Zhou, Q. Liu, Y. Liu, and J. Wang. 2020. "Protein tyrosine phosphatase 1B (PTP1B) inhibitorsfrom the deep-sea fungus Penicillium chrysogenum SCSIO 07007." *Bioorg Chem* 96:103646. doi: 10.1016/j.bioorg.2020.103646.

Hitora, Y., A. Sejiyama, K. Honda, Y. Ise, F. Losung, R. E. P. Mangindaan, and S. Tsukamoto. 2021. "Fluorescent image-based high-content screening of extracts of natural resources for cell cycle inhibitors and identification of a new sesquiterpene quinone from the sponge, Dactylospongia metachromia." *Bioorg Med Chem* 31:115968. doi: 10.1016/j.bmc.2020.115968.

Hitora, Y., K. Takada, Y. Ise, S. P. Woo, S. Inoue, N. Mori, H. Takikawa, S. Nakamukai, S. Okada, and S. Matsunaga. 2020. "Metachromins X and Y from a marine sponge Spongia sp. and their effects on cell cycle progression." *Bioorg Med Chem* 28(2):115233. doi: 10.1016/j.bmc.2019.115233.

Hong, S. W., H. S. Lee, K. H. Jung, H. Lee, and S. S. Hong. 2012. "Protective effect of fucoidan against acetaminophen-induced liver injury." *Arch Pharm Res* 35(6):1099–105. doi: 10.1007/s12272-012-0618-5.

Jiao, W. H., B. H. Cheng, G. D. Chen, G. H. Shi, J. Li, T. Y. Hu, and H. W. Lin. 2018. "Dysiarenone, a dimeric C(21) meroterpenoid with inhibition of COX-2 expression from the marine sponge Dysidea arenaria." *Org Lett* 20(10):3092–5. doi: 10.1021/acs.orglett.8b01148.

Jiao, W. H., B. H. Cheng, G. H. Shi, G. D. Chen, B. B. Gu, Y. J. Zhou, L. L. Hong, et al. 2017. "Dysivillosins A-D, unusual anti-allergic meroterpenoids from the marine sponge Dysidea villosa." *Sci Rep* 7(1):8947. doi: 10.1038/s41598-017-04021-z.

Kolesnikova, S. A., E. G. Lyakhova, A. I. Kalinovsky, R. S. Popov, E. A. Yurchenko, and V. A. Stonik. 2018. "Oxysterols from a Marine Sponge Inflatella sp. and Their Action in 6-hydroxydopamine-induced cell model of Parkinson's disease." *Mar Drugs* 16(11). doi: 10.3390/md16110458.

Lazarus, J. V., E. Roel, and A. M. Elsharkawy. 2020. "Hepatitis C virus epidemiology and the impact of interferon-free hepatitis C virus therapy." *Cold Spring Harb Perspect Med* 10(3). doi: 10.1101/cshperspect.a036913.

Lee, J. Y., G. J. Kim, J. K. Choi, Y. A. Choi, N. H. Jeong, P. H. Park, H. Choi, and S. H. Kim. 2018. "4-(hydroxymethyl)catechol extracted from fungi in marine sponges attenuates rheumatoid arthritis by inhibiting PI3K/Akt/NF-kappaB signaling." *Front Pharmacol* 9:726. doi: 10.3389/fphar.2018.00726.

Li, J., W. Wu, F. Yang, L. Liu, S. P. Wang, W. H. Jiao, S. H. Xu, and H. W. Lin. 2018. "Popolohuanones G –I, dimeric sesquiterpene quinones with IL-6 inhibitory activity from the marine sponge Dactylospongia elegans." *Chem Biodivers* 15(6):e1800078. doi: 10.1002/cbdv.201800078.

Lomonaco, R., C. Ortiz-Lopez, B. Orsak, A. Webb, J. Hardies, C. Darland, J. Finch, et al. 2012. "Effect of adipose tissue insulin resistance on metabolic parameters and liver histology in obese patients with nonalcoholic fatty liver disease." *Hepatology* 55(5):1389–97. doi: 10.1002/hep.25539.

Luo, X., Q. Wang, X. Tang, J. Xu, M. Wang, P. Li, and G. Li. 2021. "Cytotoxic manoalide-type sesterterpenes from the sponge Luffariella variabilis collected in the South China Sea." *J Nat Prod* 84(1):61–70. doi: 10.1021/acs.jnatprod.0c01026.

MacLachlan, J. H., and B. C. Cowie. 2015. "Hepatitis B virus epidemiology." *Cold Spring Harb Perspect Med* 5(5):a021410. doi: 10.1101/cshperspect.a021410.

Makhmoor, T., S. Naheed, S. Shujaat, S. Jalil, S. Hayat, M. I. Choudhary, K. M. Khan, J. M. Alam, and S. Nazir. 2013. "Hepatoprotection by chemical constituents of the marine brown alga Spatoglossum variabile: A relation to free radical scavenging potential." *Pharm Biol* 51(3):383–90. doi: 10.3109/13880209.2012.732582.

Milic, S., D. Lulic, and D. Stimac. 2014. "Non-alcoholic fatty liver disease and obesity: Biochemical, metabolic and clinical presentations." *World J Gastroenterol* 20(28):9330–7. doi: 10.3748/wjg.v20.i28.9330.

Mohibbullah, M., J. S. Choi, M. M. H. Bhuiyan, M. N. Haque, M. K. Rahman, I. S. Moon, and Y. K. Hong. 2018. "The red alga Gracilariopsis chorda and its active constituent arachidonic acid promote spine dynamics via dendritic filopodia and potentiate functional synaptic plasticity in hippocampal neurons." *J Med Food* 21(5):481–8. doi: 10.1089/jmf.2017.4026.

Mondal, D., K. Das, and A. Chowdhury. 2022. "Epidemiology of liver diseases in India." *Clin Liver Dis (Hoboken)* 19(3):114–7. doi: 10.1002/cld.1177.

Nair, D. G., R. Weiskirchen, and S. K. Al-Musharafi. 2015. "The use of marine-derived bioactive compounds as potential hepatoprotective agents." *Acta Pharmacol Sin* 36(2):158–70. doi: 10.1038/aps.2014.114.

Oh, J. H., J. S. Choi, and T. J. Nam. 2018. "Fucosterol from an edible brown alga Ecklonia stolonifera prevents soluble amyloid beta-induced cognitive dysfunction in aging rats." *Mar Drugs* 16(10). doi: 10.3390/md16100368.

Oh, J. H., E. Y. Kim, and T. J. Nam. 2018. "Phycoerythrin-derived tryptic peptide of a red alga Pyropia yezoensis attenuates glutamate-induced ER stress and neuronal senescence in primary rat hippocampal neurons." *Mol Nutr Food Res* 62(8):e1700469. doi: 10.1002/mnfr.201700469.

Park, C., G. Y. Kim, W. I. Kim, S. H. Hong, D. I. Park, N. D. Kim, S. J. Bae, J. H. Jung, and Y. H. Choi. 2007. "Induction of apoptosis by (Z)-stellettic acid C, an acetylenic acid from the sponge Stelletta sp., is associated with inhibition of telomerase activity in human leukemic U937 cells." *Chemotherapy* 53(3):160–8. doi: 10.1159/000100809.

Park, S. R., J. H. Kim, H. D. Jang, S. Y. Yang, and Y. H. Kim. 2018. "Inhibitory activity of minor phlorotannins from Ecklonia cava on alpha-glucosidase." *Food Chem* 257:128–34. doi: 10.1016/j.foodchem.2018.03.013.

Paschos, P., and K. Paletas. 2009. "Non alcoholic fatty liver disease and metabolic syndrome." *Hippokratia* 13(1):9–19.

Peng, B. R., K. H. Lai, Y. C. Chang, Y. Y. Chen, J. H. Su, Y. M. Huang, P. J. Chen, S. S. Yu, C. Y. Duh, and P. J. Sung. 2020. "Sponge-derived 24-Homoscalaranes as potent anti-inflammatory agents." *Mar Drugs* 18(9). doi: 10.3390/md18090434.

Piao, S. J., W. H. Jiao, F. Yang, Y. H. Yi, Y. T. Di, B. N. Han, and H. W. Lin. 2014. "New hippolide derivatives with protein tyrosine phosphatase 1B inhibitory activity from the marine sponge Hippospongia lachne." *Mar Drugs* 12(7):4096–109. doi: 10.3390/md12074096.

Piao, S. J., Y. L. Song, W. H. Jiao, F. Yang, X. F. Liu, W. S. Chen, B. N. Han, and H. W. Lin. 2013. "Hippolachnin A, a new antifungal polyketide from the South China Sea sponge Hippospongia lachne." *Org Lett* 15(14):3526–9. doi: 10.1021/ol400933x.

Piao, S. J., H. J. Zhang, H. Y. Lu, F. Yang, W. H. Jiao, Y. H. Yi, W. S. Chen, and H. W. Lin. 2011. "Hippolides A-H, acyclic manoalide derivatives from the marine sponge Hippospongia lachne." *J Nat Prod* 74(5):1248–54. doi: 10.1021/np200227s.

Pipitone, R. M., R. Zito, G. Lupo, A. Javed, C. La Mantia, G. Di Maria, G. Pratelli, et al. 2023. "Curcumin and andrographolide co-administration safely prevent steatosis induction and ROS Production in HepG2 cell line." *Molecules* 28(3). doi: 10.3390/molecules28031261.

Qureshi, K., and G. A. Abrams. 2007. "Metabolic liver disease of obesity and role of adipose tissue in the pathogenesis of nonalcoholic fatty liver disease." *World J Gastroenterol* 13(26):3540–53. doi: 10.3748/wjg.v13.i26.3540.

Rehm, J., and K. D. Shield. 2013. "Alcohol and mortality: Global alcohol-attributable deaths from cancer, liver cirrhosis, and injury in 2010." *Alcohol Res* 35(2):174–83.

Roerecke, M., A. Vafaei, O. S. M. Hasan, B. R. Chrystoja, M. Cruz, R. Lee, M. G. Neuman, and J. Rehm. 2019. "Alcohol consumption and risk of liver cirrhosis: A systematic review and meta-analysis." *Am J Gastroenterol* 114(10):1574–86. doi: 10.14309/ajg.0000000000000340.

Sharma, K., D. Sharma, M. Sharma, N. Sharma, P. Bidve, N. Prajapati, K. Kalia, and V. Tiwari. 2018. "Astaxanthin ameliorates behavioral and biochemical alterations in in-vitro and in-vivo model of neuropathic pain." *Neurosci Lett* 674:162–70. doi: 10.1016/j.neulet.2018.03.030.

Sugiura, Y., M. Usui, H. Katsuzaki, K. Imai, M. Kakinuma, H. Amano, and M. Miyata. 2018. "Orally administered phlorotannins from Eisenia arborea suppress chemical mediator release and cyclooxygenase-2 signaling to alleviate mouse ear swelling." *Mar Drugs* 16(8). doi: 10.3390/md16080267.

Tsermpini, E. E., A. Plemenitas Iljes, and V. Dolzan. 2022. "Alcohol-induced oxidative stress and the role of antioxidants in alcohol use disorder: A systematic review." *Antioxidants (Basel)* 11(7). doi: 10.3390/antiox11071374.

Wang, F., Y. Li, Y. J. Zhang, Y. Zhou, S. Li, and H. B. Li. 2016. "Natural products for the prevention and treatment of hangover and alcohol use disorder." *Molecules* 21(1):64. doi: 10.3390/molecules21010064.

Wong, R. J., R. Cheung, and A. Ahmed. 2014. "Nonalcoholic steatohepatitis is the most rapidly growing indication for liver transplantation in patients with hepatocellular carcinoma in the U.S.." *Hepatology* 59(6):2188–95. doi: 10.1002/hep.26986.

Xue, M., Y. Liu, R. Lyu, N. Ge, M. Liu, Y. Ma, and H. Liang. 2017. "Protective effect of aplysin on liver tissue and the gut microbiota in alcohol-fed rats." *PLOS ONE* 12(6):e0178684. doi: 10.1371/journal.pone.0178684.

Yoon, M., and S. Cho. 2018. "Triphlorethol A, a dietary polyphenol from seaweed, decreases sleep latency and increases non-rapid eye movement sleep in mice." *Mar Drugs* 16(5). doi: 10.3390/md16050139.

Yu, S., K. Wang, Q. Li, Y. Wei, Y. Li, Q. Zhang, P. Huang, et al. 2023. "Nonalcoholic steatohepatitis critically rewires the ischemia/reperfusion-induced dysregulation of cardiolipins and sphingolipids in mice." *Hepatobiliary Surg Nutr* 12(1):3–19. doi: 10.21037/hbsn-21-133.

Zheng, C. J., M. J. Sohn, S. Lee, and W. G. Kim. 2013. "Meleagrin, a new FabI inhibitor from Penicillium chryosogenum with at least one additional mode of action." *PLOS ONE* 8(11):e78922. doi: 10.1371/journal.pone.0078922.

Zheng, J., Z. Li, Y. Manabe, M. Kim, T. Goto, T. Kawada, and T. Sugawara. 2018. "Siphonaxanthin, a carotenoid from green algae, inhibits lipogenesis in hepatocytes via the suppression of liver X receptor alpha activity." *Lipids* 53(1):41–52. doi: 10.1002/lipd.12002.

Zheng, J., X. Tian, W. Zhang, P. Zheng, F. Huang, G. Ding, and Z. Yang. 2019. "Protective effects of fucoxanthin against alcoholic liver injury by activation of Nrf2-mediated antioxidant defense and inhibition of TLR4-mediated inflammation." *Mar Drugs* 17(10). doi: 10.3390/md17100552.

16 Bioactive Metabolites from Marine Greens and Their Health Benefits

Madan Kumar Arumugam

16.1 INTRODUCTION

Marine organisms are potentially prolific sources of highly bioactive secondary metabolites, which are substances manufactured by plants and marine organisms that make them competitive in their own environment and that might represent useful leads in the development of new pharmaceutical agents (Karthikeyan, Joseph, and Nair 2022). These agents are small molecules that exert a wide range of biological effects on marine organisms themselves and on other living species (Zaporozhets and Besednova 2020; Khalifa et al. 2019). These functional metabolites can be divided further into two broad categories: primary metabolites and secondary metabolites. Primary metabolites, namely, carbohydrates, amino acids, lipids, and nucleic acids, are responsible for the development and growth of organisms. Secondary metabolites, on the other hand, are a group of compounds that, while not essential, give organisms the ability to overcome various obstacles and survive, and allow them to interact and adapt to their surroundings; thus, the same organisms growing in various locations may have different concentrations of these compounds, or even different compounds in their makeup (Jimenez-Lopez et al. 2021; Montero et al. 2018; Liu, Hansen, and Lin 2011). They induce cellular signals as signaling molecules and for "defense and offense" in marine species (Barzkar et al. 2019). They act as antimicrobials and perform the role of attractants or conversely, repellents. Over 50,000 secondary metabolites have been discovered in the plant kingdom. Medicinal herbs and many modern medicines rely on secondary plant metabolites for their actions (El Gamal 2010). During the last four decades, numerous novel compounds have been isolated from marine organisms, and many of these substances have been demonstrated to possess interesting biological activities (Faulkner 2001). Marine greens consist of different biologically active components with the potential for human clinical applications. Recently, studies have reported that the microalgae metabolites are attracting enormous attention for their potential biological activity (Shimizu 1996). Among various marine greens, the microalgal phyla provide novel and diversified secondary metabolites, which have been recognized for their remarkable pharmacological activity. Moreover, microalgae are considered to be the actual producers of some highly bioactive compounds found in marine resources (Shimizu 1996), as they are the main producers of various highly active therapeutic molecules from the marine-associated green species. Other than these marine greens, there are red algae, which are considered the most important source of many biologically active metabolites in comparison to other algal classes. Also, the principal uses of seaweeds are as a source of human food and as a source of adhesives. During the last four decades, numerous novel biologically active compounds have been isolated from marine organisms, and many of these substances have been demonstrated to possess interesting therapeutic activities against various diseases. The potential health benefits of marine greens are mainly due to their chemical composition, which includes small molecules, proteins, carbohydrates, essential amino acids, minerals, essential fatty acids, vitamins, and pigments (Wells et al. 2017; Menaa et al. 2021). In this respect, several bioactive components

 DOI: 10.1201/9781003369738-20

of marine green organisms/species play significant roles in improved human body functions and therapeutic potential. Furthermore, experimental studies support other important functions, such as immunomodulation and antiviral effects of supplementation with marine greens.

16.2 MARINE GREENS

These green algae are found in marine or freshwater habitats, and some even live on land, growing on soil, trees, or rocks (Gong et al. 2021). About 6000 varied species have been reported from various studies (Palkova 2004). Many of these species live as single-cell organisms, while other species form colonies or are coenocytic, composed of one large cell without cross-walls, long filaments, or larger macroscopic seaweeds (Umen 2014). The transition from unicellular to multicellular organisms is considered one of the major innovations in eukaryotic evolution (Szathmary and Smith 1995). Green algae are also found in temperate and tropical waters. They grow attached to the bottom, but not at great depths, as they require good levels of light to photosynthesize. There are about 50–100 common green algae species on coral reefs, forming the major source of food for most herbivorous fish and many other smaller animals (e.g. crustaceans, worms, and gastropods) on reefs. Green algae follow a reproduction cycle called alternation of generations and also reproduce asexually by fission, budding, or fragmentation or by zoospores. These algae produce various secondary metabolites and chemicals, which have a wide range of biological applications in humans. Studies have reported the true origins of compounds found in marine invertebrates, and marine green algae have been a subject of discussion (El Gamal 2010). They may vary from one compound to another, with unique functional properties, but there are strong hints that dietary or symbiotic algae are among the participants in the production of these metabolites (El Gamal 2010; Wells et al. 2017).

Another important natural marine material is green seaweeds, which have been repeatedly used as a major source for extracting bioactive substances over the years because of their widespread marine distribution and large biomass (Ibanez and Cifuentes 2013). These green seaweeds act as functional foods with potential health benefits and important nutritional benefits. Many countries focus on growing these marine seaweeds, which are collected for consumption as food and are especially known for their high nutritional value and health benefits (Kellogg and Lila 2013; Sabeena Farvin and Jacobsen 2013; Yangthong, Hutadilok-Towatana, and Phromkunthong 2009; Agregan et al. 2017). Studies have reported the isolation and extraction of bioactive metabolites from marine seaweeds, which are an important natural resource because of their many active ingredients, particularly those that may be used for both medical and food supplementation purposes (Neto et al. 2018; Lopes et al. 2020; Catarino, Silva, and Cardoso 2018). Most green seaweeds are major sources of lipid components and proteins, and other micro- and macro-nutrients such as peptides, polysaccharides, carotenoids, phenolic compounds, and alkaloids. Thallus, holdfast, mucilage, and whole plants have also been reported as important ingredients (Alves et al. 2010; Lahaye and Robic 2007). Among all these active ingredients, polysaccharides are the components most intensively investigated for medical purposes (Figureueroa et al. 2022). These marine seaweeds and green algae have recently been used for various applications, especially for extracting natural bioactive substances for biomedical applications (Figureueroa et al. 2022). Wang et al. reported that marine green algae are rich sources of bioactive natural compounds with a wide range of biochemical and physiological activities, including immunomodulation, anti-inflammation, antioxidant, anticoagulant, and antitumor (Wang et al. 2014). Studies have also shown that green seaweeds are rich sources of sulfated polysaccharides, which are chemically and physicochemically different from those of land plants, and may have special physiological effects on the human body (Figureueroa et al. 2022; Lahaye and Robic 2007; Tziveleka et al. 2018). These sulfated polysaccharides have various biological and pharmacological activities, such as cell recognition, cell adhesion, and regulation of receptor functions, which are of great interest in medicine (Shi et al. 2013; Li et al. 2019; Tabarsa et al. 2012). Although more than 30,000 human diseases have been clinically proven and their clinical pathology described, fewer than one-third can be treated symptomatically, and only a few can be cured (Karthikeyan, Joseph, and Nair 2022). There is always a need for new therapeutic compounds or agents to treat medical needs that are currently unmet. Naturally derived products have played a major role in drug discovery (Veeresham 2012; Dzobo 2022; Barcenilla et al. 2022). The ecology of marine natural products reveals that many of the compounds isolated from the marine sources are chemical weapons and have evolved into highly potent inhibitors of biological processes in the prey, predators, or competitors of the marine organisms that utilize them for survival (Lahaye and Robic 2007; Karthikeyan, Joseph, and Nair 2022). These chemicals are recognized as having important medicinal value as well as their biological functions. These metabolites are in two different categories, primary and secondary metabolites, which are used as new therapeutic agents, especially to treat a large variety of diseases for which there are no effective therapies (Seca and Pinto 2019). Several secondary metabolites isolated from the same seaweed, *Sargassum incisifolium*, and some semisynthetic derivatives were tested to evaluate their potential as modulators of inflammatory bowel diseases, such as Crohn's disease and ulcerative colitis, using various in vitro assays (Nyambe et al. 2019). In fact, inflammatory bowel diseases have become a global health challenge, since conventional treatments exhibit moderate efficacy and have significant side effects (Nyambe et al. 2019; Ardizzone and Bianchi Porro 2005). So, far various methods have been created to cure different types of cancers, different mutant types of viral and fungal infections, inflammatory conditions, and neurodegenerative diseases that cannot be treated successfully (Malve 2016). The development of multi-resistance of harmful pathogenic microorganisms to a variety of antibiotics and of cancer cells to anticancer and antitumor drugs necessitates the synthesis of new replacement drugs (Mansoori et al. 2017). The pathogens responsible for malaria and tuberculosis have developed resistance to most available drugs. Thus, heavy investment in the development of new antibiotics will be necessary in the next millennium. Similarly, exploring chemical entities from natural marine sources, especially from marine greens, can fulfill the requirements for human health benefits. This chapter will discuss the various biological metabolites isolated from the marine natural greens and their application to human health benefits as new drug discovery, as it is necessary to identify sources of new biological activities and chemical structural diversity.

16.3 ANTIBACTERIAL ACTIVITY

Antibacterial activity or antimicrobial activity is the most important characteristic studied in the medical field; it provides significant protection against microorganisms, biological fluids, and aerosols, as well as disease transmission (Bhatwalkar et al. 2021; Magrys, Olender, and Tchorzewska 2021). In the past decades, considerable efforts have been directed toward the isolation of biologically active chemicals from marine sources, and currently, more than 10,000 natural products of potential biotechnological interest have been isolated (Barzkar et al. 2019). Sesquiterpenes are colorless lipophilic compounds isolated from marine green algae. The sesquiterpenes have strong antibacterial activity against various human bacterial pathogens. One important sesquiterpene is Cycloeudesmol (Figure 16.1A), an antibiotic cyclopropane containing sesquiterpene; it was isolated from the marine alga *Chondria oppositiclada* Dawson (El Gamal 2010). Cycloeudesmol was found to be a potent antibiotic against bacterial pathogens such as *Staphylococcus aureus* and *Candida albicans* (Perez, Falque, and Dominguez 2016). Ali et al. reported the antibacterial activity of Lyengaroside A, which was isolated from the green alga *Codium iyengarii* and displayed moderate antibacterial activity (Ali et al. 2002). Similarly, an extract of the green alga *Caulerpa prolifera* exhibited moderate to significant activity against unidentified strains of marine bacteria (Smyrniotopoulos et al. 2003). A further study has shown that the relevant targets of algal extracts include oral microbiota. Stearidonic acid (Figure 16.1B) and gamma-linolenic acid (Figure 16.1C) are bioactive compounds extracted from *Ulva linza* that showed efficient antibacterial activity against *Prevotella intermedia* and *Porphyromonas gingivalis* (Ahmed et al. 2020). *Avrainvillea* sp. contains bromophenols that showed inhibitory activity against both *Bacillus subtilis* and *Staphylococcus aureus*; they were

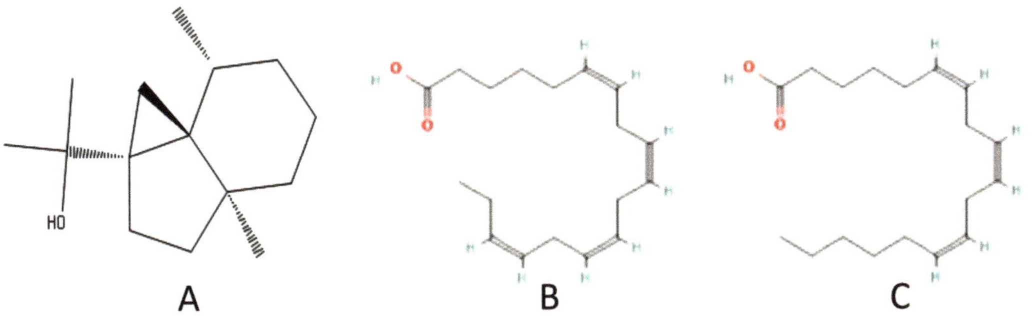

FIGURE 16.1 Antibacterial activity. (A) Cycloeudesmol, (B) stearidonic acid, (C) gamma-linolenic acid compounds isolated from marine greens.

also active against *Pseudomonas aeruginosa* as well as *Escherichia coli*, *Serratia marcescens*, and *Candida albicans* (Xiao et al. 2019; Ozcelik et al. 2009).

16.4 ANTICANCER ACTIVITY

Cancer is a dreadful human disease that is increasing with changing lifestyles, nutrition, environmental factors, industrial harm, and global warming (Sithranga Boopathy and Kathiresan 2010). Cancer cells cease to respond correctly to chemical signals sent by other cells. By entering and manipulating normal tissues, these aberrant cells have the potential to kill. It is critical to hunt for chemicals that can cause "apoptosis" (programmed cell death) to occur (Green and Llambi 2015; Zhang and Yu 2013). The development of cancer registries throughout the world has led to a search for novel drugs that are toxic to the cancer cells while having no harmful effect on normal cells (Greenwell and Rahman 2015) (Figure 16.2) The anticancer drugs used previously exhibited relatively high toxicity not only to the tumor cells, but also to the normal cells of the body part in which the cancer had developed (Saadaoui et al. 2020). Cell deregulation, which occurs when pro-apoptotic signals are lost or anti-apoptotic signals are increased, can result in a range of clinical diseases, including cancer. As a result, promoting apoptosis is a significant anticancer therapeutic target. Cancer treatments do not have potent medicines, as the currently available drugs cause side effects in some instances (Gurib-Fakim 2006). Several of these molecules derived from marine green sources and associated organisms have proven to be beneficial in combating cancer by either preventing the proliferation of cancerous cells (Karanam and Arumugam 2022) or enhancing apoptosis in cancerous cell lines present in humans (Karanam and Arumugam 2022; Sirpu Natesh, Arumugam, and Karanam 2018). Similarly, other marine species and their associated microorganisms are also responsible for producing a variety of bioactive compounds. Important compounds are those that mediate the essential ecological functions in the development and growth of algae species, including quorum sensing signaling molecules, compounds with biological activities, substances that promote growth, and other effective molecules (Singh and Reddy 2014). These compounds have been demonstrated to possess pharmacological, nutritional, and biofunctional values along with their potential therapeutic properties in the biomedical area (Penalver et al. 2020). Therefore, the screening, isolation, and production of anticancer drugs developed from marine green algae-derived compounds can be an interesting approach to potentiate anticancer properties. In addition, these anticancer drugs can be useful to overcome some limitations that arise from specific characteristics of each compound, such as systemic toxicity, weak therapeutic effect, and limited bioavailability (Alves et al. 2018). Highly sulfated polysaccharides possess tremendous antiviral activity. Figure 16.3A shows an active anticancer compound isolated from the Curaso strain of the marine Cyanobacterium *Lyngbya majuscula* by Gerwick's group in 1994. It is an important lead compound for a new type of potential anticancer drug. It also plays a role as an antimitotic agent that

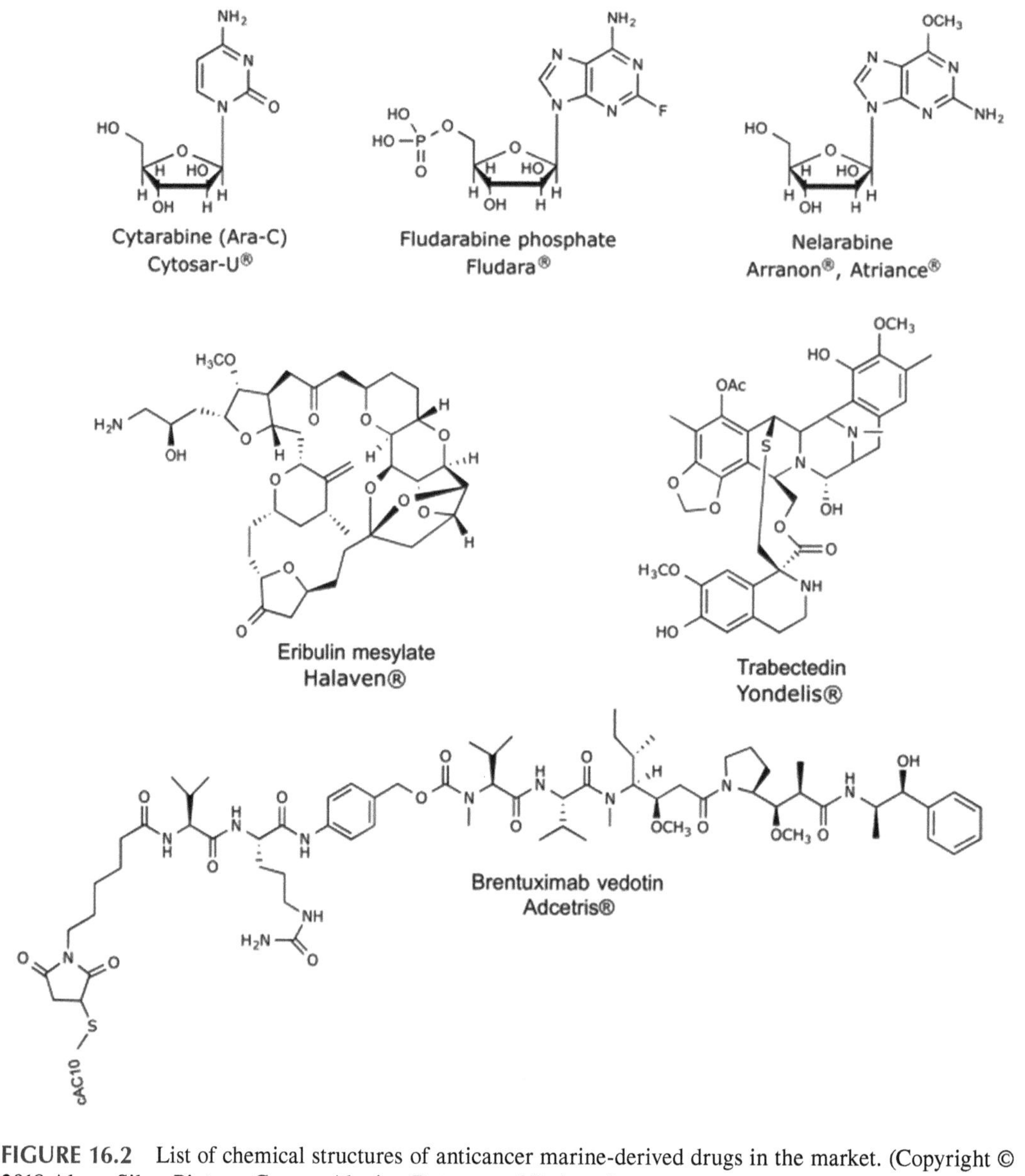

FIGURE 16.2 List of chemical structures of anticancer marine-derived drugs in the market. (Copyright © 2018 Alves, Silva, Pinteus, Gaspar, Alpoim, Botana, and Pedrosa.)

inhibits microtubule assembly and the binding of colchicine to tubulin (Shalaby 2011). A metabolite from the blue-green alga *Lyngbya majuscula* that showed an anticancer effect was later identified as Lyngbyatoxin-A (Figure 16.3B), the most thoroughly investigated metabolite. Lyngbyatoxin shows cytotoxicity against P388 leukemia, but also, a few studies have reported that it acts as a co-carcinogen. It has been suggested that this substance may play a role in human stomach cancer among Hawaiians, who consume large quantities of edible seaweed on which *Lyngbya majuscula* grows epiphytically (El Gamal 2010). Another promising source of anti-carcinogenic components is marine herbs, which are widely accepted for their anticancer activity. Apart from being an effective source of primary metabolites of vitamins, minerals, and protein, they contain a good amount of (2R,3R)-3′,4′,5,5′,7-pentahydroxyflavan-3-yl gallate and 3,4,5-trihydroxybenzoic acid (Farasat et al. 2014). Similarly, marine algae were reported as an important source of anticancer molecules. One such example is a sulfated polysaccharide derivative of brown algae "fucoidan" which was isolated

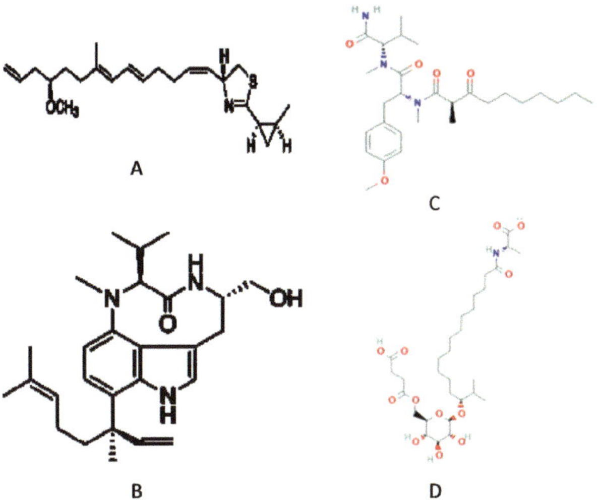

FIGURE 16.3 Anticancer (A) Curacin A, (B) Lyngbyatoxin-A, and antifungal (C) Majuscuiamide C, (D) Ieodoglucomide C compounds isolated from marine greens.

and extracted. It was found to be effective against apoptosis in a human lymphoma cell line (Aisa et al. 2005). An aqueous microalgal extract can also reduce cancer cell metastasis in vivo, which appears to be related to the preferential killing of suspended cancer cells and the anti-colony-forming properties of the microalgal extract (Somasekharan et al. 2016). There are other remarkable marine drugs available on the market. It is particularly interesting to see that a few, such as Cytarabine (Cytosar-U®), Vidarabine (Vira-A®) (US discontinued), Ziconotide (Prialt®), Brentuximab Vedotin (Adcetris®), Eribulin Mesylate (Halaven®), Omega-3-acid ethyl esters (Lovaza®), Trabectedin (Yondelis®), Fludarabine Phosphate (Fludara®), and Nelarabine (Arranon®), were approved by the Food and Drug Administration (FDA) in the US Pharmacopeia and/or by the European Agency for the Evaluation of Medicinal Products (EMA). Iota-carrageenan (Carragelose®), an over-the-counter product, was approved by the EMA (Alves et al. 2018). Drugs that are used in cancer therapies are shown in Figure 16.3, and the majority of the anticancer drugs that are in clinical trials are also for application in cancer therapy, which reveals the great potential of marine compounds as anticancer drugs (Alves et al. 2018; Alves et al. 2016).

16.5 ANTIFUNGAL ACTIVITY

Fungal infections are potentially fatal complications associated with the use of popular immunosuppressive drugs, including anticancer drugs (Ascioglu et al. 2002), and the development of safe antifungal drugs for clinical use would greatly benefit patients (Roemer and Krysan 2014). The rise in the incidence of fungal infections has exacerbated the need for next-generation antifungal agents, as many of the currently available drugs have undesirable side effects, are ineffective against new or re-emerging fungal strains, or lead to rapid development of resistance. Some algal extracts have also been shown to inhibit fungal enzyme activity due to the presence of bioactive metabolites. Previous studies have reported that isolauraldehyde isolated from the organic extract of the red alga *Laurencia obtusa* had significant antifungal activity against *Candida albicans* with a minimum inhibitory concentration (MIC) of 70mg/mL, and showed medium activity against *Aspergillus fumigatus* and *Aspergillus flavus* with a MIC of 100 and 1000 mg/mL, respectively (Alarif et al. 2016). Ieodoglucomide C (Figure 16.3A–D) and ieodoglycolipid are two glycolipids that are both isolated from the aquatic bacterium *Bacillus licheniformis*. It was found that they both have potent antifungal activity, with MIC values of 0.02–0.03 μM against the human pathogens *Candida albicans*, *Colletotrichum acutatum*, *Botrytis cinerea*, *Rhizoctonia solani*, and *Aspergillus niger* (Choudhary et al. 2017). Mickymaray and Alturaiki reported that marine macroalgae contained bioactive compounds that an excellent inhibitory activity against a variety of fungal pathogens and thus, may be useful for combating fungal infections and recovering from chronic asthmatic states (Mickymaray and Alturaiki 2018). Aristolene and 3,7,11,15-tetramethyl-2-hexadecen-1-ol from ethanolic fractions of *Laurencia paniculata* are reported to have antifungal and antimicrobial activity *Laurencia paniculata* (Mickymaray and Alturaiki 2018).

16.6 ANTIMUTAGENIC ACTIVITY

The term "antimutagen" refers to chemical agents possessing the ability to diminish the rate or frequency of induced or spontaneous gene mutations and DNA adduct formation, thereby preventing cell damage (Kada and Shimoi 1987; Fodor 1971). This group of agents includes both natural and synthetic compounds. Two different types of antimutagens, i.e. desmutagens and bioantimutagens, can be distinguished. Desmutagens that function extracellularly are able to inactivate mutagenic agents before they reach DNA. On the other hand, bioantimutagens act within the cell and participate in mutation suppression after DNA damage. These compounds are able to influence genome repair and replication. Cymo-barbatol and 4-isocymobarbatol are two major compounds having an antimutagenic effect, which were isolated from the marine green alga *Cymopolia barbata* (Budzalek et al. 2021). Both compounds were found to be nontoxic over a broad concentration range against *Salmonella typhimurium* strains and exhibited strong inhibition of the mutagenicity of 2-aminoanthracene and ethylmethanesulfonate.

16.7 ANTI-INFLAMMATORY ACTIVITY

Inflammation usually occurs when infectious microorganisms such as bacteria, viruses, or fungi invade the body, reside in particular tissues, and/or circulate in the blood (Artis and Spits 2015). Thus, the inflammation may happen in response to protective processes such as tissue injury, cell death, cancer, ischemia, and degeneration (Vosshenrich and Di Santo 2013). Mostly, both the innate immune response as well as the adaptive immune response are involved in the formation of inflammation (Gasteiger and Rudensky 2014). There are many investigations ongoing to isolate chemical entities that possess strong anti-inflammatory activity from marine red algae. In one such investigation, a marine red alga, *Ceratodictyon spongiosum*, containing the symbiotic sponge *Sigmadocia symbiotica*, collected from Indonesia, afforded two isomers of a new bioactive thiazole containing cyclic heptapeptide: cis,cis-ceratospongamide and trans,trans-ceratospongamide (Tan et al. 2000). The trans-ceratospongamide shows anti-inflammation activity by potent inhibition of sPLA2 expression, whereas the cis isomer is inactive. Trans-ceratospongamide was also shown to inhibit the expression of a human-sPLA2 promotor-based reporter by 90%. The anti-inflammatory activity of compounds was also studied to show the inhibition of phospholipase A2 by secreted hepatocellular carcinoma cells, which is stimulated by interleukins (Kehraus et al. 2002; Bugni et al. 2002; Doi, Asano, and Yoza 2003). Similarly, bromophenolic metabolites such as vidalols isolated from the Caribbean red alga *Vidalia obtusaloba* showed strong anti-inflammatory activity through the inhibition of phospholipase enzyme (Wiemer, Idler, and Fenical 1991). The new compounds were discovered as part of an organized effort to isolate new naturally occurring anti-inflammatory agents with a focus on those that may function through inhibition of phospholipase A2 (Dennis et al. 2011). Marine alkaloids originating from marine organisms are also described in various scientific literatures, presenting different biological and pharmacological activities (Wu et al. 2019). Several studies describe the presence of alkaloids in Alcyonacea coral species, with antiviral, antitumor, antibiotic, and immunomodulatory activities (Kubota et al. 2016). Similarly, alkaloids isolated from marine sources and associated sponge symbionts have potential anti-inflammatory activity. Asperversiamides, an indole alkaloid derived from the marine fungus *Aspergillus versicolor*, showed anti-inflammatory activity by reducing NO levels, inhibiting induced nitric oxide synthase (iNOS) activity, as reported by Huaqiang et al. (Li et al. 2018). Similarly, other alkaloids, such as penicamide A, extracted from the fungus *Penicillium canescens* isolated from the ascidian *Styela plicata*, also acted synergistically with lipopolysaccharide (LPS)-stimulated macrophages. Chaetoglobosin Fex, the chytocalasan-based alkaloid extracted from the fungus *Chaetomium globosum*, was able to reduce the production of tumor necrosis factor (TNF)-α, interleukin (IL)-6, and MCP-1 and also inhibited IκB-α degradation, the translocation of the NF-κB p65 subunit to the cell nucleus, and the phosphorylation of p38, ERK1/2, and JNK MAPKs by LPS-stimulated macrophages (Dou et al. 2011). The indolic alkaloids caulerpin and racemosin showed protective activity on SH-SY5Y cell viability; the latter can give rise to alkylamide derivatives capable of inducing cell death in the breast cancer cell line MDA-MB-231, and showed inhibitory activity and anti-inflammatory activity (Lunagariya et al. 2019; Liu et al. 2013).

16.8 ANTIVIRAL ACTIVITY

Viral diseases, caused by various pathogenic virus infections, are still one of the leading causes of death in humans worldwide (Richards et al. 1978). The risk of virus infection and/or disease in a human population is determined by the characteristics of the virus, of susceptible individuals, and of the host population, such as innate and acquired resistance. In addition, virus transmission is affected by behavioral, environmental, and ecological factors (Burrell, Howard, and Murphy 2017). Although many antiviral agents have been developed and are used for the treatment of viral infections, the emergence of drug resistance, side effects, and the necessity for extensive clinical use are the main reasons for failure of antiviral therapy (Witvrouw and De Clercq 1997; Richards et al. 1978). Sulfated poly-mannuroguluronate isolated from brown algae has recently entered a Phase II clinical trial in China as the first anti–acquired immune deficiency syndrome compound (Meiyu et al. 2003); similarly, Witvrouw and De Clercq have reported that fucoidans and sulfated polysaccharides extracted from some marine brown seaweeds show antiviral activity against infectious diseases, such as HIV, herpes simplex virus types, and cytomegalovirus (Witvrouw and De Clercq 1997).

16.9 ANTIMALARIAL ACTIVITY

Malaria is a life-threatening disease spread to humans by some types of mosquitoes. It is caused by plasmodia, ancient, single-celled protozoans transmitted to humans by the bites of female Anopheles mosquitoes (Abdalla et al. 1980; Weatherall and Abdalla 1982). Antimalarial drug screening based on phenotypic methods has placed numerous preclinical candidates from marine greens and organisms into antimalarial drug discovery (Guiguemde et al. 2012). Snyderol, a sesquiterpene derivative isolated from the red alga *Laurencia obtusa*, was active against different clones of the malarial parasite *Plasmodium falciparum* (Budzalek et al. 2021) (Figure 16.4).

16.10 CONCLUSION

Marine greens contain a great variety of bioactive molecules with potential health benefits. Several types of marine micro- and macro algae are already consumed as food additives and nutritional supplements, potentially exerting their beneficial effects through diet. There is, however, a compelling necessity to consider the algal bioactive compounds (Figure 16.4), which include anticancer, antibacterial, antimicrobial, antifungal, and antimalarial drug candidates for new drug discovery programs, and to investigate their potential biological effects in order to find new pharmaceuticals with preventive and therapeutic efficacy.

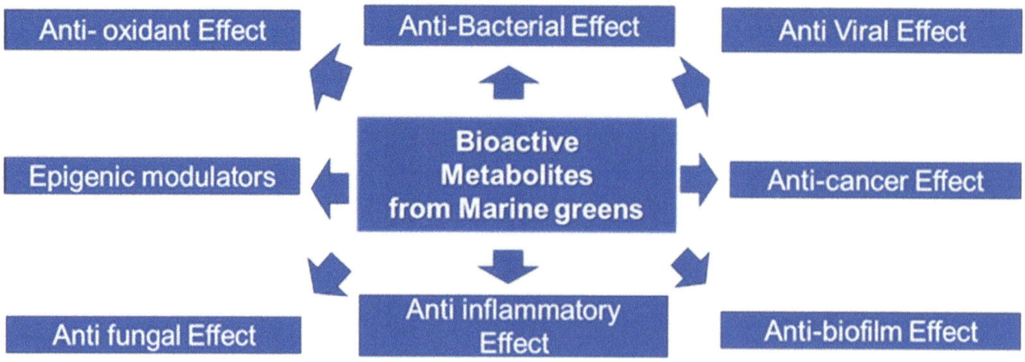

FIGURE 16.4 Bioactive metabolites and their various biological activities.

REFERENCES

Abdalla, S., D. J. Weatherall, S. N. Wickramasinghe, and M. Hughes. 1980. "The anaemia of P. falciparum malaria." *Br J Haematol* 46(2):171–83. doi: 10.1111/j.1365-2141.1980.tb05956.x.

Agregan, R., P. E. Munekata, R. Dominguez, J. Carballo, D. Franco, and J. M. Lorenzo. 2017. "Proximate composition, phenolic content and in vitro antioxidant activity of aqueous extracts of the seaweeds Ascophyllum nodosum, Bifurcaria bifurcata and Fucus vesiculosus. Effect of addition of the extracts on the oxidative stability of canola oil under accelerated storage conditions." *Food Res Int* 99(Pt 3):986–94. doi: 10.1016/j.foodres.2016.11.009.

Ahmed, S. A., D. A. Abdelrheem, H. R. A. El-Mageed, H. S. Mohamed, A. A. Rahman, K. N. M. Elsayed, and S. A. Ahmed. 2020. "Destabilizing the structural integrity of COVID-19 by caulerpin and its derivatives along with some antiviral drugs: An in silico approaches for a combination therapy." *Struct Chem* 31(6):2391–412. doi: 10.1007/s11224-020-01586-w.

Aisa, Y., Y. Miyakawa, T. Nakazato, H. Shibata, K. Saito, Y. Ikeda, and M. Kizaki. 2005. "Fucoidan induces apoptosis of human HS-sultan cells accompanied by activation of caspase-3 and down-regulation of ERK pathways." *Am J Hematol* 78(1):7–14. doi: 10.1002/ajh.20182.

Alarif, W. M., K. O. Al-Footy, M. S. Zubair, M. Halid Ph, M. A. Ghandourah, S. A. Basaif, S. S. Al-Lihaibi, S. E. Ayyad, and F. A. Badria. 2016. "The role of new eudesmane-type sesquiterpenoid and known eudesmane derivatives from the red alga Laurencia obtusa as potential antifungal-antitumour agents." *Nat Prod Res* 30(10):1150–5. doi: 10.1080/14786419.2015.1046378.

Ali, M. S., M. Saleem, R. Yamdagni, and M. A. Ali. 2002. "Steroid and antibacterial steroidal glycosides from marine green alga Codium iyengarii Borgesen." *Nat Prod Lett* 16(6):407–13. doi: 10.1080/10575630290034249.

Alves, A., S. G. Caridade, J. F. Mano, R. A. Sousa, and R. L. Reis. 2010. "Extraction and physico-chemical characterization of a versatile biodegradable polysaccharide obtained from green algae." *Carbohydr Res* 345(15):2194–200. doi: 10.1016/j.carres.2010.07.039.

Alves, C., S. Pinteus, A. Horta, and R. Pedrosa. 2016. "High cytotoxicity and anti-proliferative activity of algae extracts on an in vitro model of human hepatocellular carcinoma." *Springerplus* 5(1):1339. doi: 10.1186/s40064-016-2938-2.

Alves, C., J. Silva, S. Pinteus, H. Gaspar, M. C. Alpoim, L. M. Botana, and R. Pedrosa. 2018. "From marine origin to therapeutics: The antitumor potential of marine algae-derived compounds." *Front Pharmacol* 9:777. doi: 10.3389/fphar.2018.00777.

Ardizzone, S., and G. Bianchi Porro. 2005. "Biologic therapy for inflammatory bowel disease." *Drugs* 65(16):2253–86. doi: 10.2165/00003495-200565160-00002.

Artis, D., and H. Spits. 2015. "The biology of innate lymphoid cells." *Nature* 517(7534):293–301. doi: 10.1038/nature14189.

Ascioglu, S., J. H. Rex, B. de Pauw, J. E. Bennett, J. Bille, F. Crokaert, D. W. Denning, et al. 2002. "Defining opportunistic invasive fungal infections in immunocompromised patients with cancer and hematopoietic stem cell transplants: An international consensus." *Clin Infect Dis* 34(1):7–14. doi: 10.1086/323335.

Barcenilla, C., A. Alvarez-Ordonez, M. Lopez, O. Alvseike, and M. Prieto. 2022. "Microbiological safety and shelf-life of low-salt meat products–A review." *Foods* 11(15). doi: 10.3390/foods11152331.

Barzkar, N., S. Tamadoni Jahromi, H. B. Poorsaheli, and F. Vianello. 2019. "Metabolites from marine microorganisms, micro, and macroalgae: Immense scope for pharmacology." *Mar Drugs* 17(8). doi: 10.3390/md17080464.

Bhatwalkar, S. B., R. Mondal, S. B. N. Krishna, J. K. Adam, P. Govender, and R. Anupam. 2021. "Antibacterial properties of organosulfur compounds of garlic (Allium sativum)." *Front Microbiol* 12:613077. doi: 10.3389/fmicb.2021.613077.

Budzalek, G., S. Sliwinska-Wilczewska, K. Wisniewska, A. Wochna, I. Bubak, A. Latala, and J. M. Wiktor. 2021. "Macroalgal defense against competitors and herbivores." *Int J Mol Sci* 22(15). doi: 10.3390/ijms22157865.

Bugni, T. S., G. P. Concepcion, G. C. Mangalindan, M. K. Harper, R. D. James, and C. M. Ireland. 2002. "p-Sulfooxyphenylpyruvic acid from the red macro alga Ceratodictyon spongiosum and its sponge symbiont Haliclona cymaeformis." *Phytochemistry* 60(4):361–3. doi: 10.1016/s0031-9422(02)00098-5.

Burrell, C. J., C. R. Howard, and F. A. Murphy. 2017. "Epidemiology of viral infections." *Fenner White's Med Virol* 2017:185–203. doi: 10.1016/B978-0-12-375156-0.00013-8. Epub 2016 Nov 11.

Catarino, M. D., A. M. S. Silva, and S. M. Cardoso. 2018. "Phycochemical constituents and biological activities of Fucus spp.." *Mar Drugs* 16(8). doi: 10.3390/md16080249.

Choudhary, A., L. M. Naughton, I. Montanchez, A. D. W. Dobson, and D. K. Rai. 2017. "Current status and future prospects of marine natural products (MNPs) as antimicrobials." *Mar Drugs* 15(9). doi: 10.3390/md15090272.

Dennis, E. A., J. Cao, Y. H. Hsu, V. Magrioti, and G. Kokotos. 2011. "Phospholipase A2 enzymes: Physical structure, biological function, disease implication, chemical inhibition, and therapeutic intervention." *Chem Rev* 111(10):6130–85. doi: 10.1021/cr200085w.

Doi, M., A. Asano, and K. Yoza. 2003. "cis,cis-Ceratospongamide n,n-dimethylacetamide hemisolvate in the presence of twinning." *Acta Crystallogr C* 59(6):o323–5. doi: 10.1107/s010827010300369x.

Dou, H., Y. Song, X. Liu, W. Gong, E. Li, R. Tan, and Y. Hou. 2011. "Chaetoglobosin Fex from the marine-derived endophytic fungus inhibits induction of inflammatory mediators via toll-like receptor 4 signaling in macrophages." *Biol Pharm Bull* 34(12):1864–73. doi: 10.1248/bpb.34.1864.

Dzobo, K. 2022. "The role of natural products as sources of therapeutic agents for innovative drug discovery." *Comprehensive Pharmacology* 2022:408–22. doi: 10.1016/B978-0-12-820472-6.00041-4. Epub 2022 Jun 9.

El Gamal, A. A. 2010. "Biological importance of marine algae." *Saudi Pharm J* 18(1):1–25. doi: 10.1016/j.jsps.2009.12.001.

Farasat, M., Khavari-Nejad, R. A., Nabavi, S. M., F. Namjooyan. 2014. "Antioxidant activity, total phenolics and flavonoid contents of some edible green seaweeds from northern coasts of the persian gulf." *Iran J Pharm Res* 13(1):163–70.

Faulkner, D. J. 2001. "Marine natural products." *Nat Prod Rep* 18(1):1–49. doi: 10.1039/b006897g.

Figureueroa, F. A., R. T. Abdala-Diaz, C. Perez, V. Casas-Arrojo, A. Nesic, C. Tapia, C. Duran, et al. 2022. "Sulfated polysaccharide extracted from the green algae *Codium bernabei*: Physicochemical characterization and antioxidant, anticoagulant and antitumor activity." *Mar Drugs* 20(7). doi: 10.3390/md20070458.

Fodor, I. 1971. "DNA-repair enzymes." *Usp Sovrem Biol* 71(3):331–52.

Gasteiger, G., and A. Y. Rudensky. 2014. "Interactions between innate and adaptive lymphocytes." *Nat Rev Immunol* 14(9):631–9. doi: 10.1038/nri3726.

Gong, N., K. Shao, K. Shen, Y. Gu, Y. Liu, J. Ye, C. Hu, et al. 2021. "Chemical control of overwintering green algae to mitigate green tide in the Yellow Sea." *Mar Pollut Bull* 168:112424. doi: 10.1016/j.marpolbul.2021.112424.

Green, D. R., and F. Llambi. 2015. "Cell death signaling." *Cold Spring Harb Perspect Biol* 7(12). doi: 10.1101/cshperspect.a006080.

Greenwell, M., and P. K. Rahman. 2015. "Medicinal plants: Their use in anticancer treatment." *Int J Pharm Sci Res* 6(10):4103–12. doi: 10.13040/IJPSR.0975-8232.6(10).4103-12.

Guiguemde, W. A., A. A. Shelat, J. F. Garcia-Bustos, T. T. Diagana, F. J. Gamo, and R. K. Guy. 2012. "Global phenotypic screening for antimalarials." *Chem Biol* 19(1):116–29. doi: 10.1016/j.chembiol.2012.01.004.

Gurib-Fakim, A. 2006. "Medicinal plants: Traditions of yesterday and drugs of tomorrow." *Mol Aspects Med* 27(1):1–93. doi: 10.1016/j.mam.2005.07.008.

Ibanez, E., and A. Cifuentes. 2013. "Benefits of using algae as natural sources of functional ingredients." *J Sci Food Agric* 93(4):703–9. doi: 10.1002/jsfa.6023.

Jimenez-Lopez, C., A. G. Pereira, C. Lourenco-Lopes, P. Garcia-Oliveira, L. Cassani, M. Fraga-Corral, M. A. Prieto, and J. Simal-Gandara. 2021. "Main bioactive phenolic compounds in marine algae and their mechanisms of action supporting potential health benefits." *Food Chem* 341(2):128262. doi: 10.1016/j.foodchem.2020.128262.

Kada, T., and K. Shimoi. 1987. "Desmutagens and bio-antimutagens--Their modes of action." *BioEssays* 7(3):113–6. doi: 10.1002/bies.950070305.

Karanam, G., and M. K. Arumugam. 2022. "Potential anticancer effects of cyclo(-Pro-Tyr) against N-diethyl nitrosamine induced hepatocellular carcinoma in mouse through PI3K/AKT signaling." *Environ Toxicol* 37(2):256–69. doi: 10.1002/tox.23395.

Karthikeyan, A., A. Joseph, and B. G. Nair. 2022. "Promising bioactive compounds from the marine environment and their potential effects on various diseases." *J Genet Eng Biotechnol* 20(1):14. doi: 10.1186/s43141-021-00290-4.

Kehraus, S., G. M. Konig, A. D. Wright, and G. Woerheide. 2002. "Leucamide A: A new cytotoxic heptapeptide from the Australian sponge Leucetta microraphis." *J Org Chem* 67(14):4989–92. doi: 10.1021/jo020058r.

Kellogg, J., and M. A. Lila. 2013. "Chemical and in vitro assessment of Alaskan coastal vegetation antioxidant capacity." *J Agric Food Chem* 61(46):11025–32. doi: 10.1021/jf403697z.

Khalifa, S. A. M., N. Elias, M. A. Farag, L. Chen, A. Saeed, M. F. Hegazy, M. S. Moustafa, et al. 2019. "Marine natural products: A source of novel anticancer drugs." *Mar Drugs* 17(9). doi: 10.3390/md17090491.

Kubota, T., K. Nakamura, K. Sakai, J. Fromont, T. Gonoi, and J. Kobayashi. 2016. "Hyrtinadines C and D, new azepinoindole-type alkaloids from a marine sponge Hyrtios sp." *Chem Pharm Bull (Tokyo)* 64(7):975–8. doi: 10.1248/cpb.c16-00201.

Lahaye, M., and A. Robic. 2007. "Structure and functional properties of ulvan, a polysaccharide from green seaweeds." *Biomacromolecules* 8(6):1765–74. doi: 10.1021/bm061185q.

Li, H., W. Sun, M. Deng, Q. Zhou, J. Wang, J. Liu, C. Chen, et al. 2018. "Asperversiamides, linearly fused prenylated indole alkaloids from the marine-derived fungus Aspergillus versicolor." *J Org Chem* 83(15):8483–92. doi: 10.1021/acs.joc.8b01087.

Li, Y., X. Wang, Y. Jiang, J. Wang, H. Hwang, X. Yang, and P. Wang. 2019. "Structure characterization of low molecular weight sulfate Ulva polysaccharide and the effect of its derivative on iron deficiency anemia." *Int J Biol Macromol* 126:747–54. doi: 10.1016/j.ijbiomac.2018.12.214.

Liu, D. Q., S. C. Mao, H. Y. Zhang, X. Q. Yu, M. T. Feng, B. Wang, L. H. Feng, and Y. W. Guo. 2013. "Racemosins A and B, two novel bisindole alkaloids from the green alga Caulerpa racemosa." *Fitoterapia* 91:15–20. doi: 10.1016/j.fitote.2013.08.014.

Liu, M., P. E. Hansen, and X. Lin. 2011. "Bromophenols in marine algae and their bioactivities." *Mar Drugs* 9(7):1273–92. doi: 10.3390/md9071273.

Lopes, D., T. Melo, F. Rey, J. Meneses, F. L. Monteiro, L. A. Helguero, M. H. Abreu, A. I. Lillebo, R. Calado, and M. R. Domingues. 2020. "Valuing bioactive lipids from green, red and brown macroalgae from aquaculture, to foster functionality and biotechnological applications." *Molecules* 25(17). doi: 10.3390/molecules25173883.

Lunagariya, J., P. Bhadja, S. Zhong, R. Vekariya, and S. Xu. 2019. "Marine natural product bis-indole alkaloid caulerpin: Chemistry and biology." *Mini Rev Med Chem* 19(9):751–61. doi: 10.2174/1389557517666170927154231.

Magrys, A., A. Olender, and D. Tchorzewska. 2021. "Antibacterial properties of Allium sativum L. against the most emerging multidrug-resistant bacteria and its synergy with antibiotics." *Arch Microbiol* 203(5):2257–68. doi: 10.1007/s00203-021-02248-z.

Malve, H. 2016. "Exploring the ocean for new drug developments: Marine pharmacology." *J Pharm Bioallied Sci* 8(2):83–91. doi: 10.4103/0975-7406.171700.

Mansoori, B., A. Mohammadi, S. Davudian, S. Shirjang, and B. Baradaran. 2017. "The different mechanisms of cancer drug resistance: A brief review." *Adv Pharm Bull* 7(3):339–48. doi: 10.15171/apb.2017.041.

Meiyu, G., L. Fuchuan, X. Xianliang, L. Jing, Y. Zuowei, and G. Huashi. 2003. "The potential molecular targets of marine sulfated polymannuroguluronate interfering with HIV-1 entry. Interaction between SPMG and HIV-1 rgp120 and CD4 molecule." *Antiviral Res* 59(2):127–35. doi: 10.1016/s0166-3542(03)00068-8.

Menaa, F., U. Wijesinghe, G. Thiripuranathar, N. A. Althobaiti, A. E. Albalawi, B. A. Khan, and B. Menaa. 2021. "Marine algae-derived bioactive compounds: A new wave of nanodrugs?." *Mar Drugs* 19(9). doi: 10.3390/md19090484.

Mickymaray, S., and W. Alturaiki. 2018. "Antifungal efficacy of marine macroalgae against fungal isolates from bronchial asthmatic cases." *Molecules* 23(11). doi: 10.3390/molecules23113032.

Montero, L., A. Del Pilar Sanchez-Camargo, E. Ibanez, and B. Gilbert-Lopez. 2018. "Phenolic compounds from edible algae: Bioactivity and health benefits." *Curr Med Chem* 25(37):4808–26. doi: 10.2174/0929867324666170523120101.

Neto, R. T., C. Marcal, A. S. Queiros, H. Abreu, A. M. S. Silva, and S. M. Cardoso. 2018. "Screening of Ulva rigida, Gracilaria sp., Fucus vesiculosus and Saccharina latissima as functional ingredients." *Int J Mol Sci* 19(10). doi: 10.3390/ijms19102987.

Nyambe, M. N., T. C. Koekemoer, M. van de Venter, E. D. Goosen, and D. R. Beukes. 2019. "In vitro evaluation of the phytopharmacological potential of Sargassum incisifolium for the treatment of inflammatory bowel diseases." *Medicines (Basel)* 6(2). doi: 10.3390/medicines6020049.

Ozcelik, B., I. Gurbuz, T. Karaoglu, and E. Yesilada. 2009. "Antiviral and antimicrobial activities of three sesquiterpene lactones from Centaurea solstitialis L. ssp. solstitialis." *Microbiol Res* 164(5):545–52. doi: 10.1016/j.micres.2007.05.006.

Palkova, Z. 2004. "Multicellular microorganisms: Laboratory versus nature." *EMBO Rep* 5(5):470–6. doi: 10.1038/sj.embor.7400145.

Penalver, R., J. M. Lorenzo, G. Ros, R. Amarowicz, M. Pateiro, and G. Nieto. 2020. "Seaweeds as a functional ingredient for a healthy diet." *Mar Drugs* 18(6). doi: 10.3390/md18060301.

Perez, M. J., E. Falque, and H. Dominguez. 2016. "Antimicrobial action of compounds from marine seaweed." *Mar Drugs* 14(3). doi: 10.3390/md14030052.

Richards, J. T., E. R. Kern, L. A. Glasgow, J. C. Overall, Jr., E. F. Deign, and M. T. Hatch. 1978. "Antiviral activity of extracts from marine algae." *Antimicrob Agents Chemother* 14(1):24–30. doi: 10.1128/AAC.14.1.24.

Roemer, T., and D. J. Krysan. 2014. "Antifungal drug development: Challenges, unmet clinical needs, and new approaches." *Cold Spring Harb Perspect Med* 4(5). doi: 10.1101/cshperspect.a019703.

Saadaoui, I., R. Rasheed, N. Abdulrahman, T. Bounnit, M. Cherif, H. Al Jabri, and F. Mraiche. 2020. "Algae-derived bioactive compounds with anti-lung cancer potential." *Mar Drugs* 18(4). doi: 10.3390/md18040197.

Sabeena Farvin, K. H., and C. Jacobsen. 2013. "Phenolic compounds and antioxidant activities of selected species of seaweeds from Danish coast." *Food Chem* 138(2–3):1670–81. doi: 10.1016/j.foodchem.2012.10.078.

Seca, A. M. L., and D. C. G. A. Pinto. 2019. "Biological potential and medical use of secondary metabolites." *Medicines (Basel)* 6(2). doi: 10.3390/medicines6020066.

Shalaby, E. A. 2011. "Algae as promising organisms for environment and health." *Plant Signal Behav* 6(9):1338–50. doi: 10.4161/psb.6.9.16779.

Shi, J., C. Cheng, H. Zhao, J. Jing, N. Gong, and W. Lu. 2013. "In vivo anti-radiation activities of the Ulva pertusa polysaccharides and polysaccharide-iron(III) complex." *Int J Biol Macromol* 60:341–6. doi: 10.1016/j.ijbiomac.2013.06.001.

Shimizu, Y. 1996. "Microalgal metabolites: A new perspective." *Annu Rev Microbiol* 50:431–65. doi: 10.1146/annurev.micro.50.1.431.

Singh, R. P., and C. R. Reddy. 2014. "Seaweed-microbial interactions: Key functions of seaweed-associated bacteria." *FEMS Microbiol Ecol* 88(2):213–30. doi: 10.1111/1574-6941.12297.

Sirpu Natesh, N., M. Arumugam, and G. Karanam. 2018. "Apoptotic role of marine sponge symbiont Bacillus subtilis NMK17 through the activation of caspase-3 in human breast cancer cell line." *Mol Biol Rep* 45(6):2641–51. doi: 10.1007/s11033-018-4434-y.

Sithranga Boopathy, N., and K. Kathiresan. 2010. "Anticancer drugs from marine flora: An overview." *J Oncol* 2010:214186. doi: 10.1155/2010/214186.

Smyrniotopoulos, V., D. Abatis, L. A. Tziveleka, C. Tsitsimpikou, V. Roussis, A. Loukis, and C. Vagias. 2003. "Acetylene sesquiterpenoid esters from the green alga Caulerpa prolifera." *J Nat Prod* 66(1):21–4. doi: 10.1021/np0202529.

Somasekharan, S. P., A. El-Naggar, P. H. Sorensen, Y. Wang, and H. Cheng. 2016. "An aqueous extract of marine microalgae exhibits antimetastatic activity through preferential killing of suspended cancer cells and anticolony forming activity." *Evid Based Complement Alternat Med* 2016:9730654. doi: 10.1155/2016/9730654.

Szathmary, E., and J. M. Smith. 1995. "The major evolutionary transitions." *Nature* 374(6519):227–32. doi: 10.1038/374227a0.

Tabarsa, M., J. H. Han, C. Y. Kim, and S. G. You. 2012. "Molecular characteristics and immunomodulatory activities of water-soluble sulfated polysaccharides from Ulva pertusa." *J Med Food* 15(2):135–44. doi: 10.1089/jmf.2011.1716.

Tan, L. T., R. T. Williamson, W. H. Gerwick, K. S. Watts, K. McGough, and R. Jacobs. 2000. "cis,cis- and trans,trans-ceratospongamide, new bioactive cyclic heptapeptides from the Indonesian red alga Ceratodictyon spongiosum and symbiotic sponge Sigmadocia symbiotica." *J Org Chem* 65(2):419–25. doi: 10.1021/jo991165x.

Tziveleka, L. A., N. Pippa, P. Georgantea, E. Ioannou, C. Demetzos, and V. Roussis. 2018. "Marine sulfated polysaccharides as versatile polyelectrolytes for the development of drug delivery nanoplatforms: Complexation of ulvan with lysozyme." *Int J Biol Macromol* 118(A):69–75. doi: 10.1016/j.ijbiomac.2018.06.050.

Umen, J. G. 2014. "Green algae and the origins of multicellularity in the plant kingdom." *Cold Spring Harb Perspect Biol* 6(11):a016170. doi: 10.1101/cshperspect.a016170.

Veeresham, C. 2012. "Natural products derived from plants as a source of drugs." *J Adv Pharm Technol Res* 3(4):200–1. doi: 10.4103/2231-4040.104709.

Vosshenrich, C. A., and J. P. Di Santo. 2013. "Developmental programming of natural killer and innate lymphoid cells." *Curr Opin Immunol* 25(2):130–8. doi: 10.1016/j.coi.2013.02.002.

Wang, L., X. Wang, H. Wu, and R. Liu. 2014. "Overview on biological activities and molecular characteristics of sulfated polysaccharides from marine green algae in recent years." *Mar Drugs* 12(9):4984–5020.

Weatherall, D. J., and S. Abdalla. 1982. "The anaemia of Plasmodium falciparum malaria." *Br Med Bull* 38(2):147–51. doi: 10.1093/oxfordjournals.bmb.a071751.

Wells, M. L., P. Potin, J. S. Craigie, J. A. Raven, S. S. Merchant, K. E. Helliwell, A. G. Smith, M. E. Camire, and S. H. Brawley. 2017. "Algae as nutritional and functional food sources: Revisiting our understanding." *J Appl Phycol* 29(2):949–82. doi: 10.1007/s10811-016-0974-5.

Wiemer, D. F., D. D. Idler, and W. Fenical. 1991. "Vidalols A and B, new anti-inflammatory bromophenols from the Caribbean marine red alga Vidalia obtusaloba." *Experientia* 47(8):851–3. doi: 10.1007/BF01922471.

Witvrouw, M., and E. De Clercq. 1997. "Sulfated polysaccharides extracted from sea algae as potential antiviral drugs." *Gen Pharmacol* 29(4):497–511. doi: 10.1016/s0306-3623(96)00563-0.

Wu, Q., B. Nay, M. Yang, Y. Ni, H. Wang, L. Yao, and X. Li. 2019. "Marine sponges of the genus Stelletta as promising drug sources: Chemical and biological aspects." *Acta Pharm Sin B* 9(2):237–57. doi: 10.1016/j.apsb.2018.10.003.

Xiao, X. N., F. Wang, Y. T. Yuan, J. Liu, Y. Z. Liu, and X. Yi. 2019. "Antibacterial activity and mode of action of dihydromyricetin from Ampelopsis grossedentata leaves against food-borne bacteria." *Molecules* 24(15). doi: 10.3390/molecules24152831.

Yangthong, M., N. Hutadilok-Towatana, and W. Phromkunthong. 2009. "Antioxidant activities of four edible seaweeds from the southern coast of Thailand." *Plant Foods Hum Nutr* 64(3):218–23. doi: 10.1007/s11130-009-0127-y.

Zaporozhets, T. S., and N. N. Besednova. 2020. "Biologically active compounds from marine organisms in the strategies for combating coronaviruses." *AIMS Microbiol* 6(4):470–94. doi: 10.3934/microbiol.2020028.

Zhang, L., and J. Yu. 2013. "Role of apoptosis in colon cancer biology, therapy, and prevention." *Curr Colorectal Cancer Rep* 9(4). doi: 10.1007/s11888-013-0188-z.

17 Modulatory and Neuroprotective Roles of Marine Green Spirulina on the Brain and Neurodegeneration

Abraham Olufemi Asuku, Maryam Tayo Ayinla, Ayodeji Johnson Ajibare, and R. Krishnamurthy

17.1 MARINE GREEN SPIRULINA

According to Kulshreshtha *et al.* (2008), marine green spirulina is an undivided, tiny, elongated, thin, spiral cyanobacterium that can develop in both salty and alkaline conditions and doubles its biomass between 48 and 120 hours. The cyanobacteria species *Spirulina platensis* and *Spirulina maxima* are primarily referred to as spirulina (Sinha *et al.*, 2018). Marine green spirulina was a staple food for the Aztec and Maya civilizations for many years (López-Romero *et al.*, 2018); today, it is regarded as safe for human consumption due to its extremely low toxicity (Gutiérrez-Salmeán *et al.*, 2015) and accounts for over 30% of the global microalgae biomass production. Marine green spirulina is grown all over the world as food or as an essential component in many nutraceutical compositions (Costa *et al.*, 2019). Thanks to a scientific study that revealed its high protein and natural vitamin content, the World Health Organization (WHO) named marine green spirulina the future's best food in 1996 (López-Romero *et al.*, 2018).

Marine green spirulina (Figure 17.1) appears to be able to stimulate the entire human physiology and demonstrates therapeutic properties by fighting cancer, allergy, oxidative stress, inflammation, bacterial and viral diseases, and diabetes as well as having a wide range of other advantageous properties (Kulshreshtha *et al.*, 2008).

17.2 PRODUCTION

Today, specialized algae farms in outdoor tanks and bioreactors are where the majority of marine green spirulina used as a food supplement is produced worldwide (Belay, 2008).

The majority of industrial production methods use paddle wheels to mix spirulina cultures in shallow raceways (Habib *et al.*, 2008).

The seven main steps in the manufacture of marine green spirulina are cleaning and filtration, concentration, neutralization, disintegration, dehydration, packing, and storage. Marine green spirulina is concentrated, its water and salt content are removed, and it is made ready for distribution by these procedures. Chemical composition tests, microbiological testing, and tests for pollutants such as pesticides, heavy metals, and extraneous elements (rat hair, insect fragments, etc.) are all part of the quality control process for marine green spirulina that meets food standards (Habib *et al.*, 2008).

DOI: 10.1201/9781003369738-21

FIGURE 17.1 Marine green spirulina.

17.3 COMPOSITION, NUTRIENTS, AND PHYTO-DERIVATIVES

Because marine green spirulina contains great quantities of proteins, vitamins, fatty acids (essential), minerals, lipids, and other nutrients (Liestianty, 2019), it is frequently used as a nutritious dietary item, which when consumed, enhances physical and mental performance and benefits human health. Marine green spirulina is among the most important super-edibles on earth, according to the WHO, and NASA employs it for space flight because a small amount may offer a variety of nutrients (Karkos *et al.*, 2011). In reality, marine green spirulina has a high concentration of minerals like iron, calcium, zinc, magnesium, manganese, and potassium as well as B vitamins, especially vitamin B12 (Kennedy, 2016). Gamma-linolenic acid (GLA), one of the necessary fatty acids, is also found in spirulina. Its phyto complex is abundant in pigments like allophycocyanin, phycocyanin, and chlorophyll (Makkar *et al.*, 2020). It's essential to remember that the body swiftly and effectively absorbs marine green spirulina nutrients, thereby quickly restoring deficient nutrients to normal ranges (Wells *et al.*, 2017).

17.4 THE BRAIN AND NEURODEGENERATION/ NEURODEGENERATIVE DISEASES

The brain serves as the body's primary control center for the majority of physiological processes. It processes, integrates, and coordinates signals from different sense organs and then gives feedbacks that are adequately relayed to the peripherals in different parts of the body. Thus, preserving brain health is essential to ensure proper function in all spheres of mental well-being, physiology, and societal living.

Maintaining a disease-free condition and consuming dietary sources can help the body to avoid or treat brain diseases, which can then have an impact on all bodily processes (Gómez-Pinilla, 2008). Increasing malfunction and loss of the structure and function of the neurons, leading to their death, are known as neurodegeneration (Yildiz-Unal *et al.*, 2015). The central nervous system (CNS) is affected by many disorders that cause neurodegeneration; the symptoms of acute as well as chronic neurodegenerative diseases are determined by the loss of particular numbers of neurons connected to functional neural networks.

The term "neurodegenerative disease" refers to a broad category of neurological conditions that predominantly impact CNS neurons and lead to a gradual loss of CNS neurons that eventually results in deficiencies in particular brain processes (memory, movement, and cognition). Examples of neurodegenerative disorders are PD, amyotrophic lateral sclerosis, MS, AD, and Huntington's disease.

Numerous biological processes (neuroinflammation, mitochondrial malfunction, oxidative stress, aberrant protein misfolding and aggregation, excitotoxicity, and apoptosis) have been linked to neurodegeneration (Singh *et al.*, 2019; Cenini *et al.*, 2019). These biological mechanisms are thought to be responsible for the development and pathophysiology of neurodegenerative disorders. A plethora of studies have been performed to clarify the mechanism and possible therapeutic targets for treating neurodegenerative disorders. Therefore, by interacting with the pathophysiological change process, neuroprotection techniques and related processes function to stop or prevent the process of neurodegeneration.

17.5 EFFECT OF MARINE GREEN SPIRULINA ON THE BRAIN

The usage of marine green spirulina has been linked to many neuroprotective effects in various brain regions, according to numerous in vivo studies utilizing various animal models.

Neuroprotection has been examined using experimental models under a variety of conditions, including neuroinflammation, neurodegeneration, neurotoxicity, and more. In a neuroinflammation model, an example of neuroprotection in vivo was discovered. For instance, it was discovered that administering marine green spirulina to young rats reduced the serious inflammatory problem caused by lipopolysaccharide (LPS), which resulted in a drop in neural stem cell proliferation. It was discovered that consuming marine green spirulina as a dietary supplement dramatically boosted the ability of neural precursor cells of the hippocampus to multiply rapidly (Bachstetter *et al.*, 2010).

In a model using LPS administered to young rats, marine green spirulina was given to nursing mothers to further demonstrate the neuroprotective effect of the algae against neuroinflammatory problems. In this instance, LPS-induced inflammatory interleukin (IL)-1β was reduced, while the expression of brain-derived neurotrophic factor (BDNF) and the antioxidants Nrf2 and γ-glutamylcysteine ligase catalytic subunit (GCLC) was increased, perhaps through mechanisms that normalized phosphorylated AKT (pAKT) (Patil *et al.*, 2016). Later, a comparable model further supported this result. It has been demonstrated that feeding marine green spirulina to nursing mothers can protect them from the neuroinflammation caused by LPS, in particular by reversing the decline in antioxidant defense that occurs after inflammation, an effect that is linked to a lower amount of phosphorylated p38 and a higher amount of antioxidant defense (Patil *et al.*, 2018) (Figure 17.2).

17.6 MECHANISM

Phycocyanobilin (PCB), a component of marine green spirulina, has a structure resembling that of bilirubin, an endogenous NADPH inhibitor, and is converted to phycocyanorubin by the same enzyme (biliverdin reductase). Bilirubin is a catabolite of heme. Spirulina has also been linked to decreased NADPH complex expression (Riss *et al.*, 2007). The primary NADPH oxidase inhibitor and suppressor mechanisms of marine green spirulina appear to perform a function in neurology.

The effect of a protein extract of marine green spirulina on iron-induced oxidative stress, which is described by the collapse of the glutathione (GSH) pathway's enzyme activities, a decrease in reduced glutathione, and a rise in the ratio of oxidized glutathione (GSSG) and GSH, was examined by Bermejo-Bescós *et al.* (2008) in a neuroblastoma cell line (SH-SY5Y). By way of the antioxidant enzyme glutathione peroxidase (GPx), GSH reduces H_2O_2 to H_2O and is oxidized to GSSG, which is then reduced to GSH by the enzyme glutathione reductase (GR).

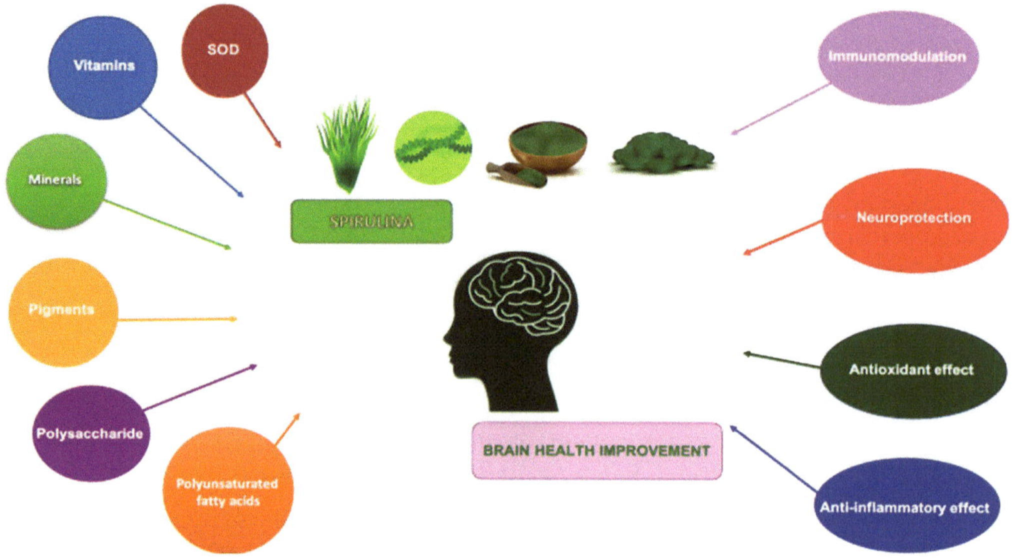

FIGURE 17.2 Constituents of marine green spirulina and effects on brain health. (From Trotta, T. *et al.*, *Nutrients*, 14, 676, 2022.)

Fe^{2+} can produce hydroxyl or alkoxy radicals, which have been demonstrated to be destroyed by the marine green spirulina's protein extract and phycocyanin, which preserve the action of antioxidant enzymes.

Additionally, it was demonstrated that marine green spirulina functions as an iron chelator, delaying the interaction of iron with oxygen or peroxides (Bermejo-Bescós *et al.*, 2008).

17.7 EFFECT OF MARINE GREEN SPIRULINA ON NEURODEGENERATIVE DISEASES

A collection of abnormal protein folding events that result in the production of amyloid fibrils or shapeless aggregates are defined as neurodegenerative disorders (Sweeney *et al.*, 2017). According to Godoy *et al.* (2008), there is ample scientific proof that oxidative stress and inflammation are primarily involved in the development and progression of neurodegenerative disorders. However, marine green spirulina has been proven to have positive effects on AD, PD, and MS.

17.7.1 Effect of Marine Green Spirulina on Multiple Sclerosis

Several research studies have been carried out to study the use of marine green spirulina as therapy for MS, demonstrating its potential in this area. It appears to be able to shield axons from demyelination in particular.

Rats who were given C-phycocyanin (C-PC) treatment had myelin that was solidified, crushed, and compressed but did not exhibit any symptoms of axonal disintegration (Pentón-Rol *et al.*, 2011).

Experimental auto-immune encephalomyelitis (EAE), which has the ability to simulate the primary neuroimmunological as well as histological features of this disorder, is the most popular model utilized to research this disease (Bittner *et al.*, 2014).

The ability of C-PC to induce the Treg subset—a heterogeneous T cell population responsible for sustaining self-antigen tolerance—was shown by Pentón-Rol *et al.* in their 2011 study of MS patients.

Furthermore, C-PC significantly decreased the expression of PPAR and CD44, which are responsible for the demyelination process, and significantly increased the expression of a group of genes associated with remyelination, gliogenesis, and axon–glia processes. More recently, oral C-PC therapy in rat and mouse models of EAE improved EAE clinical development and restored motor function and motor coordination. IL-6 and interferon (IFN)-γ, which are known as pro-inflammatory cytokines, were significantly reduced in the brain after Phycocyanobilin (PCB) treatment, and oxidative stress markers were positively controlled. Myelin sheath integrity was also conserved (Cervantes-Llanos *et al.*, 2018).

C-PC selectively blocks cyclooxygenase 2 (COX-2). It has been shown that blocking COX-2 protects oligodendrocyte precursor cells from various types of injury (Shiow *et al.*, 2017).

Medications for MS may hold promise if they can combat inflammation and oxidative stress, stimulate Treg, and shield axons from myelin loss, as was the case with marine green spirulina therapy.

17.7.2 Effect of Marine Green Spirulina on Parkinson's Disease

Neuroinflammation can be controlled by foods that can effectively fight oxidative stress and inflammation, and marine green spirulina has largely been proven to protect the neurons.

Studies conducted in vivo and in vitro revealed that marine green spirulina's antioxidant components, such as phycocyanin, have the ability to prevent or delay damage caused by oxidative stress by reducing the build-up of free radicals and stimulating the catalase (CAT), superoxide dismutase (SOD), and glutathione peroxidase (GPx) antioxidant enzyme systems (Thaakur and Jyothi, 2007).

Dopamine neurons are thought to be particularly vulnerable to oxidative stress and inflammation, according to research on experimental PD models (McGeer and McGeer, 2008), and stimulation of microglial cells is recognized to be responsible for the etiology of neurodegenerative disorders (Pabon *et al.*, 2012).

One month after a 6-hydroxydopamine (6-OHDA) lesion, restoration of SNpc-positive TH neurons and striatal dopamine-positive TH nerve fibers was noticed in rats given a marine green spirulina–rich diet. A marker linked to the stimulation of M1 microglia, major histocompatibility complex II (MHC II), was well expressed, and this also revealed a reduction in the number of stimulated microglia (M1 phenotype, pro-inflammatory) (Strömberg *et al.*, 2005).

In an α-Syn model of PD, scientists demonstrated that marine green spirulina was neuroprotective, noting significant quantities of NeuN+ and TH+ cells as well as a decline in the quantities of stimulated microglial cells as measured by MHC II expression (Pabon *et al.*, 2011).

In neurotoxicity induced by 6-OHDA in the rat, marine green spirulina had a protective effect by causing a decrease in the brain's free radicals and lipoperoxidation levels (Tobón-Velasco *et al.*, 2013).

In the brain of mice administered MPTP, pre-treatment with marine green spirulina polysaccharide dramatically boosted immunoreactive staining and the expression of mRNA of tyrosine hydroxylase and dopamine transporter (DAT). Additionally, the activities of GPx and SOD in the midbrain and serum were much higher than those of monoamine oxidase B (MAO B), indicating that this polysaccharide's ability to fight oxidative stress may be responsible for its neuroprotective impact (Zhang *et al.*, 2015).

In one study, Kumar *et al.* (2012) discovered that adding marine green spirulina to the diet significantly reduces SOD and CAT activity in DJ-193 flies, a *Drosophila melanogaster* model of PD. This increases longevity and locomotor activity.

17.7.3 Effect of Marine Green Spirulina on Alzheimer's Disease

Researchers are working to produce natural bioactive substances like marine green spirulina to treat neurodegenerative diseases such as AD (Kim *et al.*, 2010). Substances that can prevent the build-up of Aβ deposits in the CNS are believed to be possible treatment agents for AD (Guo *et al.*, 2010). The favorable impact of marine green spirulina in AD has been proven by scientific research. Recently, Luo and Jing showed that C-PC can prevent bovine serum albumin (BSA) from forming amyloid by preventing BSA's structural conversion (to α-helices and β-sheets) (Luo and Jing, 2020).

Singh *et al.* demonstrated in a worm model (*Caenorhabditis elegans*) of AD that phycocyanin might carry out its neuroprotective effect in AD by inhibiting the activity of β-secretase (Singh *et al.*, 2014), which together with γ-secretase creates Aβ via cleavage of APP (Das and Yan, 2019).

The development of hyperphosphorylated Tau's neurofibrillary tangles, aggregated Aβ plaques, and soluble Aβ oligomeric species in AD causes microglia to become activated, which in turn causes neuroinflammation (Ozben and Ozben, 2019).

In response to the neurotoxicity caused by oligomeric Aβ in mice, several investigations found that C-PC reduced inflammation and neuronal death (Li *et al.*, 2020).

In another study, it was shown that when C-PC was administered, it increased the levels of miRNA-335, decreased programmed cell death, and reduced HDAC3 expression in the hippocampus of rats given oligomeric Aβ peptides (Li *et al.*, 2020), which showed a promising target for treating AD, as blockage of hippocampal tissue HDAC3 decreased Aβ and amyloid plaque status and reduced microglia stimulation in the brains of APP/PS1 rats (Zhu *et al.*, 2017).

Furthermore, HDAC3 failed to impact the phosphorylated Tau protein levels in the hippocampus of 24- and 36-week-old APP/PS1 rats (Zhu *et al.*, 2017). RGFP-966, an agent that selectively blocked and silenced HDAC3, on the other hand, is claimed to reduce both hyperphosphorylated Tau and Aβ protein burden (Janczura *et al.*, 2018), indicating a potential impact of C-PC on Tau protein.

Also, findings in male Wistar rats revealed that when they are administered nicotine for 60 days, marine green spirulina lipopolysaccharides can suppress the phosphorylation of Tau protein (Elsonbaty and Ismail, 2020).

ICAM-1, a biomarker of endothelial dysfunction, is elevated in AD; astrocytes, endothelial cells, and microglia exhibit an elevated amount of ICAM-1 in diseased conditions, which contributes to neuritic plaques on neurite growth and neurodegeneration, according to experimental evidence (Walker *et al.*, 2017; Guha *et al.*, 2021). Additionally, chemokines like CXCL2 function in neuronal cells through the chemokine receptor (CXCR2), where they are responsible for the pathophysiology of AD (De Paola *et al.*, 2007).

The rates of expression of mRNA of the two markers were well attenuated in ischemia rats treated with marine green spirulina (Marn-Prida *et al.*, 2013), indicating that a similar mitigating impact on AD may lessen CXCL2 and ICAM-induced damage in this condition.

Rats with cognitive impairment induced by intra-cerebroventricular Streptozotocin were used to study the effect of phycocyanin in improving cognitive deficit. In this AD model, the administration of phycocyanin significantly reduced neuroinflammation, and elevated levels of BDNF and insulin-like growth factor-1 (IGF-1) (Agrawal *et al.*, 2020).

17.8 CONCLUSION

A plethora of evidence shows that marine green spirulina exhibits a unique neuroprotective mechanism, such as anti-inflammatory and antioxidant actions in the brain tissues, that could make it a useful pharmaceutical for avoiding and providing a remedy for the problem of neurodegenerative diseases. Nevertheless, more findings are required to fully understand the mechanisms of action of marine green spirulina, despite the abundance of positive scientific evidence that has been demonstrated both in vivo and in vitro.

REFERENCES

Adam DB, Jennifer J, Andrea S, Jennifer LV, Charles H, Michael JC, Douglas RL, Juan T, Paul RS, Cyndy DS, Cesario B, Yuji K, Naoki T, Caemelina G, Paula CB (2010) Spirulina promotes stem cell genesis and protects against Lipopolysaccharides induced declines in Neural stem cell proliferation. *PLOS ONE* 5(5), e10496. DOI: 10.1371/journal.pone0010496.

Agrawal M, Perumal Y, Bansal S, Arora S, Chopra K (2020) Phycocyanin alleviates ICV-STZ induced cognitive and molecular deficits via PI3-kinase dependent pathway. *Food Chem. Toxicol.* 145, 111684.

Al-Dhabi NA (2013) Heavy metal analysis in commercial Spirulina products for human consumption. *Saudi J. Biol. Sci.* 20(4), 383–388. DOI: 10.1016/j.sjbs.2013.04.006.

Ascherio A, Schwarzschild MA (2016) The epidemiology of Parkinson's disease: Risk factors and prevention. *Lancet Neurol.* 15(12). DOI: 10.1016/s1474-4422(16)30230-7.

Bachstetter AD, Jernberg J, Schlunk A, Vila JL, Hudson C, Cole MJ, Shytle RD, Tan J, Sanberg PR, Sanberg CD (2010) Spirulina promotes stem cell genesis and protects against LPS induced declines in neural stem cell proliferation. *PLOS ONE* 5(5), e10496.

Balestrino R, Schapira AHV (2020) Parkinson disease. *Eur. J. Neurol.* 27(1), 27–42.

Bedard K, Krause KH (2007) The NOX family of ROS-generating NADPH oxidases: Physiology and pathophysiology. *Physiol. Rev.* DOI: 101152/physrev.00044.2005.

Belay A (2008) Spirulina (Arthrospira): Production and quality assurance. In *Spirulina in Human Nutrition and Health*. CRC Press, Boca Raton, FL. 7(7): 1–26. DOI: 10.5772/31848.

Bermejo-Bescós P, Piñero-Estrada E, del Fresno AMV (2008) Neuroprotection by Spirulina platensis protean extract and phycocyanin against iron-induced toxicity in SH-SY5Y neuroblastoma cells. *Toxicol. Vitro* 22(6), 1496–1502.

Bittner S, Afzali AM, Wiendl H, Meuth SG (2014) Myelin oligodendrocyte glycoprotein (MOG35-55) induced experimental autoimmune encephalomyelitis (EAE) in C57BL/6 mice. *J. Vis. Exp.* 15, 51275.

Bolós M, Perea JR, Avila J (2017) Alzheimer's disease as an inflammatory disease. *Biomol. Concepts* 8(1), 37–43.

Bras JM, Singleton A (2009) Genetic susceptibility in Parkinson's disease. *Biochim. Biophys. Acta* 1792(7), 597–603.

Cenini G, Lloret A, Cascella R (2019) Oxidative stress in neurodegenerative diseases: From a mitochondrial point of view. *Oxid. Med. Cell. Longev.* Article ID 2105607, 18.

Cervantes-Llanos M, Lagumersindez-Denis N, Marín-Prida J, Pavón-Fuentes N, Falcon-Cama V, Piniella-Matamoros B, Camacho-Rodríguez H, Fernández-Massó JR, Valenzuela-Silva C, Raíces-Cruz I, et al. (2018) Beneficial effects of oral administration of C Phycocyanin and phycocyanobilin in rodent models of experimental autoimmune encephalomyelitis. *Life Sci.* 194, 130–138.

Chamorro G, Pérez-Albiter M, Serrano-García N, Mares-Sámano JJ, Rojas P (2006) Spirulina maxima pretreatment partially protects against 1-methyl-4-phenyl-1,2,3,6-tetrahydropyridine neurotoxicity. *Nutr. Neurosci.* 9(5–6), 207–212. DOI: 10.1080/10284150600929748.

Ciferri O (1983) Spirulina, the edible microorganism. *Microbiol. Rev.*, 47(4): 551–578. DOI: 10.1128/mr.47.4.551-578.1983

Correale J, Marrodan M, Ysrraelit MC (2019) Mechanisms of neurodegeneration and axonal dysfunction in progressive multiple sclerosis. *Biomedicines* 7(1), 14. DOI: 10.3390/biomedicines7010014.

Costa JAV, Freitas BCB, Rosa GM, Moraes L, Morais MG, Mitchell BG (2019) Operational and economic aspects of Spirulina based biorefinery. *Bioresour. Technol.* 292, 121946.

Colonna M, Butovsky O (2017) Microglia function in the central nervous system during health and neurodegeneration. *Annu. Rev. Immunol.* 35(1), 441–468.

Das B, Yan RA (2019) A Close look at BACE1 inhibitors for Alzheimer's disease treatment. *CNS Drugs* 33(3), 251–263.

De la Monte SM (2012) Contributions of brain insulin resistance and deficiency in amyloid-related neurodegeneration in Alzheimer's disease. *Drugs* 72(1), 49–66.

Deng H, Wang P, Jankovic J (2018) The genetics of Parkinson disease. *Ageing Res. Rev.* 42, 72–85.

De Paola M, Buanne P, Biordi L, Bertini R, Ghezzi P, Mennini T (2007) Chemokine MIP-2/CXCL2, acting on CXCR2, induces motor neuron death in primary cultures. *Neuroimmunomodulation* 14(6), 310–331.

Diaz-Aparicio I, Beccari S, Abiega O, Sierra A (2016) Clearing the corpses: Regulatory mechanisms, novel tools, and therapeutic potential of harnessing microglial phagocytosis in the diseased brain. *Neural Regen. Res.* 11(10), 1533–1539.

Dietrich D, Hoeger S (2005) Guidance values for microcystins in water and cyanobacterial supplement products (blue-green algal supplements): A reasonable or misguided approach? *Toxicol. Appl. Pharmacol.* 203(3), 273–289. DOI:10.1016/j.taap.2004.09.005.

Elbaz A, Carcaillon L, Kab S, Moisan F (2016) Epidemiology of Parkinson's disease. *Rev. Neurol.* 172(1), 14–26.

Elsonbaty SM, Ismail AFM (2020) Nicotine encourages oxidative stress and impairment of rats' brain mitigated by Spirulina platensis lipopolysaccharides and low-dose ionizing radiation. *Arch. Biochem. Biophys.* 689, 108382.

Friese MA, Schattling B, Fugger L (2014) Mechanisms of neurodegeneration and axonal dysfunction in multiple sclerosis. *Nat. Rev. Neurol.* 10(4), 225–238.

Gitte SJ, Cassandra D, Miki L, Kathleen FB (2016) Clinical safety of a high dose of phycocyanin-enriched aqueous extract from Arthrospira (Spirulina) platensis: Results from a randomized, double-blind, placebo-controlled study with a focus on anticoagulant activity and platelet activation. *J. Med. Food* 19(7), 645–653. DOI: 10.1089/jmf.2015.0143.

Godoy MCP, Tarelli R, Ferrari CC, Sarchi MI, Pitossi FJ (2008) Central and systemic IL-1 exacerbates neurodegeneration and motor symptoms in a model of Parkinson's disease. *Brain* 131(7), 1880–1894.

Gutiérrez-Salmeán G, Fabila-Castillo L, Chamorro-Cevallos G (2015) Nutritional and toxicological aspects of Spirulina (Arthrospira). *Nutr. Hosp.* 32, 34–40.

Guha S, Paidi RK, Goswami S, Saha P, Biswas SC (2021) ICAM-1 protects neurons against Amyloid-β and improves cognitive behaviors in 5xFAD mice by inhibiting NF-kB. *Brain Behav. Immun.* 100, 194–210.

Guo JP, Yu S, McGeer PL (2010) Simple in vitro assays to identify amyloid-βaggregation blockers for Alzheimer's disease therapy. *J. Alzheimers Dis.* 19(4), 1359–1370.

Gómez-Pinilla F (2008) Brain foods: The effects of nutrients on brain function. *Nat. Rev. Neurosci.* 9(7), 568–578.

Hinds TD, Jr, Stec DE (2018) Bilirubin, a cardiometabolic signaling molecule. *Hypertension* 72(4), 788–795. DOI: 10.1161/hypertensionaha.118.11130.

Habib MA, Parvin M, Huntington TC, Hasan MR (2008) A review on culture, production and use of Spirulina as food for humans and feeds for domestic animals and fish. FAO Fisheries and Aquaculture Circular. 978-92-5-106106-0.

Heussner AH, Mazija L, Fastner J, Dietrich DR (2012) Toxin content and cytotoxicity of algal dietary supplements. *Toxicol. Appl. Pharmacol.* 265(2), 263–271. DOI: 10.1016/J.taap.2012.10.005.

Huang Z, Guo BJ, Wong RNS, Jiang Y (2007) Characterization and antioxidant activity of selenium-containing phycocyanin isolated from Spirulina platensis. *Food Chem.* 100(3), 1137–1143.

Hutadilok-Towatana N, Reanmongkol W, Panichayupakaranant P (2010) Evaluation of the toxicity of Arthrospira (Spirulina) platensis extract. *J. Appl. Phycol.* 22(5), 599–605.

Hwang JH, Lee IT, Jeng KC, Wang MF, Hou RC, Wu SM, Chan YC (2011) Spirulina prevents memory dysfunction, reduces oxidative stress damage and augments antioxidant activity in senescence-accelerated mice. *J. Nutr. Sci. Vitaminol.* 57(2), 186–191. DOI: 10.3177/jnsv.57.186.

Janczura KJ, Volmar CH, Sartor GC, Rao SJ, Ricciardi NR, Lambert G, Brothers SP, Wahlestedt C (2018) Inhibition of HDAC3 reverses Alzheimer's disease-related pathologies in vitro and in the 3xTg-AD mouse model. *Proc. Natl. Acad. Sci. U. S. A.* 115(47), E11148–E11157. DOI: 10.1073/pnas.1805436115.

Karkos P, Leong S, Karkos C, Sivaji N, Assimakopoulos D (2011) Spirulina in clinical practice: Evidence-based human applications. *Evid. Based Complement. Altern. Med.* 2011, 531053.

Kaur R, Mehan S, Singh S (2019) Understanding multifactorial architecture of Parkinson's disease: Pathophysiology to management. *Neurol. Sci.* 40(1), 13–23.

Kennedy DOB (2016) B vitamins and the brain: Mechanisms, dose and efficacy—A review. *Nutrients* 8(2), 68.

Kim J, Lee HJ, Lee KW (2010) Naturally occurring phytochemicals for the prevention of Alzheimer's disease. *J. Neurochem.* 112(6), 1415–1430.

Kremer D, Göttle P, Hartung HP, Küry P (2016) Pushing forward: Remyelination as the new frontier in CNS diseases. *Trends Neurosci.* 39(4), 246–263.

Kumar H, Lim HW, More SV, Kim BW, Koppula S, Kim IS, Choi DK (2012) The role of free radicals in the aging brain and Parkinson's disease: Convergence and parallelism. *Int. J. Mol. Sci.* 13(8), 10478–10504.

Kulshreshtha A, Zacharia AJ, Jarouliya U, Bhadauriya P, Prasad GBKS, Bisen PS (2008) Spirulina in health care management. *Curr. Pharm. Biotechnol.* 9(5), 400–405.

Li Z, Gan L, Yan S, Yan Y, Huang W (2020) Effect of C-phycocyanin on HDAC3 and miRNA-335 in Alzheimer's disease. *Transl. Neurosci.* 11(1), 161–172.

Liddelow SA, Guttenplan KA, Clarke LE, Bennett FC, Bohlen CJ, Schirmer L, Bennett ML, Münch AE, Chung WS, Peterson TC et al. (2017) Neurotoxic reactive astrocytes are induced by activated microglia. *Nature* 541(7638), 481–487.

Liestianty D, Rodianawati I, Arfah RA, Assa A (2019) Nutritional analysis of Spirulina sp to promote as superfood candidate. *IOP Conf. Ser. Mater. Sci. Eng.* 509, 012031.

Liguori I, Russo G, Curcio F, Bulli G, Aran L, Della-Morte D, Gargiulo G, Testa G, Cacciatore F, Bonaduce D, Abete P (2018) Oxidative stress, aging, and diseases. *Clin. Interv. Aging.* 13, 757–772.

López-Romero D, Izquierdo-Vega JA, Morales-González JA, Madrigal-Bujaidar E, Chamorro-Cevallos G, Sánchez Gutiérrez M, Betanzos-Cabrera G, Alvarez-Gonzalez I, Morales-González Á, Madrigal-Santillán E (2018) Evidence of some natural products with antigenotoxic effects. Part 2: Plants, vegetables, and natural resin. *Nutrients* 10(12), 1954.

Luo YC, Jing P (2020) Molecular interaction of protein-pigment C-phycocyanin with bovine serum albumin in a gomphosis structure inhibiting amyloid formation. *Int. J. Mol. Sci.* 21(21), 8207.

Magalingam KB, Radhakrishnan A, Ping NS, Haleagrahara N (2018) Current concepts of neurodegenerative mechanisms in Alzheimer's disease. *BioMed Res. Int.* 2018, 3740461.

Makkar R, Behl T, Bungau S, Zengin G, Mehta V, Kumar A, Uddin M, Ashraf GM, Abdel-Daim MM, Arora S (2020) Nutraceuticals in neurological disorders. *Int. J. Mol. Sci.* 21(12), 4424.

Marín-Prida J, Pavón-Fuentes N, Llópiz-Arzuaga A, Fernández-Massó JR, Delgado-Roche L, Mendoza-Marí Y, San-tana SP, Cruz-Ramírez A, Valenzuela-Silva C, Nazábal-Gálvez M, et al. (2013) Phycocyanobilin promotes PC12 cell survival and modulates immune and inflammatory genes and oxidative stress markers in acute cerebral hypoperfusion in rats. *Toxicol. Appl. Pharmacol.* 272(1), 49–60.

Marles RJ, Barrett ML, Barnes J, Chavez ML, Gardiner P, Ko R, Mahady GB, Dog TL, Sarma ND, Giancaspro GI. et al. (2011) United States Pharmacopeia safety evaluation of Spirulina. *Crit. Rev. Food Sci. Nutr.* 51(7), 593–604.

Marsan DW, Conrad SM, Stutts WL, Parker CH, Deeds JR (2018) Evaluation of microcystin contamination in blue-green algal dietary supplements using a protein phosphatase inhibition-based test kit. *Heliyon* 4(3). DOI: 10.1016/j.heliyon.2018.e00573.

Matejuk A, Ransohoff RM (2020) Crosstalk between astrocytes and microglia: An overview. *Front. Immunol.* 11, 1416.

McCarty MF, Barroso-Aranda J, Contreras F (2010) NADPH oxidase mediates glucolipotoxicity-induced beta cell dysfunction- clinical implications. *Med. Hypotheses* 74(3), 596–600. DOI: 10.1016/j.mehy.2008.09.062.

McCarty MF (2007) Clinical potential of Spirulina as a source of phycocyanobilin. *J. Med. Food* 10(4), 566–570. DOI: 10.1089/jmf.2007.621.

McGeer PL, McGeer EG (2008) The α-synuclein burden hypothesis of Parkinson disease and its relationship to Alzheimer disease. *Exp. Neurol.* 212(2), 235–238.

Mosser CA, Baptista S, Arnoux I, Audinat E (2017) Microglia in CNS development: Shaping the brain for the future. *Prog. Neurobiol.* 150, 1–20.

Muys M, Sui Y, Schwaiger B, Lesueur C, Vandenheuvel D, Vermeir P, Vlaeminck SE (2018) High variability in nutritional value and safety of commercially available Chlorella and Spirulina biomass indicates the need for smart production strategies. *Bioresour. Technol.* 275, 247–257. DOI: 10.1016/j.biortech.2018.12.059.

Napoli I, Neumann H (2009) Microglial clearance function in health and disease. *Neuroscience* 158(3), 1030–1038.

Nylander A, Hafler DA (2012) Multiple sclerosis. *J. Clin. Investig.* 122(4), 1180–1188.

Ozben T, Ozben S (2019) Neuro-inflammation and anti-inflammatory treatment options for Alzheimer's disease. *Clin. Biochem.* 72, 87–89.

Pan RY, Ma J, Kong XX, Wang XF, Li SS, Qi XL, Yan YH, Cheng J, Liu Q, Jin W et al. (2019) Sodium rutin ameliorates Alzheimer's disease-like pathology by enhancing microglial amyloid-β clearance. *Sci. Adv.* 5(2), 6328.

Pabon MM, Jernberg JN, Morganti J, Contreras J, Hudson CE, Klein RL, Bickford PCA (2012) Spirulina-enhanced diet provides neuroprotection in an α-synuclein model of Parkinson's disease. *PLOS ONE* 7(9), e45256. DOI: 10.1371/journal.pone.0045256.

Pabon MM, Bachstetter AD, Hudson CE, Gemma C, Bickford PC (2011) CX3CL1 reduces neurotoxicity and microglial activation in a rat model of Parkinson's disease. *J. Neuroinflammation* 8(1), 9. DOI: 10.1186/1742-2094-8-9.

Panday A, Sahoo MK, Osorio D, Batra S (2015) NADPH oxidases: An overview from structure to innate immunity-associated pathologies. *Cell Mol. Immunol.* 12(1), 5–23. DOI: 10.1038/cmi.2014.89.

Park JC, Han SH, Byun MS, Yi D, Lee JH, Park K, Lee DY, Mook-Jung I (2017) Low serum phosphorus correlates with cerebral Aβ deposition in cognitively impaired subjects: Results from the KBASE study. *Front. Aging Neurosci.* 9, 362.

Patil J, Matte A, Nissbrandt H, Mallard C, Sandberg M (2016) Sustained effects of neonatal systemic lipopolysaccharide on IL-1β and Nrf2 in adult rat substantia nigra are partly normalized by a Spirulina-enriched diet. *Neuroimmunomodulation* 23(4), 250–259.

Patil J, Matte A, Mallard C, Sandberg M (2018) Spirulina diet to lactating mothers protects the antioxidant system and reduces inflammation in post-natal brain after systemic inflammation. *Nutr. Neurosci.* 21(1), 59–69.

Pérez-Juárez A, Chamorro G, Alva-Sánchez C, Paniagua-Castro N, Pacheco-Rosado J (2016) Neuroprotective effect of Arthrospira (Spirulina) platensis against kainic acid-neuronal death. *Pharm. Biol.* 54(8), 1408–1412.

Petrus M, Culerrier R, Campistron M, Barre A, Rougé P (2010) First case report of anaphylaxis to spirulin: Identification of phycocyanin as responsible allergen. *Allergy* 65(7), 924–925. DOI: 10.1111/j.1398-9995.2009.02257.x.

Pentón-Rol G, Martínez-Sánchez G, Cervantes-Llanos M, Lagumersindez-Denis N, Acosta-Medina EF, Falcón-Cama V, Alonso-Ramírez R, Valenzuela-Silva C, Rodríguez-Jiménez E, Llópiz-Arzuaga A. et al. (2011) C-Phycocyanin ameliorates experimental autoimmune encephalomyelitis and induces regulatory T cells. *Int. Immunopharmacol.* 11(1), 29–38.

Piovan A, Battaglia J, Filippini R, Dalla Costa V, Facci L, Argentini C, Pagetta A, Giusti P, Zusso M (2021) Pre- and early post-treatment with Arthrospira platensis (Spirulina) extract impedes lipopolysaccharide-triggered neuroinflammation in microglia. *Front. Pharmacol.* 12, 724993.

Piniella-Matamoros B, Marin-Prida J, Penton-Rol G (2021) Nutraceutical and therapeutic potential of phycocyanobilin for treating Alzheimer's disease. *J. Bio Sci.* 46(2), 42.

Riss J, Décordé K, Sutra T, Delage M, Baccou JC, Jouy N, Brune JP, Oréal H, Cristol JP, Rouanet JM (2007) Phycobiliprotein C-phycocyanin from Spirulina platensis is powerfully responsible for reducing oxidative stress and NADPH oxidase expression induced by an atherogenic diet in hamsters. *J. Agric. Food Chem.* 55(19), 7962–7967. DOI: 10.1021/jf070529g.

Roy-Lachapelle A, Solliec M, Bouchard MF, Sauvé S (2017) Detection of cyanotoxins in algae dietary supplements. *Toxins (Basel).* 9(3), 76. DOI: 10.3390/toxins9030076.

Shiow LR, Favrais G, Schirmer L, Schang AL, Cipriani S, Andres C, Wright JN, Nobuta H, Fleiss B, Gressens P, Rowitch DH (2017) Reactive astrocyte COX2-PGE2 production inhibits oligodendrocyte maturation in neonatal white matter injury. *Glia* 65(12), 2024–2037.

Singh A, Kukreti R, Saso L, Kukreti S (2019) Oxidative stress: A key modulator in neurodegenerative diseases. *Molecules* 24(8), 8.

Singh NK, Hasan SS, Kumar J, Raj I, Pathan AA, Parmar A, Shakil S, Gourinath S, Madamwar D (2014) Crystal structure and interaction of phycocyanin with β-secretase: A putative therapy for Alzheimer's disease. *CNS Neurol. Disord. Drug Targets* 13, 691–698.

Sinha S, Patro N, Patro IK (2018) Maternal protein malnutrition: Current and future perspectives of Spirulina supplementation in neuroprotection. *Front. Neurosci.* 12, 966.

Skaper SD, Facci L, Zusso M, Giusti P (2020) Corrigendum: An inflammation-centric view of neurological disease: Beyond the neuron. *Front. Cell. Neurosci.* 13, 578.

Sorrenti V, Castagna DA, Fortinguerra S, Buriani A, Scapagnini G, Willcox DC (2021) Spirulina microalgae and brain health: A scoping review of experimental and clinical evidence. *Mar. Drugs* 19(6), 293. DOI: 10.3390/md19060293.

Strömberg I, Gemma C, Vila J, Bickford PC (2005) Blueberry- and Spirulina-enriched diets enhance striatal dopamine recovery and induce a rapid, transient microglia activation after injury of the rat nigrostriatal dopamine system. *Exp. Neurol.* 196(2), 298–307.

Sveinbjornsdottir S (2016) The clinical symptoms of Parkinson's disease. *J. Neurochem.* 139(S1), 318–324.

Sweeney P, Park H, Baumann M, Dunlop J, Frydman J, Kopito R, McCampbell A, Leblanc G, Venkateswaran A, Nurmi A, Hodgson R (2017) Protein misfolding in neurodegenerative diseases: Implications and strategies. *Transl. Neurodegener.* 6(1), 6. DOI: 10.1186/s40035-017-0077-5.

Teravskis PJ, Covelo A, Miller EC, Singh B, Martell-Martínez HA, Benneyworth MA, Gallardo C, Oxnard BR, Araque A, Lee MK, Liao D (2018) A53T mutant α-synuclein induces tau-dependent postsynaptic impairment independently of neurodegenerative changes. *J. Neurosci.* 38(45), 9754–9767.

Tobón-Velasco JC, Palafox-Sánchez V, Mendieta L, García E, Santamaría A, Chamorro-Cevallos G, Limón ID (2013) Anti-oxidant effect of Spirulina (Arthrospira) maxima in a neurotoxic model caused by 6-OHDA in the rat striatum. *J. Neural Transm.* 120(8), 1179–1189.

Thaakur SR, Jyothi B (2007) Effect of Spirulina maxima on the haloperidol induced tardive dyskinesia and oxidative stress in rats. *J. Neural Transm.* 114(9), 1217–1225. DOI: 10.1007/s00702-007-0744-2.

Tocher DR, Betancor MB, Sprague M, Olsen RE, Napier JA (2019) Omega-3 long-chain polyunsaturated fatty acids, EPA and DHA: Bridging the gap between supply and demand. *Nutrients* 11(1), 89.

Trotta T, Porro C, Cianciulli A, Panaro MA (2022) Beneficial effects of Spirulina consumption on brain health. *Nutrients* 14(3), 676. DOI: 10.3390/nu14030676.

Vichi S, Lavorini P, Funari E, Scardala S, Testai E (2012) Contamination by Microcystis and microcystins of blue-green algae food supplements (BGAS) on the Italian market and possible risk for the exposed population. *Food Chem. Toxicol.* 50(12), 4493–4499. DOI: 10.1016/j.fct.2012.09.029.

Wakabayashi K, Tanji K, Odagiri S, Miki Y, Mori F, Takahashi H (2013) The Lewy body in Parkinson's disease and related neurodegenerative disorders. *Mol. Neurobiol.* 47(2), 495–508.

Walker DG, Lue LF, Tang TM, Adler CH, Caviness JN, Sabbagh MN, Serrano GE, Sue LI, Beach TG (2017) Changes in CD200 and intercellular adhesion molecule-1 (ICAM-1) levels in brains of Lewy body disorder cases are associated with amounts of Alzheimer's pathology not α-synuclein pathology. *Neurobiol. Aging* 54, 175–186.

Wells ML, Potin P, Craigie JS, Raven JA, Merchant SS, Helliwell KE, Smith AG, Camire ME, Brawley SH (2017) Algae as nutritional and functional food sources: Revisiting our understanding. *J. Appl. Phycol.* 29(2), 949–982.

Yang Y, Park Y, Cassada DA, Snow DD, Rogers DG, Lee J (2011) In vitro and in vivo safety assessment of edible blue-green algae, Nostoc commune var. sphaeroides Kützing and Spirulina plantensis. *Food Chem. Toxicol.* 49(7), 1560–1564. DOI: 10.1016/j.fct.2011.03.052.

Yildiz-Unal A, Korulu S, Karabay A (2015) Neuroprotective strategies against calpain-mediated neurodegeneration. *Neuropsychiatr. Dis. Treat.* 11, 297–310.

Zhang F, Lu J, Zhang JG, Xie JX (2015) Protective effects of a polysaccharide from Spirulina platensis on dopaminergic neu-rons in an MPTP-induced Parkinson's disease model in C57BL/6J mice. *Neural Regen. Res.* 10(2), 308–313.

Zhu X, Wang S, Yu L, Jin J, Ye X, Liu Y, Xu Y (2017) HDAC3 negatively regulates spatial memory in a mouse model of Alzheimer's disease. *Aging Cell* 16(5), 1073–1082.

18 Marine Green Biotechnology Advances as Bioactive Compounds for Medical Use

Odeseye Abdulhameed Adewale, Olasore Abiodun Amos, and Israel Emiezi Agarry

18.1 INTRODUCTION

The basic source of life is water, and its sustainability depends on water. As such, it is a known fact that the surface of the earth is composed of about 70% extant water bodies such as oceans and seas (Thakur and Thakur, 2006; Trincone, 2017, Khalifa *et al.*, 2019). Likewise, these extant water bodies are inhabited by diverse forms of living organisms. The existing living organisms range from micro to macro species, both plants and animals. Despite scientific investigations about and in water bodies, Trincone (2017) noted that more than 90% of marine species of living organisms are yet to be identified, classed, described, and taxonomized. Thus, a deeper understanding of water bodies and the organisms they contain has enhanced and elicited their potential as well as that of the inhabiting organisms for sustainable utilization in the course of human endeavors. Typical applications have begun to emerge, and their advances are being found beneficial in various fields of study. These advances via the application of science and technology have created a special study of water bodies and the living organisms that occupy them, termed Marine Biotechnology.

18.2 MARINE BIOACTIVE SUBSTANCES

Marine bodies are hosts to a plethora of living organisms that are classified on the basis of diverse features such as size, color, morphology, mode of nutrition, etc. These diverse organisms existing in the marine bodies constitute microorganisms, plants, and sponges, which are the main components of water body environments. The marine organisms, despite their diversity, possess biologically potent and active substances, often existing naturally or as metabolites. These active substances from living organisms are referred to as bioactive compounds, and they have been characterized by their numerous biological health benefits for people. The biologically active substances can be obtained from plant and animal sources existing in the water bodies. They can be isolated, extracted, and purified using specialized technologies to obtain distinctive bioactive substances such as polysaccharides, peptides, fatty acids, enzymes, vitamins, minerals, as well as metabolites of some organisms for applications in improving human health via the development of promising pharmaceutical and functional foods.

With respect to the features of emerging or existing bioactive substances, studies have reiterated their potential use for human health purposes across the globe. The achievement of these purposes can be viewed in different ways: mainly through the use of diets (food) or (pharmaceuticals) medicinal drugs. The former plays the role of improving consumer health, and it has led to the development of a special class of foods called functional foods. Functional foods are special foods that contain bioactive substances either as natural components or as fortifying agents (fortificants) with the potential to enhance health beyond the provision of basic nutrients in food (Lordan *et al.*, 2011). In

DOI: 10.1201/9781003369738-22

the latter, the application of bioactive substances has also been found beneficial in the development of medicinal drugs in the medical world to enhance human health. This often accompanies exploitation of the potential of natural products of organisms of marine origin in drug development. Khalifa *et al.* (2019) noted that 99% of all marine bacteria possess the ability to synthesize many interesting natural biologically active substances that are potential medicinal drug leads. Sometimes, the synthesis of these substances is attributed to the production of secondary metabolites by some microorganisms as defense mechanisms during or when subjected to unfavorable or extreme environmental conditions of temperature, pressure, pH, or salinity as well as when resisting predation. Thus, these secondary metabolites, which form part of the bioactive substances whose impacts on health have been evaluated in clinical trials, include the ones obtained from marine organisms such as microflora (bacteria, cyanobacteria, actinobacteria, and fungi), algae (microalgae and macroalgae), invertebrates, sponges, etc. (Khalifa *et al.*, 2019). Meanwhile, Huang *et al.* (2016) and Khalifa *et al.* (2019) in their studies estimated that over 22000 secondary metabolites are known, with 10% of the bioactive metabolites that are natural products coming from a microbial source.

18.2.1 Bioactive Compounds in Microalgae

Modern research demonstrates that microalgae include a variety of chemical compounds (Figure 18.1) that function effectively as anti-inflammatory, antioxidant, antimicrobial, and anticancer agents. Researchers have paid close attention to microalgae as a potential alternative source of new bioactive compounds.

18.2.1.1 Proteins and Amino Acids

Microalgae, being principally organized as members of the plant kingdom, are capable of utilizing the inorganic sources of nitrogen in marine water, such as ammonia, nitrates, and nitrites, to produce protein. Proteins are composed of amino acids joined together by peptide linkages. In plants and animals, proteins are high–molecular weight compounds that are built up of amino acids and contain some reactive groups, such as carbonyl and amino, at their surfaces. The large molecular size and the presence of the reactive groups account for the sensitivity of protein molecules to different conditions and reagents, such that they readily undergo transformation. Promising functional

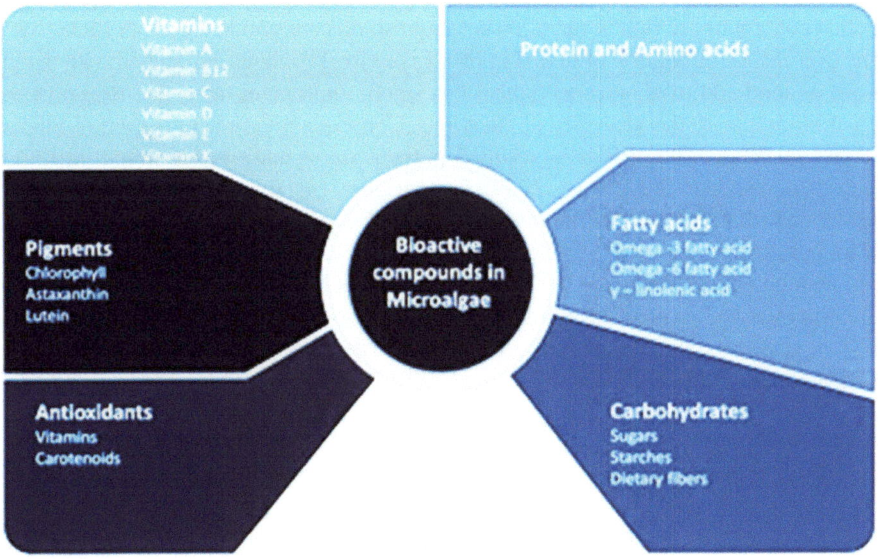

FIGURE 18.1 Bioactive compounds in microalgae.

activities have been attributed to proteins of marine origin due to their distinctive properties, such as foaming property, gel-forming ability, and microbial activity (Rasmussen and Morrissey, 2007). These enable the protein content of various microalgae species to be compared with food protein, thus forming a good alternative for human consumption. More so, the amino acids of microalgal proteins are characterized by excellent bioavailability and high productivity. For instance, Spirulina is reportedly high in protein (60–70% depending on the strain), and not only does this protein possess all of the essential amino acids, but these amino acids are characterized with excellent bioavailability (Plaza *et al.*, 2009). The industrial growth of *Dunaliella* is similar, with a protein extract yield about 100 times greater than that of agriculture and 50 times greater than that in fish production (Rasmussen and Morrissey, 2007; Lordan *et al.*, 2011). With respect to cell protection, Sheih *et al.* (2009) noted significant protective effects of linear molecules of two or more amino acids (called peptides) from the microalga *Chlorella vulgaris* against cellular damage as one of the biotechnological advances of bioactive substances in marine microalgae.

18.2.1.2 Fatty Acids

Fatty acids are the building blocks of lipids (fat-soluble constituents of living cells) in plants, animals, and microorganisms. They can be straight, branched, or ringed chains but commonly exist in combined states. In microalgae, fat exists in combination with alcohols or sugars. Palmitic and stearic acids are the most widely distributed in the lipids of organisms, with stearic acid being abundant in plants. Some fatty acids are essentially required because the required quantity cannot be synthesized in the animal body but must be taken in the diet. Such essential fatty acids include omega-3 (alpha-linolenic acid) and omega-6 (linoleic acid) fatty acids. For instance, human infants require docosahexaenoic (DHA) and arachidonic acids, of which the former can be derived from Omega-3 fatty acid and the latter from Omega-6 fatty acid. Being components of cell membranes that contribute to the regulation of blood, hormone, blood clotting, artery wall relaxation, and others, omega-3 fatty acids play vital roles in membrane structure and function, gene expression, cell signaling, and communication channels (Calder, 2016). Furthermore, omega-3 fatty acids like DHA help to increase the production of bioactive lipid mediators formed via the cyclooxygenase and lipoxygenase pathways. These mediators are recognized for their roles in anti-inflammation, immunity enhancement, and smooth muscle contraction (Freitas and Campos, 2019). As stated by Lordan *et al.* (2011), higher animals and plants are limited in the requisite enzymes to synthesize polyunsaturated fatty acids (PUFAs) of more than 18 carbons; thus, they have to be obtained via dietary sources that contain or are rich in them. Fishes and fish oils are good sources of PFAs, but there are concerns over the accumulation of toxins in fish. Likewise, problems associated with the utilization of fish oils as food additives, such as typical fish smell, unpleasant taste, and oxidative instability, are a great hindrance to the identified applications. Consequently, the cultivation of microalgae for the commercial production of PUFAs resulted in their use as additives in infant milk formulation and dietary supplements (Sheih *et al.*, 2009). Furthermore, certain amino acids can be produced by altering the growth conditions of microalgae such as *Chlorella minutissima* and Spirulina, which provide an interesting source of γ-linolenic acid (20–25% of the total lipid fraction), which is a precursor to prostaglandins, leukotrienes, and thromboxanes involved in the modulation of immunological, inflammatory, and cardiovascular responses. Also, Spirulina was found to contain sterols, including clionasterol, which has been shown to increase the production of plasminogen-activating factors in vascular endothelial cells (Mata *et al.*, 2010).

18.2.1.3 Carbohydrates

Carbohydrates are macronutrients (alongside fat and protein) and the major known source of energy in diets. They exist in varying forms, such as sugars, starches, and dietary fibers. In microalgae, carbohydrates are found in the form of sugar, starch, glucose, and other polysaccharides with high overall digestibility (Lordan *et al.*, 2011). The linkage between the high digestibility of microalgal carbohydrates and their bioactive nature is related to the reason why the use of dried whole microalgae in food and feeds is unlimited. Essentially, polysaccharide complexes from *Chlorella pyrenoidosa*, and possibly *Chlorella ellipsoidea*, contain glucose in combination with any of galactose, rhamnose, mannose, arabinose, N-acetylglucosamide, and N-acetylgalactosamine. These complexes are believed to have immune-stimulating properties and can inhibit the proliferation of *Listeria monocytogenes* and *Candida albicans* (Mata *et al.*, 2010). Chlorella has also been described as a good food additive due to the taste and flavor-adjusting potential of its coloring agent. The most important constituent in Chlorella is beta-1,3-glucan, which is an active immune stimulator, a free radical scavenger, and a reducer of blood lipids (Spolaore *et al.*, 2006).

18.2.1.4 Antioxidants

Antioxidants are compounds or substances that inhibit the action of free radicals (unstable molecules that bring about cell damage). These substances either exist naturally or are synthesized to protect body cells against free radicals that contribute to the development of chronic diseases, such as heart disease and cancer. This is often achieved by donating an electron to free radicals, thereby reducing their reactivity with body cells. Among the existing thousands of antioxidants, carotenoids, and vitamins are the common compounds of nutritional and therapeutic importance. Dietary carotenoids are vitamin A precursors, or provitamin A—they can be converted to vitamin A in the human body, thereby acting as antioxidants by inhibiting the formation of reactive oxygen (Yeum and Russell, 2022). According to Lordan *et al.* (2011), the essential bioactive substances of interest in microalgae carotenoid production are beta-carotene and astaxanthin, as they have the potential for yields in high concentration. Rasmussen and Morrissey (2007) listed *Dunaliella salina*, *Sarcina maxima*, *Chlorella protothecoides*, *Chlorella vulgaris*, and *Haematococcus pluvialis* as common strains of microalgae under study for natural production of carotenoids. Metting (1996) asserted that *Dunaliella salina* is the most suitable organism for the mass production of beta-carotene, as it can produce up to 14% of its dry weight. beta-carotene is characterized by strong antioxidant potential that helps to combat the harmful effects of free radicals as well as suppress the effects of premature aging induced by UV rays. On the whole, Lordan *et al.* (2011) reported that beta-carotene derived from microalgae is more biologically active than synthetically produced beta-carotene, and contains numerous carotenoids as well as essential nutrients that the synthetic version lacks. Furthermore, *Dunaliella salina* under irradiation stress has the potential to accrue a significant amount of xanthophylls, particularly zeaxanthin, which possesses unique biological properties with the potential for disease prevention (Yokthongwattana *et al.*, 2005). The most interesting aspect of microalgae is their vast diversity and the possibility of growing them in different conditions. This suggests they are a rich source of nutritious and bioactive compounds, including phycobilins, fatty acids, polysaccharides, vitamins, sterols, as well as essential vitamins (A, B_1, B_6, B_{12}, C, E, biotin, and pantothenic and folic acids).

18.2.1.5 Pigments

Since microalgae can produce a higher chlorophyll content than plants, they are used to make chlorophyll compounds because of their significant and conspicuous biological functions in human health (Fernandes *et al.*, 2021). Microalgae contain chlorophyll, which has profoundly important health benefits. Chlorophyll and its constituent parts are frequently employed in pharmaceutical and healthcare products. *Cladophora* sp., which contains chlorophylls a and b, *U. intestinalis*, which contains chlorophyll a, and *Sargassum ilicifolium*, which has the lowest chlorophyll content, have all been shown to accelerate the healing of wounds by more than 25% (Ibrahim *et al.*, 2023). In a prior instance, it was discovered that the chlorophyll in microalgae might speed up the healing of a persistent ulcer. By using ointments containing chlorophyll derivatives, not only is the pain relieved within a few days, but the damaged tissues' appearance is also improved (Hosikian *et al.*, 2010).

Lutein, which can be extracted from different microalgae species such as *Chlorella vulgaris* and *Haematococcus pluvialis*, has proven to be beneficial in preventing age-related macular degeneration (AMD), Alzheimer's disease (AD), and free radical scavenging for healthy skin. Lutein has been shown to act as a blue light filter, preventing eye tissue from deterioration due ti sunlight, as well as preventing build-up of oxygen-related free radicals. (Ren *et al.*, 2021).

Astaxanthin, which can be synthesized from *Haematococcus pluvialis* and *Chlorella zoofingiensis*, possesses powerful antioxidant potential. The polar nature of astaxanthin makes it valuable in treating inflammation, preventing central nervous system disorders, and mitigating Parkinson's disease (Barkia *et al.*, 2019; Villaro *et al.*, 2021).

18.2.1.6 Vitamins

Microalgae are fast-dividing and photosynthetic unicellular organisms with the capacity to synthesize vitamins, including vitamins A, B_{12}, C, D, E, and K, which are rare in plants. These vitamins can reduce fatigue, safeguard against heart disease, prevent eye diseases, and protect against premenstrual syndrome (Sathasivam *et al.*, 2019). *Dunaliella tertiolecta* and *Tetraselmis suecica* are excellent sources of vitamin B_{12}, which plays several functions during one's lifespan. Microalgae including *Chaetoceros gracilis*, *Thalassiosira pseudonana*, *Dunaliella tertiolecta*, and *Nannochloropsis oculata* have been shown to be good sources of vitamin C, all of which are vital for the production of several compounds in the human body (Ibrahim *et al.*, 2023).

18.3 MACROALGAE

Macroalgae are simple chlorophyll-containing marine organisms commonly known as seaweeds. They exist in varying forms, such as cellular (one cell), colonies (a group of cells), or tissues (an intricate structure of cells). Being a natural bounty of the sea, macroalgae grow in both shallow and deep water bodies (Sing *et al.*, 2020). The existence of seaweeds in the marine environment is symbiotic with marine animals. As with microalgae, the classification of seaweeds hinges on their color (green, red, blue, and brown). Meanwhile, Lordan *et al.* (2011) attributed the variation in the color of seaweeds to their habitat compared with microalgae. The color or phyto-pigmentation observed in seaweeds is related to energy obtained from sunlight, considering the fact that the light intensity requirement of most macroalgae for photosynthesis varies with their color. It was asserted by Bocanegra *et al.* (2009) that green macroalgae, which have a high capacity to absorb a large quantity of sunlight energy, are commonly found in coastal waters, while brown and red seaweeds are found at greater marine depths where the penetration of sunlight energy is limited. They require carbon dioxide released by marine animals during respiration for food formation during photosynthesis and release oxygen for the respiration and survival of marine animals. The uniqueness of seaweeds lies in their high adaptability, mostly in extreme conditions, and their richness in biologically active compounds, making them good sources of nutrients and medicinal cures (Sing *et al.*, 2020). Polysaccharides, proteins, fibers, minerals, vitamins, and other functional compounds are among the range of bioactive compounds provided by seaweeds, as enumerated by Cardoso *et al.* (2015). Studies have established the activity and functionality of these compounds for utilization as nutrients as well as in the pharmaceutical, nutraceutical, and cosmetic industries. For instance, Sing *et al.* (2020) noted that this potential is attributed to the ability of polysaccharides (which are one of the major components of seaweeds) to disperse in water, forming colloidal solutions. Similarly, the biocidal compounds derived from some species of seaweeds have been described by Selvin (2004), with high efficacy in killing parasitic vectors in both their larval and adult stages.

18.3.1 Bioactive Compounds in Macroalgae

18.3.1.1 Polysaccharides

Polysaccharides are complex carbohydrates such as starch or cellulose that are made up of sugar molecules linked in either a straight or a branched chain structure. Extracts from macroalgae can be used to prepare a powdered extract used as a gelling agent or thickener (agar), complex carbohydrates used for commercial preparation of food and drink (carrageenan), and a viscous liquid used as thickener or emulsifier (alginates), due to their rich polysaccharide component. These components constitute the most commercially exploited components in seaweeds due to their textural and stabilizing qualities. Thus, they are widely applied in food production for aqueous solution thickening, gel formation, water-soluble film formation, and as stabilizers in ice cream production (Rasmussen and Morrissey, 2007). Although macroalgae are relatively rich in carbohydrate content, they cannot be considered an energy-rich food because of their low digestibility (Bocanegra *et al.*, 2009).

In red seaweed varieties, the distinctive polysaccharides are starch, cellulose, xylan, and mannan as well as water-soluble fiber fractions. In brown macroalgae, the normal polysaccharides are fucoidan, laminarin, cellulose, alginates, and mannitol, whereas the fibers are mainly cellulose and insoluble alginates. However, most of these polysaccharides are not digestible by the human gastrointestinal tract; as such, they are described as dietary fibers (Lordan *et al.*, 2011). Thus, algae fiber has been established to promote human health, mostly through nutrition. Specifically, fucoidans are reported to display numerous physiological and biological properties, including anticoagulant, antiviral, antitumor, antithrombotic, and antioxidant activities. Also, they have been reported to have effects on the inflammatory and immune systems (Berteau and Mulloy, 2003; Pomin and Mourão, 2008). It was further added that the therapeutic potential of fucoidans increases with their degree of sulfation. Alternatively, laminarin, the second-largest polysaccharide found in brown algae, is said to be effective in modulating intestinal metabolic activities. This is often done through its effects on mucus composition, intestinal pH, and short-chain fatty acid production (Devillé *et al.*, 2007). Porphyran is another sulfated polysaccharide found in red macroalgae and is used as a gelling agent, a nutritional supplement, and an antioxidant (O'Sullivan *et al.*, 2010). Oligosaccharides are other carbohydrate derivatives that are described as having a low degree of polymerization. Most of them are identified as performing activities beneficial to health, including immuno-stimulation, antioxidant, and antitumor. They are beneficial to human health by enhancing the growth of prebiotic bacteria when they act as a substrate for the growth of health-promoting anaerobic bacteria such as bifidobacteria and lactobacilli. Overall, algal-derived polysaccharides are described as having great potential for valuable bioactive functions because of their ease of isolation and numerous health as well as industrial benefits.

18.3.1.2 Proteins and Amino Acids

Proteins from sea sources are considered complete or wholesome because they contain all the essential amino acids. In seaweeds, the protein composition contains all the essential amino acids, although they vary in concentration from phylum to phylum. Essential amino acids including leucine, valine, and methionine are among the abundant amino acids found in seaweeds like *Palmaria palmata* in concentrations comparable to those reported for ovalbumin, whereas the concentrations of isoleucine and threonine are comparable to those of legume proteins (Lordan *et al.*, 2011). Furthermore, *Ulva rigida* is rich in leucine, phenylalanine, valine, and histidine, and the concentration of histidine, which is an essential amino acid important for children's growth, is comparable to that found in eggs and legumes. Thus, edible seaweeds that are rich in bioactive compounds like amino acids act like a novel food and help not only in preventing diseases but also in other therapeutic actions such as antihypertension, immunomodulatory, antithrombotic, antioxidant, anticancer, antimicrobial, anti-obesity, antiviral, antifungal, antibacterial, antidiabetic, antihypertensive, immunomodulatory, anticoagulant, anti-inflammatory, anti-estrogenic, thyroid stimulating, neuroprotective, antioxidant, and tissue healing properties.

According to Lordan et al (2011), these characteristics make seaweeds a remarkable natural resource with a diverse range of biological functions attributed to their proteins. When proteins are digested, linear molecules of linked amino acids called peptides result. These bioactive peptides are produced by various methods such as solvent extraction, enzymatic hydrolysis, and microbial fermentation of marine proteins (Lordan *et al.*, 2011). The bioactive peptides contain 3–20 amino acids and are responsible for the biological functionality of the protein. Furthermore, reports have shown little digestibility of algal protein *in vivo*, whereas a high rate of *in vitro* digestibility has been recorded using proteolytic enzymes such as pepsin, pancreatin, and pronase. For instance, the *in vitro* digestibility of proteins from the red seaweed *Porphyra tenera* is approximately 70% (Mabeau and Fleurence, 1993).

18.3.1.3 Fatty Acids

Fatty acids are composed of triglycerides and can be categorized as saturated and unsaturated fatty acids. The most common saturated fatty acid is palmitic, while palmitoleic, oleic, and linoleic are among the most common unsaturated fatty acids. From the foregoing, linoleic acid and α-linolenic acid are critical for use by body cells. α-Linolenic acid is a precursor to omega-3 fatty acids—eicosa-pentaenoic acid (EPA) and docosahexaenoic acid (DHA)—as linoleic acid is a precursor to omega-6 fatty acids (Wiley-Blackwell, 2020). Omega-3 and omega-6 are essential fatty acids; they are best consumed in the diet for effectiveness. Studies have shown that EPA and DHA are easily and effectively produced by marine algae. Thus, seaweeds are judged as a good source of essential fatty acids due to the high proportion in their composition. Generally, unsaturated fatty acids play important roles in regulatory functions such as blood clotting, blood pressure, brain development, and inflammatory responses as well as good functioning of the nervous system (Wall *et al.*, 2010). For instance, Dembitsky (1991) characterized green algae such as *Ulva pertusa* with a good quality of hexadecatetraenoic, oleic, and palmitic acids whereas n-3 fatty acids and octadecatetraenoic acids were found prominent in *Laminaria* sp. and *Undaria pinnatifida.*

18.3.1.4 Vitamins and Minerals

Vitamins are organic substances that are essential for metabolic activities in minute quantities in human nutrition. In seaweeds, the amount and composition of vitamins depend on the type of processing or unit operation they are subjected to, such as dehydration. For example, the quantity of dehydrated beta-tocopherol in *Himanthalia elongata* (33.3 μg/g dry weight) was reportedly higher than in canned *Himanthalia elongata* (12 μg/g dry weight) (Sánchez-Machado *et al.*, 2002). Table 18.1 gives the list of dietary vitamins of algal origin. Seaweeds are also characterized by a rich and unparalleled deposit of minerals and trace elements because of their capacity to retain inorganic substances. This wealth of mineral deposits, as noted by Burtin (2003), was attributed to the features of their cell membrane, and the fraction of the mineral content in some species accounts for up to 36% dry matter.

18.4 CHRONIC DISEASES OF IMPORTANCE IN MEDICAL USE OF MARINE GREEN BIOACTIVE SUBSTANCES

Research over the years has successfully demonstrated the potential of bioactive substances, most importantly as functional food ingredients in improving human health. The findings of the researchers have revealed that the marine world is a deposit or reserve of various under-tapped or untapped functional bioactive substances. This awareness has triggered the interest of both the food and pharmaceutical industries in the development of products of health importance to consumers. Nutrition targets functional marine substances like omega-3, omega-6, carotenoids, vitamins, minerals, bioactive peptides, amino acids, polysaccharides, etc. with the aim of either preventing diseases or slowing down the rate of disease progression in the human body. Similar overlapping functions encapsulate the targets of pharmaceutical industries as well, with a focus on substances such as enzymes, receptors, and transcription factors (Kadam and Prabhasankar, 2010). However, Biesalski *et al.* (2009) stated explicitly that the health-promoting features of any naturally occurring bioactive substance to be exploited should be preventive rather than curative.

18.4.1 Diabetes

Diabetes is a metabolic disorder that is associated with loss of control of sugar in the body. This disorder occurs due to inadequate production of insulin, thereby affecting the blood sugar level. One

TABLE 18.1
Algal Sources of Dietary Vitamins

Vitamins	Algal species	Reported level (µg/g dry wt)
Vitamin C	*Ulva* sp.	94.20–1250
	Monostroma undulatum	1590–4550
	Undaria pinnatifida	1847.38
	Ascophyllum nodosum	81.75
	Laminaria digitata	355.25
	Porphyra umbilicalis	1610.63
	Palmaria palmata	690
	Thalassiosira pseudonana	1100
	Chaetoceros muelleri	16000
	Gracilaria changgi	285
Vitamin E	*Ulva rigida*	19.70
	Ascophyllum nodosum	3.63
	Dunaliella tertiolecta	200–500
	Undaria pinnatifida	145–174
	Laminaria digitata	34.38
	Porphyra umbilicalis	14.25
	Palmaria palmata	162
α-Tocopherol	*Porphyridium cruentum*	55.2
	Laminaria ochroleuca	8.9 ± 2.1
	Saccorhiza polychides	5.7 ± 1.3
	Himanthalia elongata	12.0–33.3
	Tetraselmis suecica	190–1080
γ-Tocopherol	*Porphyridium cruentum*	51.3
Carotenoids	*Porphyridium cruentum*	1020 ± 140
α-Carotene	*Chlorella pyrenoidosa*	4232.50
	Dunaliella salina	*2410–2690*
β-Carotene	*Ascophyllum nodosum*	4314.3
	Chlorella pyrenoidosa	80–500
	Chlorella vulgaris	–
	Chlorococcum	4950–138250
	Dunaliella salina	–
	Fucus serratus	–
	Fucus vesiculosus	52 ± 4
	Gracilaria changgi	80 ± 30
	Haematococcus pluvialis	–
	Laminaria digitata	–
	Laminaria saccharina	–
	Pelvetia canaliculata	–
	Phormidium sp.	–
	Porphyra tenera	2040
	Synechocystis sp.	

Source: Lordan, S.R. *et al.*, *Mar. Drugs*, 9, 1056, 2011.

of the approaches to managing diabetes is dietary management, which involves the use of diets to maintain blood glucose and lipid concentrations. Diabetes is classified as either Type 1, when there is decreased production of insulin because of the autoimmune process that destroys the pancreatic beta cells; or Type 2, which causes damage to the eyes, nerves, and kidneys. Type 1 diabetes is often seen in childhood, and its management requires lifelong insulin injections, whereas Type 2 is typically found in middle-aged persons, and its management often requires the use of both diet and drugs. The target of diabetes management is to mitigate the development of associated complications such as diabetic neuropathy, retinopathy, and nephropathy (Lunn and Theobald, 2006). To this effect, some marine bioactive extracts have been characterized as providing improved control of glycemic index (GI). *In vitro*, Pascual *et al.* (2007) reported *Xetospongia muta* sponge and *Bunodosoma granulifera* and *Bartholomea annulata* sea anemones to be effective in inhibiting dipeptidyl peptidase IV activity. Also, when experimenting with the efficacy of certain bioactive extracts, such as antidiabetic agents from *Ulva alga* and *Posidonia oceanica phanerogam* on diabetic Wistar rats, Gokce and Haznedaroglu (2008) and Celikler *et al.* (2009) were able to realize a decrease in concentration of the blood glucose of the subjects. Delarue *et al.* (1996) examined the antidiabetic properties of fish oil in healthy human subjects and found a reduction in glucose oxidation and increased fat oxidation and glycogen storage.

18.4.2 Cardiovascular Disease

Cardiovascular diseases (CVDs) are diseases that are connected with the heart and blood vessels. CVD is termed a "rich man's disease" and is a leading cause of mortality and morbidity around the globe. It affects the heart, arteries, and veins (blood vessels) as well as the blood circulation, which often results in ailments such as heart failure, stroke, atherosclerosis, deep vein thrombosis, etc. In nutritional terms, studies have ascertained that there is a strong inverse relationship between the consumption of fiber-rich foods and chronic diseases like CVDs. The possibility of this is linked to the hypocholesterolemic effect of the soluble fraction of fiber. Thus, considering the high soluble dietary fiber composition of seaweeds, it follows that the content of soluble polysaccharides can induce hypocholesterolemic and antihypertensive effects in consumers of seaweeds (Lordan *et al.*, 2011). Similarly, omega-3-rich oil extracted from marine organisms has the potential to offer cardio-protective ability, or the ability to influence the likelihood of CVD risk factors. Lordan *et al.* (2011) affirmed that numerous epidemiological and experimental studies have shown and proven that diets rich in fish and fish oils are associated with the potential to reduce the risks of CVDs. Furthermore, modification of the levels of high-density lipoprotein (HDL) and low-density lipoprotein (LDL) cholesterol in the bloodstream is another role played by marine bioactive substances in food, and it reduces the incidence of hypertension. Omega-3 PUFAs and astaxanthin were recorded to be effective against CVDs. When used as dietary supplements, they either prevent or alleviate certain CVDs by reducing inflammation, associated with the development of coronary heart disease (CHD) (Lordan *et al.*, 2011).

18.4.3 Inflammation

Inflammation is a protective response to injury or infection of tissue in any part of the body. The response may result in swelling, redness, heat, or pain in the injured or infected part of the body. When the response is exaggerated, misdirected, or prolonged, health may be affected adversely, while conditions such as inflammatory bowel disease, arthritis, and asthma may result (Miyashita, 2009). In arthritis, for instance, which is a description of an inflammatory condition of body joints, the ailment is a pain-associated disease that is common in elderly people. Dietary PUFAs (omega-3) have been proven to be effective as a therapy with 3–4 months of supplementation (Lordan *et al.*, 2011). Similarly, EPA and DHA were noted by Laye (2010) as anti-inflammatory agents that could minimize neuro-inflammation when the diet is rich in them. Thus, among the numerous anti-inflammatory compounds with potential pharmacological applications that have been isolated from marine sources, omega-3 PUFAs appear to be the most prominent and promising agents. Thus, increased consumption of marine-sourced omega-3 fatty acids may help in offering a complementary role to drug therapy management of inflammation. Baked foods, margarine, dairy products, and meat-based and fish-based meals could be used as nutrient vehicles to increase the use of PUFAs.

18.4.4 Cancer

Cancer is a condition that is caused by the presence of a growth due to uncontrollable cell multiplication, thereby destroying healthy tissues. Cancer remains a contemporary disease threatening human life, as millions of new cases are constantly emerging globally. This rising trend in cancer cases is attributed to unhealthy lifestyle, environmental changes, and modernization. White *et al.* (2014) related the predisposing factors of cancer to internal factors (immune conditions, genetic predisposition) and external factors (tobacco, chemicals, radiation, and infectious organisms). With the emergence of biotechnology in the 1980s, various efforts have been directed toward exploiting marine resources with the aim of preventing and treating the menace of cancer. Consequently, various natural products from marine sources that have displayed different anticancer activities are under investigation. In the early 1990s, marine-derived anticancer bioactive compounds began to undergo clinical trials, and the successes over time have given rise to some bioactive substances that have been approved for use against cancer. Instances include trabectedin, which was isolated from the Caribbean marine tunicate *Ecteinascidia turbinata*, and aplidine, a bioactive extract from the Mediterranean marine tunicate *Aplidium albicans.* Rinehart (2000) and Mayer and Gustafson (2008) agreed that a marine-derived bioactive extract from *Aplidium albicans* has powerful antitumor potential with possible applications in prostate, gastric, breast, and colon cancer treatment.

Meanwhile, the mode of action of marine-derived anticancer compounds varies; some bioactive compounds exhibit cytotoxic effects by attacking macromolecules expressed by cancer cells, while a meaningful number act as antitumor agents by inhibiting the growth of tumor cells in humans (Newman and Cragg, 2014). Others, like biologically active peptides and polyketide metabolites, are effective at killing cancer cells by inducing apoptotic death by affecting cell signals via activation of the protein kinase family (Sithranga Boopathy and Kathiresa, 2010). Some metabolites also function in cancer therapy by inhibiting cell division by disrupting the polymerization of actin (Bubb *et al.*, 1995), inhibiting cyclin-dependent kinase 4 (Soni *et al.*, 2000), inhibiting protein synthesis (Fukuoka *et al.*, 2000), or blocking topoisomerase II (Marshall *et al.*, 2003). Generally, for cancer treatment, 17 clinically approved drugs of marine origin were listed by Rotter *et al.* (2021), with the prospect of addenda in years to come, as more are undergoing clinical trials. These drugs include cytarabine (Cytosar-U), nelarabine (Arranon), Fludarabine (Fludara®), trabectedin (Yondelis®), eribulin mesylate (Halaven®), brentuximab vedotin (Adcetris®), plitidepsin (Aplidin), polatuzumab vedotin (Polivy®), and enfortumab vedotin-ejfv (PADCEV®), and currently lurbinectedin (Zepzelca) and belantamab mafodotin (BLENREP) are being tried for cancer treatment (Rotter et al., 2021).

18.5 CONCLUSION

The marine green environment is a unique habitat, harboring diverse organisms with great potential to produce bioactive substances that are characterized by distinctive biological and physicochemical properties. Although most marine green bioactive substances are still in the research and trial stages, a few have displayed great potential to improve human well-being in the diet and as pharmaceuticals. Understanding that diet plays a critical role in human health, consumers are not only concerned with tasty and convenient food but also with nutrition. This is really great in connection with marine algae and their composition of carbohydrates, proteins, lipids, and fibers that occur naturally, as well as their metabolites, whose extraction is simple and economical.

REFERENCES

Aransiola, S. A., Victor-Ekwebelem, M. O., Ikhumetse, A. A., Abioye, O. P. (2021). Challenges and future prospects of biotechnology. In N. R. Maddela, L. C. García (eds.), *Innovations in Biotechnology for a Sustainable Future.* Springer Nature Switzerland AG 2021. DOI: 10.1007/978-3-030-80108-3_20.

Aransiola, S. A., Victor-Ekwebelem, M. O., Leh-Togi Zobeashia, S. S., Maddela, N. R. (2023). Sources and techniques for biofuel generation. In M. P. Shah (ed.), *Green Approach to Alternative Fuel for a Sustainable Future.* Elsevier Inc. DOI: 10.1016/B978-0-12-824318-3.00026-6.

Barkia, I., Saari, N., Manning, S. R. (2019). Microalgae for high-value products towards human health and nutrition. *Mar. Drugs*, 17(5), 304.

Biesalski, H.-K., Dragsted, L. O., Elmadfa, I., Grossklaus, R., Müller, M., Schrenk, D., Walter, P., Weber, P. (2009). Bioactive compounds: Definition and assessment of activity. *Nutrition*, 25(11–12), 1202–1205.

Berteau, O., Mulloy, B. (2003). Sulfated fucans, fresh perspectives: Structures, functions, and biological properties of sulfated fucans and an overview of enzymes active towards this class of polysaccharide. *Glycobiology*, 13(6), 29R–40R.

Bocanegra, A., Bastida, S., Benedí, J., Ródenas, S., Sánchez-Muniz, F. J. (2009). Characteristics and nutritional and cardiovascular-health properties of seaweeds. *J. Med. Food*, 12(2), 236–258.

Bubb, M., Spector, I., Bershadsky, A., Korn, E. D. (1995). Swinholide is a microfilament disrupting marine toxin that stabilizes actin dimers and severs actin filaments. *J. Biol. Chem.*, 270(8), 3463–3466.

Burtin, P. (2003). Nutritional value of seaweeds. *EJEAFChe*, 2, 498–503.

Calder, P. C. (2016). Docosahexaenoic acid. *Ann. Nutr. Metab.*, 69(Suppl 1), 7–21.

Carvalho, W., Silva, D.-D. V., Canilha, L. (2005). Aditivos alimentares produzidos por via fermentativa part I: ácidos orgânicos. *Rev. Anal.*, 18, 70–76.

Celikler, S., Tas, S., Vatan, O., Ziyanok-Ayvalik, S., Yildiz, G., Bilaloglu, R. (2009). Anti-hyperglycemic and antigenotoxic potential of Ulva rigida ethanolic extract in the experimental diabetes mellitus. *Food Chem. Toxicol.*, 47(8), 1837–1840.

Da Silveira, J. M. F. J., Borges, I. D. C., Buainain, A. M. (2005). Biotecnologia e agricultura: Da ciencia e tecnologia aos impactos da inovação. *São Paulo em Perspect*, 19, 101–114.

Delarue, J., Couet, C., Cohen, R., Bréchot, J. F., Antoine, J. M., Lamisse, F. (1996). Effects of fish oil on metabolic responses to oral fructose and glucose loads in healthy humans. *Am. J. Physiol.*, 270(2), E353–E362.

Dembitsky, V. M., Pechenkina-Shubina, E. E., Rozentsvet, O. A. (1995). Glycolipids and fatty acids of some seaweeds and marine grasses from the black sea. *Phytochemistry*, 30(7), 2279–2283.

Devillé, C., Gharbi, M., Dandrifosse, G., Peulen, O. (2007). Study on the effects of laminarin, a polysaccharide from seaweed, on gut characteristics. *J. Sci. Food Agric.*, 87(9), 1717–1725.

Fernandes, A. S., Nascimento, T. C. D., Pinheiro, P. N., Jacob-Lopes, E., Zepka, L. Q. (2021). Determination of profile of chlorophyll compounds in microalgae species/determinação do perfil de compostos de clorofila em espécies de microalgas. *Braz. J. Dev.*, 7(1), 4381–4399.

Forti, L., Di Mauro, S., Cramarossa, M. R., Filippucci, S., Turchetti, B., Buzzini, P. (2015). Non-conventional yeasts whole cells as efficient biocatalysts for the production of flavors and fragrances. *Molecules*, 20(6), 10377–10398.

Freitas, R., Campos, M. M. (2019). Protective effects of omega-3 fatty acids in cancer-related complications. *Nutrients*, 11(5), 945.

Fukuoka, K., Yamagishi, T., Ichihara, T., Nakaike, S., Iguchi, K., Yamada, Y., Fukumoto, H., Yoneda, T., Samata, K., Ikeya, H., Nanaumi, K., Hirayama, N., Narita, N., Saijo, N., Nishio, K. (2000). Mechanism of action of aragusterol a (yta0040), a potent anti-tumor marine steroid targeting the g(1) phase of the cell cycle. *Int. J. Cancer*, 88(5), 810–819.

García-Casal, M. N., Pereira, A. C., Leets, I., Ramírez, J., Quiroga, M. F. (2007). High iron content and bioavailability in humans from four species of marine algae. *J. Nutr.*, 137(12), 2691–2695.

Gokce, G., Haznedaroglu, M. Z. (2008). Evaluation of antidiabetic, antioxidant and vasoprotective effects of Posidonia oceanica extract. *J. Ethnopharmacol.*, 115(1), 122–130.

Hosikian, A., Lim, S. u., Halim, R., Danquah, M. K. (2010). Chlorophyll extraction frommicroalgae: A review on the process engineering aspects. *Int. J. Chem. Eng.*, 2010, 1–11.

Huang, C., Leung, R. K. K., Guo, M., Tuo, L., Guo, L., Yew, W. W., Lou, I., Lee, S. M. Y., Sun, C. (2016). Genome-guided investigation of antibiotic substances produced by Allosalinactinospora lopnorensis CA15-2 T from Lop nor region, China. *Sci. Rep.*, 6(1), 20667.

Ibrahim, T. N. B. T., Feisal, N. A. S., Kamaludin, N. H., Cheah, W. Y., How, V., Bhatnagar, A., Ma, Z., Show, P. L. (2023). Biological active metabolites from microalgae for healthcare and pharmaceutical industries: A comprehensive review. *Bioresour. Technol.*, 372, 128661.

James, C. (2014). ISAAA Briefs brief 49 global status of commercialized biotech. *GM Crops*, 2014, 1–270.

Kadam, S., Prabhasankar, P. (2010). Marine foods as functional ingredients in bakery and pasta products. *Food Res. Int.*, 43(8), 1975–1980.

Kafarski, P. (2012). Rainbow code of biotechnology. *Chemik*, 66, 814–816.

Khalid, S., Abbas, M., Saeed, F., Bader-Ul-Ain, H., Suleria, H. (2018). Therapeutic potential of seaweed bioactive compounds. Retrieved January 14, 2021, from https://www.intechopen.com/books/seaweedbiomaterials/therapeutic-potential-of-seaweed-bioactivecompounds.

Khalifa, S. A. M., Elias, N., Farag, M. A., Chen, L., Saeed, A., Hegazy, M.-E. F., Moustafa, M. S., Abd El-Wahed, A., Al-Mousawi, S. M., Musharraf, S. G., Chang, F.-R., Iwasaki, A., Suenaga, K., Alajlani, M., Göransson, U., El-Seedi, H. R. (2019). Marine natural products: A source of novel anticancer drugs. *Mar. Drugs*, 17(9), 491. DOI: 10.3390/md17090491.

Layé, S. (2010). Polyunsaturated fatty acids, neuroinflammation and well-being. *Prostaglandins Leukot. Essent. Fatty Acids*, 82(4–6), 295–303.

Lordan, S. R., Ross, P., Stanton, C. (2011). Marine bioactives as functional food ingredients: Potential to reduce the incidence of chronic diseases. *Mar. Drugs*, 9(6), 1056–1100. DOI: 10.3390/md9061056.

Lunn, J., Theobald, H. (2006). The health effects of dietary unsaturated fatty acids. *Nutr. Bull.*, 31(3), 178–224.

Mabeau, S., Fleurence, J. (2006). Seaweed in food products: Biochemical and nutritional aspects. *Trends Food Sci. Technol.*, 4(4), 103–107.

MacArtain, P., Gill, C. I. R., Brooks, M., Campbell, R., Rowland, I. R. (2007). Nutritional value of edible seaweeds. *Nutr. Rev.*, 65(12), 535–543.

Marshall, K. M., Matsumoto, S. S., Holden, J. A., Concepción, G. P., Tasdemir, D., Ireland, C. M., Barrows, L. R. (2003). The anti-neoplastic and novel topoisomerase ii-mediated cytotoxicity of neoamphimedine, a marine pyridoacridine. *Biochem. Pharmacol.*, 66(3), 447–458.

Mata, T. M., Martins, A. A., Caetano, N. S. (2010). Microalgae for biodiesel production and other applications: A review. *Renew. Sustain. Energ. Rev.*, 14(1), 217–232.

Mayara, C. S., Barcelos, F. B., Lupki, G. A. C., David, L. N., Molina, G. (2018). The colors of biotechnology: General overview and developments of white, green and blue areas. Federation of European microbiology journals. *FEMS Microbiol. Lett.*, 365(21), fny239. DOI: 10.1093/femsle/fny239.

Mayer, A. M. S., Gustafson, K. R. (2008). Marine pharmacology in 2005–2006: Antitumour and cytotoxic compounds. *Eur. J. Cancer*, 44(16), 2357–2387.

Merriam-Webster. Biotecnology. 2016. https://www.merriam-webster.com/dictionary/biotechnology.

Metting, F. B. (1996). Biodiversity and application of microalgae. *J. Ind. Microbiol.*, 17(5–6), 477–489.

Miyashita, K. (2009). Function of marine carotenoids. *Forum Nutr.*, 61, 136–146.

Newman, D., Cragg, G. (2014). Marine-sourced anti-cancer and cancer pain control agents in clinical and late preclinical development. *Mar. Drugs*, 12(1), 255–278.

O'Sullivan, L., Murphy, B., McLoughlin, P., Duggan, P., Lawlor, P. G., Hughes, H., Gardiner, G. E. (2010). Prebiotics from marine macroalgae for human and animal health applications. *Mar. Drugs*, 8(7), 2038–2064.

Pascual, I., Lopéz, A., Gómez, H., Chappé, M., Saroyán, A., González, Y., Cisneros, M., Charli, J. L., Chávez, M. (2007). Screening of inhibitors of porcine dipeptidyl peptidase IV activity in aqueous extracts from marine organisms. *Enzyme Microb. Technol.*, 40(3), 414–419.

Plaza, M., Herrero, M., Cifuentes, A., Ibáñez, E. (2009). Innovative natural functional ingredients from microalgae. *J. Agric. Food Chem.*, 57(16), 7159–7170.

Pomin, V. H., Mourão, P. A. S. (2008). Structure, biology, evolution, and medical importance of sulfated fucans and galactans. *Glycobiology*, 18(12), 1016–1027.

Ramos, M. V., Melo, D. F., Silva, A. L. C. (2016). Biotecnologia: A ciência, o bacharelado, a demanda socioeconômica. UFC.

Rasmussen, R. S., Morrissey, M. T. (2007). Marine biotechnology for production of food ingredients. *Adv. Food Nutr. Res.*, 52, 237–292.

Ren, Y., Sun, H., Deng, J., Huang, J., Chen, F. (2021). Carotenoid production from microalgae: Biosynthesis, salinity responses and novel biotechnologies. *Mar. Drugs* 19(12), 713.

Rinehart, K. (2000). Antitumor compounds from tunicates. *Med. Res. Rev.*, 20(1), 1–27.

Rotter, A., et al. (2021). The essentials of marine biotechnology. *Front. Mar. Sci.*, 8, 629629. DOI: 10.3389/fmars.2021.62962.

Sato, T., Yamada, Y., Ohtani, Y., Mitsui, N., Murasawa, H., Araki, S. (2001). Production of Menaquinone (vitamin K2)-7 by Bacillus subtilis. *J. Biosci. Bioeng.*, 91(1), 16–20.

Sánchez-Machado, D., López-Hernández, J., Paseiro-Losada, P. (2002). High-performance liquid chromatographic determination of [alpha]-tocopherol in macroalgae. *J. Chromatogr. A*, 976(1–2), 277–284.

Schenberg, A. C. G. (2010). Biotecnologia e desenvolvimento sustentavel. *Rev. Estud. Avancados*, 24, 7–17.

Sheih, I. C., Wu, T.-K., Fang, T. J. (2009). Antioxidant properties of a new antioxidative peptide from algae protein waste hydrolysate in different oxidation systems. *Bioresour. Technol.*, 100(13), 3419–3425.

Singh, H., Parida, A., Debbarma, K., Ray, D., Banerjee, P. (2020). Common marine organisms: A novel source of medicinal compounds. *Int. J. Bioresour. Sci. IJBS*, 7(2), 39–49. DOI: 10.30954/2347-9655.02.2020.1.
Sithranga Boopathy, N., Kathiresan, K. (2010). Anticancer drugs from marine flora: An overview. *J. Oncol.*, 2010, 18.
Soni, R., Muller, L., Furet, P., Schoepfer, J., Stephan, C., Zumstein-Mecker, S., Fretz, H., Chaudhuri, B. (2000). Inhibition of cyclin-dependent kinase 4 (cdk4) by fascaplysin, a marine natural product. *Biochem. Biophys. Res. Commun.*, 275(3), 877–884.
Thakur, N. L., Thakur, A. N. (2006). Marine Biotechnology: An overview. *Indian J. Biotechnol.*, 5, 263–268.
Trincone, A. (2017). Enzymatic processes in marine biotechnology. *Mar. Drugs MPDI*, 15(93), 3–30.
Villadsen, J. (2007). Innovative technology to meet the demands of the white biotechnology revolution of chemical production. *Chem. Eng. Sci.*, 62(24), 6957–6968.
Villaro, S., Ciardi, M., Morillas-Espana, A., Sanchez-Zurano, A., Acien-Fernandez, G., Lafarga, T. (2021). Microalgae derived astaxanthin: Research and consumer trends and industrial use as food. *Foods*, 10(10), 2303.
Wall, R., Ross, R. P., Fitzgerald, G. F., Stanton, C. (2010). Fatty acids from fish: The anti-inflammatory potential of long-chain omega-3 fatty acids. *Nutr. Rev.*, 68(5), 280–289.
White, M. C., Holman, D. M., Boehm, J. E., Peipins, L. A., Grossman, M., Jane Henley, S. (2014). Age and cancer risk: A potentially modifiable relationship. *Am. J. Prev. Med.*, 46(3), S7–S15.
Yeum, K.-J., Russell, R. M. (2002). Carotenoid bioavailability and bioconversion. *Annu. Rev. Nutr.*, 22(1), 483–504.

19 Potentially Toxic Metals in Africa, Fresh and Marine Environment

Marine Green Contamination, Ecological and Human Health Risk

A.J. Adewumi, O.D. Ogundele, and I.S. Emumejakpor

19.1 INTRODUCTION

The marine environment makes up more than 70% of the Earth's surface (USGS 2019). It contains about 97% of the world's total water (USGS 2019). It plays important roles in the hydrogeological and biogeochemical cycles (de Moura *et al.* 2012). About 78% of global precipitation happens around the world's oceans, and about 86% of global evaporation originates from the sea (NASA 2023). In many African countries, marine environments are important sources of revenue via tourism (FAO 2017). The rapid expansion of coastal cities such as Lagos, Cairo, and Cape Town and an increase in seaports, coastal fisheries, industries, and agriculture contributes around 56% to the West African gross domestic product (GDP) (FAO 2017).

Metals with a relatively high density in comparison to water (H_2O) are referred to as heavy metals. They are able to induce toxicity even at low exposure. Due to their use for several purposes, such as agricultural, domestic, and industrial applications, heavy metals have become threats to human health (Tchounwou *et al.* 2012; Aransiola *et al.* 2021). Heavy metals are reported to originate from different geogenic and anthropogenic sources (Adewumi *et al.* 2022; Ogundele *et al.* 2023; Aransiola *et al.* 2013). Industrial, agricultural, pharmaceutical, and home effluents, as well as other metal-based industrial operations, are examples of anthropogenic sources (Tchounwou *et al.* 2012). It has also been claimed that natural events like weathering and volcanic activity contribute immensely to heavy metal contamination (Adewumi *et al.* 2022).

19.2 GEOLOGY OF AFRICA'S MARINE ENVIRONMENT

The coastal environments of Africa are characterized by diverse geology due to the continent's varied topography, tectonic history, and climate. In the north, the Mediterranean coast is marked by rugged mountains, and the Red Sea coast is a narrow plain bordered by steep cliffs (Stern & Johnson 2019). The East African coast is home to low-lying sandy beaches, coral reefs, and mangrove swamps (Stern & Johnson 2019). The geology of these environments is largely shaped by plate tectonics (Stern & Johnson 2019). The East African Rift System, a geological feature that runs from Ethiopia to Mozambique, has uplifted large parts of the East African coast, creating steep cliffs and rocky headlands. The rift system has also caused extensive volcanic activity, resulting in the formation of offshore islands such as Zanzibar and Pemba. The West African coast is dominated by ancient Precambrian rocks, which have been subjected to extensive erosion and

DOI: 10.1201/9781003369738-23

deposition over millions of years (Dillon & Sougy 1974). The Niger and Congo Rivers have deposited vast amounts of sediment into the Atlantic Ocean, creating large deltas and estuaries (Bonne 2014). The coastal plain is characterized by sandy beaches, lagoons, and mangrove swamps. The Mediterranean coast is marked by rugged mountains, including the Atlas Mountains in Morocco and the Aures Mountains in Algeria (van Vleck 1936). These mountains form a barrier between the coast and the Sahara Desert, which has resulted in a semi-arid climate along the coast. Overall, the coastal environments of Africa are characterized by diverse geology shaped by plate tectonics, erosion, and climate. These environments are important ecosystems, providing habitat for a wide range of marine and terrestrial species and supporting the livelihoods of millions of people.

19.3 HEAVY METAL CONTAMINATION OF MARINE ENVIRONMENTS IN AFRICA

Heavy metal contamination of the marine environment is a major issue of serious concern across the globe (Appiah-Opong *et al.* 2021; Ridwan & Ethadal 2022). Heavy metals affect both the biotic and abiotic components of the marine environment. In Africa, heavy metals affect the water, sediments, animals, and plants within the marine environment. Recent studies revealed that marine water in Egypt (Nour & El-Sorogy 2020), Ghana (Appiah-Opong *et al.* 2021), Ivory Coast (Tuo *et al.* 2012), Libya (Saleh *et al.* 2017), Mauritania (Yedih *et al.* 2022), Morocco (Jounaid *et al.* 2022), Nigeria (Yahaya *et al.* 2022), Senegal (Diop *et al.* 2014), and South Africa (Sparks *et al.* 2019) has high concentrations of heavy metals such as As, Cd, Co, Cr, Cu, Ni, Pb, and Zn that are above the world surface water standard (Table 19.1). The concentration of Cr in oceanic sediments of Africa is between 4.56 mg/kg (Senegal) and 1102.36 mg/kg (Ghana), with the amount in DR Congo and Ghana above the average crustal value (ACV). The amount of Cu in marine sediments of Africa is between 1.90 mg/kg (South Africa) and 120.14 mg/kg (Cameroun), with its concentration in those of Cameroun, Ghana, Egypt, and Sudan above the ACV. Nickel in the marine sediments of Africa is between 2.15 mg/kg (Mauritania) and 319.67 mg/kg (Ghana), with its concentration in those of Cameroon, Ghana, and Sudan also above the ACV. Lead in oceanic sediments of Africa is between 1.61 mg/kg (Senegal) and 404.15 mg/kg (Ivory Coast), with its amount in those from Angola, Benin Republic, Ghana, Ivory Coast, South Africa, Tanzania, Kenya, Egypt, and Sudan above the ACV. Zinc in marine sediments of Africa is between 1.30 mg/kg (South Africa) and 480.52 mg/kg (DR Congo), with those in Cameroun, DR Congo, Ghana, Ivory Coast, Tunisia, Kenya, Egypt, and Sudan above the ACV. Fish in the sea bioaccumulate more heavy metals in the coastal areas across the world (Bosch *et al.* 2016; Jezierska & Witeska 2006). Bosch *et al.* (2016) reported high amounts of potentially toxic metals in shells collected around the African oceanic waters.

19.4 MARINE GREENS IN AFRICA

Marine greens are abundant in coastal environments across the world. In Africa, marine greens are abundant in the coastal areas around the continent. A study by Hain *et al.* (2022) in coastal areas of Benin Republic revealed that *Amphora coffaeformis*, *Chroococcus* sp., *Oscillatoria subulatum*, *Planktothrix compressa*, *Spirogyra* sp., *Tetrastrum triangulare*, *Trachelomonas intermedia*, *Trachelomonas* sp., *Ulothrix* sp., *Anabaena piscinale*, *Anabaneopsis elenkinii*, *Bacillaria paxillifer*, *Campilodiscus baleanicus*, *Campilodiscus clypens*, *Chaetoceros muelleri*, *Closterium aciculare*, *Closterium kuetzingii*, *Coscinodiscus rudolfii*, *Cyclotella kuetzingi*, *Diatoma vulgare*, *Gomphonemaoli vaceum*, *Gyrosigma accuminatum*, *Komvopheron constrictum*, *Mastogloia braunii*, *Melosira granulata*, *Microcystis aeruginosa*, *Nostoc piscinal*, *Staureneis* sp., *Surirella capronii*, *Synechocystis aeruginosus*, and *Synechocystis aqualis* are common marine greens available in the environment. In the coastal areas of Cameroon, Fonge *et al.* (2013) reported that about 125 species of Chlorophyta, Cyanophyta, Bacillariophyta, Euglenophyta, Dinophyta, Chrysophyta, and

TABLE 19.1

Average Distribution of Heavy Metals in Oceanic Water and Sediments in Africa

	As (mg/l)	Cd (mg/l)	Co (mg/l)	Cr (mg/l)	Cu (mg/l)	Ni (mg/l)	Pb (mg/l)	Zn (mg/l)	Author
Marine water									
Egypt		**0.09**			**0.43**	**0.01**	**0.51**	**0.23**	Nour and El-Sorogy 2020
Ghana	**0.0033**				**0.012**		**0.047**	**0.015**	Appiah-Opong *et al.* 2021
Ivory Coast		**0.04**				**0.26**	**0.67**	**13.22**	Tuo *et al.* 2012
Libya				**0.0022**	**0.0264**	**0.007**		**0.461**	Saleh *et al.* 2017
Mauritania			0.02			**0.02**			Yedih *et al.* 2022
Morocco		**0.065**			**0.05**		**0.086**	0.001	Jounaid *et al.* 2022
Nigeria		**0.061**		**0.087**	**2.17**	**0.76**	**1.41**		Yahaya *et al.* 2022
Senegal		**0.23**	0.12	**0.45**	**1.93**	**1.28**	**0.48**	**21.29**	Diop *et al.* 2014
South Africa					**0.01**		**0.01**	**0.16**	Sparks *et al.* 2019
World Surface Water Standard	0.0026	0.0001		0.0002	0.0009	0.006	0.00003	0.005	Saleh *et al.* 2017
Marine sediments									
	As (mg/kg)	**Cd (mg/kg)**	**Co (mg/kg)**	**Cr (mg/kg)**	**Cu (mg/kg)**	**Ni (mg/kg)**	**Pb (mg/kg)**	**Zn (mg/kg)**	**Author**
Angola	**7.55**		8.92		12.15	4.36	**30.12**	29.47	Dinis *et al.* 2020
Benin Republic		14.81					**23.40**	52.42	Yehouenou *et al.* 2013
Cameroun		0.08	130.64	79.36	**120.14**	**229.60**		**110.00**	Ekoa-Bessa *et al.* 2020
DR Congo	**15.76**	0.13	16.46	**166.53**	17.36	27.39		**480.52**	Suami *et al.* 2020
Ghana	**182.03**		554.13	**1102.36**	**91.27**	**319.67**	**62.70**	**212.34**	Kuranchie-Mensah *et al.* 2013
Ivory Coast		2.85				45.63	**404.15**	**208.77**	Tuo *et al.* 2013
Libya		0.83	5.59		17.3	22.65	11.69	26.55	Nour and El-Sorogy 2017
Mauritania	1.01	0.16	0.49	13.03	5.55	2.15	5.08	19.02	Yehdhih *et al.* 2020
Morocco	**8.50**	0.16	3.57	26.60	6.60	16.50	6.13	51.70	Tnounmi *et al.* 2020
Nigeria	**5.00**		6.44	14.22	6.43	6.69	4.49	26.44	Adeyemi *et al.* 2019
Senegal		0.24	0.30	4.56	7.29	3.06	1.61	46.74	Diop *et al.* 2014
South Africa		1.70			1.90		**251.00**	1.30	Okoro *et al.* 2014
Tunisia	**4.17**	81.40		77.22	37.00	11.08	10.71	**104.90**	Naifor *et al.* 2018
Mozambique	**3.00**	0.11		6.50	3.70	5.60	9.70	14.00	Boitsov *et al.* 2021

(Continued)

TABLE 19.1 (CONTINUED)

Average Distribution of Heavy Metals in Oceanic Water and Sediments in Africa

Tanzania	0.00048	3.60	4.70	21.60	13.50	15.50	**35.40**	58.40	Muzuka 2008
Algeria	**7.50**		2.70	33.50	7.20	10.30	8.40	31.80	Ahmed *et al.* 2018
Kenya					28.03		**230.00**	**332.50**	Mwatsahu *et al.* 2020
Egypt		1.80			**110.00**	60.00	**20.00**	**110.00**	Ridwan and Ethadal 2022
Sudan					**68.80**	**102.00**	**17.10**	**83.00**	Idris *et al.* 2007
Average Crustal Value	1.80			100	55	75	12.5	70	

Note: Metal values above the limit are represented in bold.

Haptophyta existed. In the coastal area of Gabon, Zongo *et al.* (2022) reported that there were seven species of marine green that existed in the marine environment. These are *Bryopsis plumose*, *Bryopsis pennata*, *Caulerpa sertularioides*, *Caulerpa taxifolia*, *Chaetomorpha linum*, *Cladophora prolifera*, and *Ulva flexuosa*. In Cote D'Ivoire, the coastal areas are dominated by phytoplankton such as *Aphanizomenon* sp., *Aphanizomenon* sp., *Anabaena planctonica*, *Oscillatoria princeps*, *Oscillatoria tenuis*, *Pseudanabaena limnetica*, *Lyngbya* sp., *Aulacoseira granulata*, *Anabaena constricta*, *Phormidium*, *uncinatum*, *Phormidium* sp., *Merismopedia glauca*, *Merismopedia elegans*, *Oscillatoria limosa*, *Aphanothece castagnei*, *Microcystis wesenbergii*, *Lyngbya martensiana*, *Microcystis aeruginosa*, *Anabaena* sp., and the Chlorophyte *Pediastrum duplex* var. *gracillimum* (Seu-Anoï *et al.* 2011). In the coastal area of Morocco, some of the major marine greens available are *Coscinodiscus* sp., *Hemidiscus* sp., *Leptocylindrus danicus*, *Melosira nummuloides*, *Detonula* sp., *Lauderia annulata*, *Lauderia borealis*, *Chaetoceros decipiens*, *Gymnodinium* sp., and *Protoperidinium* sp. (Natij *et al.* 2014). In the coastal area of Nigeria, Akanmu and Onyeama (2020) described 86 phytoplankton species. Examples of the major phytoplankton species described include *Chaetoceros*, *Coscinodiscus*, *Melosira*, *Odontella*, *Thalassiosira*, *Bacillaria*, *Fragillaria*, *Nitzschia*, *Thalassionema*, *Oscillatoria*, *Trichodesmium*, *Noctiluca*, *Ceratium*, *Peridinium*, and *Protoperidinium*. In this region, diatoms are the most common marine greens reported. In the coastal area of South Africa, Gama *et al.* (2005) reported the presence of *Ulva* sp., *Micromonas* sp., *Pyramimonas* sp., *Nitzschia closterium*, *Melosira nummuloides*, *Diatoma* sp., *Peridinium* sp., *Gymnodinium* sp., and *Katodinium* sp. In the marine environment of Togo, 795 species of microalgae, which include *Bacillariophyceae*, *Cyanophyceae*, *Chorophyceae*, *Conjugatophyceae*, and *Euglenophyceae*were reported by Issifou *et al.* (2014). In the coastal environments of Tunisia, Rekik *et al.* (2016) showed that some of the abundant marine greens include *Amphora* sp., *Coscinodiscus* sp., *Grammatophora* sp., *Navicula* sp., *Alexandrium* sp., *Peridinium* sp., *Prorocentrum lima*, *Prorocentrum triestinium*, *Protoperidinium* sp., and *Euglena ascusformis*. Along the Algerian coastline, Boudjenah *et al.* (2019) reported 296 species of marine green, which include *Navicula* sp.,*Gymnodium* sp., *Prorocentrum mecans*, *Peridinium* sp., *Chaetoceros* sp., *Melosira* sp., and *Chaetoceros* sp. Nihal *et al.* (2014) reported more than 207 marine greens along the Egyptian Mediterranean coastal waters, which include *Carteria* sp., *Chlorella vulgaris*, *Asterionella glacialis*, *Cylindrotheca closterium*, *Melosira granulata*, *Navicula cryptocephala*, *Pseudo-Nitzschia palea*, *Pseudo-Nitzschia sigma*, *Pseudo-Nitzschia delicatissima*, *Skeletonema costatum*, and *Gonyaulax* spp. Along the Tanzanian coastal waters, some of the marine greens observed by Sekadende *et al.* (2021) in the area include *Amphiprora alata*, *Chaetoceros decipiens*, *Coscinodiscus nodulifera*, *Guinardia striata*, *Leptocylindrus danicus*, *Rhizosolenia alata*, *Thalassiothrix* sp., and *Protoperidinium steinii*.

19.5 SOURCES OF HEAVY METALS IN MARINE GREENS IN AFRICA

In marine settings, heavy metal concentrations are low, usually in nanograms or milligrams per kilogram. Studies across the world have confirmed that oceanic waters are now mostly contaminated by toxic metals (Singh *et al.* 2022), and this has continually increased the fear of environmental degradation. There are many factors that have led to a surge in ocean pollution by heavy metals. Some of these factors include the exponential increase in human population (Halpern *et al.* 2019), increase in urban expansion (Singh *et al.* 2022), more industrial growth (Cordes *et al.* 2016), search for and use of natural resources (Grigalunas & Opaluch 1988; Aranssiola *et al.*), modern agricultural practices (Cordes *et al.* 2016), and uncontrolled wastewater discharge into the sea (Potter *et al.* 2004). In Africa, the rate of population expansion is higher than in any other continent of the world (Kröner *et al.* 2023). The increase in population has contributed greatly to the contamination of the marine environment in Africa. For example, the population of Accra, a coastal city in Ghana, increased from 1,668,000 in the year 2000 to 2,660,000 in the year 2023, with an annual percentage increase of between about 1.50% and 3.00% (Macrotrends 2023a). Dakar, a city located along the coast of Senegal, had a surge in population growth of 1,862,000 in the year 2000 and 3,430,000 in the year 2023, with a yearly percentage of between 2.99% and 3.13%, respectively (Macrotrends 2023b). The population of Lagos, a cosmopolitan megacity on the coast of Nigeria, in the years 2000 and 2023 was 7,281,000 and 15,946,000, respectively. This represents an annual increase of between 3% and 4% each (Macrotrends 2023c). For the coastal city of Abidjan in Cote D'Ivoire, the population in the years 2000 and 2023 was 3,007,000 and 5,686,000, respectively, with a percentage growth of 2.77% and 3.08% each (Macrotrends 2023d). Also, the population of Alexandria city in Egypt, which borders the Mediterranean Sea, in the years 2000 and 2010 was 3,546,000 and 5,588,000, respectively, with a percentage growth of 2.04% and 1.09% per annum. The United Nations has projected that the population of Africa will increase exponentially from 819 million in the year 2000 to 2.5 billion in the year 2050 (Statista 2023). The observed increase in the population of these cities has led to an increase in waste generation. In Nigeria, Lagos alone generates about 13,000 metric tonnes of municipal waste per day (Theconversion 2021). In Accra, Ghana, the amount of waste generated per capita is 0.70 kg (AMA 2020), while the city of Cape Town, a coastal city in South Africa, generates about 1.7 million tonnes of municipal waste yearly and amasses approximately 0.6 million tonnes of commercial and industrial wastes and about 0.07 million tonnes of agricultural waste per annum (GreenCape 2020). The wastes generated from these expanding cities host heavy metals, which are finally mobilized into the ocean, thereby increasing their potential threat to both living and non-living things in the environment.

Increased industrialization has also played a major role in the release of heavy metals into the marine environment of Africa. For example, Lagos, Nigeria is an industrial hub that hosts industries such as maritime, steel, baking, manufacturing, engineering, textile, cement, quarry, welding, and paint companies. Also in the free zone city of Alexandria, Egypt, which accounts for over 40% of the nation's industries, there exist companies such as chemicals, petrochemicals, oil refinement, iron production, steel production, and agro-allied industries. Through the release of poisonous exhausts, waste waters, and waste products, these companies contribute significantly to the release of toxic metals into the oceanic and adjoining environment.

19.6 HEAVY METAL UPTAKE BY MARINE GREENS IN AFRICA

The sea is very important to the socio-economic development of the African continent. However, pollution of the ocean by heavy metals and other forms of pollutants is posing a great threat to the ecosystem. Within this ecosystem, marine greens are among the major oceanic plants that are affected by the release of organic and inorganic pollutants, especially potentially toxic metals. Plants have the ability to take up and accumulate potentially toxic metals in their bodies (Adewumi & Lawal 2023). This process is referred to as "bioaccumulation" For plants to bioaccumulate heavy metals in their bodies, the concentration must be high in their ambient environment, such as the water, soils, and sediments on which they grow (Tshithukhe *et al.* 2021). The main part of the plant that aids bioaccumulation of heavy metals is the root (Collin *et al.* 2022). Some bottom-dwelling marine greens, such as seaweeds or marine algae, because of their non-flowering nature, use a hold-fast to absorb metals from their immediate environment (Agarwal *et al.* 2022). The rate of uptake of heavy metals by marine greens along the coastal areas of Africa is generally high (Tshithukhe *et al.* 2021; Addico & deGraft-Johnson 2016). In the coastal areas of Cape Town, South Africa, a study by Tshithukhe *et al.* (2021) revealed that marine greens such as *Pontederia crassipes* and *Salvinia molesta* bioaccumulate heavy metals. The highest concentrations of As, Cd, Cr, Cu, Fe, Hg, Pb, and Zn in these marine greens were 2.35 mg/kg, 2.67 mg/kg, 0.82 mg/kg, 7.10 mg/kg, 2.37 mg/kg, 4.19 mg/kg, 0.24 mg/kg, and 8.34 mg/kg, respectively. A report by Vlachos *et al.* (1998) showed that the range of concentrations of As, Co, Cr, Cu, Ni, and Pb in KwaZulu-Natal beaches in South Africa was 24–1428mg/kg, 10–498mg/kg, 9–168 mg/kg, 2–87mg/kg, 9–107mg/kg, and 10–75 mg/kg, respectively. Ferletta *et al.* (1996) assessed the amount of heavy metals in marine macroalgae from the Zanzibar area of Dar es Salaam, Tanzania. The study showed that in marine greens in this area, the range of concentrations of Cd, Cr, Cu, Fe, Mn, Ni, Pband Zn was 0.14–5.62 mg/kg, 2.88–33.64 mg/kg, 2.62–17.42 mg/kg, 78.76–3619.77 mg/kg, 9.76–111.78 mg/kg, 9.76–111.78 mg/kg, 3.62–22.93 mg/kg, 6.13–53.19 mg/kg and 5.59–104.58 mg/kg respectively. On the western coast of Ghana, Addico and deGraft-Johnson (2016) reported that the range of concentrations of Cu, Zn, Fe, Pb, Cd, and As in marine greens was 24–36 mg/kg, 16–100 mg/kg, 1209–5910 mg/kg, 86–335 mg/kg, 95–119 mg/kg and1–2mg/kg, respectively. Aly *et al.* (2006) determined the content of toxic metals in marine greens from the Egyptian Red Sea coast. The study showed that on the coast of Suez, the range of the amount of Cd, Cu, Ni, Pb, Mn, Co, Cr and Zn in marine greens was 2.10–10.40 mg/kg, 2.30–17.30 mg/kg, 6.30–17.10 mg/kg, 19.40–50.50 mg/kg, 8.30–66.80 mg/kg, 2.70–16.60 mg/kg, 0.60–13.40 mg/kg and 23.40–90.40 mg/kg respectively, while for those from the Mars Alam coast, the range of concentrations was 0.40–2.40 mg/kg, 0.60–4.30 mg/kg, 1.00–7.10 mg/kg, 3.10–19.20 mg/kg, 3.50–22.20 mg/kg, 1.80–9.60 mg/kg, 0.10–6.80 mg/kg and 2.00–22.10 mg/kg respectively.

19.7 POSSIBLE HEALTH RISKS OF HEAVY METALS IN MARINE GREENS

Heavy metals are dangerous not only to the environment but also to human health (Adewumi *et al.* 2020). In some parts of the world, marine greens such as seaweeds are consumed for health benefits. It is believed that marine greens are sources of dietary fiber, calcium, iron, folate, and iodine (MacArtain *et al.* 2007). According to Adewumi *et al.* (2020), oral intake of heavy metals from plants can lead to both carcinogenic and non-carcinogenic health risks.

Arsenic is a toxic element that can cause serious health problems if ingested into the human body. Consumption of As from marine greens may lead to several health risks, including cancer. Chronic exposure to As may lead to skin, lung, bladder, liver, and kidney cancers (Adewumi *et al.* 2020). It could also lead to skin lesions such as hyperkeratosis and hyperpigmentation (Hong *et al.* 2014). Also, bioaccumulation of As in the human body has been linked to an increased risk of cardiovascular disease, including hypertension and atherosclerosis (Hong *et al.* 2014). Intake of As can cause respiratory problems such as coughing and shortness of breath (Adewumi *et al.* 2020). Excessive intake of As may also lead to neurological problems, such as numbness, tingling, and peripheral neuropathy (Hong *et al.* 2014). For married couples, excessive oral ingestion of As from marine greens may cause infertility and miscarriage (Hong *et al.* 2014). It could also lead to digestive problems such as vomiting, diarrhea, and abdominal pain (Hong *et al.* 2014).

Cadmium is also a toxic heavy metal that can cause serious health problems if ingested into the human body (Genchi *et al.* 2020). Exposure to cadmium has been linked to several health risks, such as kidney damage, which can cause kidney failure (Genchi *et al.* 2020). It can also cause lung damage including chronic bronchitis and a condition known as "cadmium pneumonitis," which can be fatal (Adewumi *et al.* 2020). Cadmium is also carcinogenic, causing lung, prostate, and bladder cancer. Exposure to cadmium can cause reproductive dysfunction, including reduced fertility and birth defects (Adewumi *et al.* 2020). It could also lead to hypertension or high blood pressure, which can cause heart disease and stroke (Adewumi *et al.* 2020). Anemia has been linked with excessive intake of Cd (Genchi *et al.* 2020). Anemia is a condition in which there are not enough red blood cells to carry oxygen to the body's tissues (Genchi *et al.* 2020). Another health issue associated with exposure to Cd is osteoporosis, which is a condition that weakens bones and makes them more susceptible to fractures (Genchi *et al.* 2020).

Zinc is another major dietary metal necessary for the proper functioning of human health (Plum *et al.* 2010). Consuming too much zinc from marine greens can lead to health issues like vomiting, stomach cramps, and diarrhea (Plum *et al.* 2010). Excessive intake of Zn can lead to the suppression of the immune system, making it difficult for the body to fight off infections (Adewumi *et al.* 2020). Also, consuming too much zinc can lead to an unpleasant metallic taste in the mouth and can also cause bad breath (Plum *et al.* 2010).

Lead is a highly toxic metal that can cause significant harm to various organs and systems in the body (Adewumi *et al.* 2020). Lead in the body system can cause neurological damage.

It is highly toxic to the nervous system (Ara & Usmani 2015). Even low doses of lead can lead to behavioral problems, learning disabilities, and cognitive impairment, especially in children (Adewumi *et al.* 2020). The presence of lead in the human body can also cause interference with the production of red blood cells, leading to anemia (Ara & Usmani 2015). Lead can also damage the kidneys, leading to kidney failure (Ara & Usmani 2015). This toxic metal can also have a negative impact on the reproductive system in both males and females, leading to infertility or other problems. Another major health challenge caused by Pb is cardiovascular disease, which has been linked to an increased risk of high blood pressure in many adults. Prenatal exposure of pregnant women to lead can cause developmental delays and intellectual disability in children (Ara & Usmani 2015). Exposure to lead can also lead to stunted growth in children. Lead can accumulate in the body over time, so even low levels of exposure can be harmful (Adewumi *et al.* 2020).

19.8 CONCLUSIONS

Heavy metals in marine greens originate from wastes generated from coastal cities around the continent. Other sources include exponential population growth, exploitation of natural resources, unabated industrialization, runoff from agricultural activities, and geological processes. Marine greens have high amounts of heavy metals, which are above internationally accepted limits. Over a period of time, they bioaccumulate these toxic metals into their bodies and subsequently release them into the marine environment after their death and decay. Marine greens are basically food for fish and other organisms in the sea. However, when humans eat fish, heavy metals in them stored in the body and may cause several carcinogenic and non-carcinogenic health problems that may eventually lead to death. It is important that the government put forward policies that will promote a healthy environment.

REFERENCES

Abdallah, A. M. A., Abdallah, M. A., Beltagy, A., & Siam, E. (2006). Contents of heavy metals in marine algae from Egyptian Red Sea coast. *Toxicological and Environmental Chemistry*, *88*(1), 9–22.

Abioye, O. P., Ezegwu, B. U., Aransiola, S. A., & Ojeba, M. I. (2020). Phycoremediation of water contaminated with arsenic (As), cadmium (Cd) and lead (Pb) from a mining site in Minna, Nigeria. *European Journal of Biological Research*, *10*(1), 35–44. http://www.journals.tmkarpinski.com/index.php/ejbr/article/view/251.

Addico, G. N. D., & deGraft-Johnson, K. A. A. (2016). Preliminary investigation into the chemical composition of the invasive brown seaweed Sargassum along the West Coast of Ghana. *African Journal of Biotechnology*, *15*(39), 2184–2191.

Adeniyi, A. O., & Akinwole, A. O. (2017). Phytoplankton composition and physico-chemical parameters of lower river Niger, Agenebode, Edo state Nigeria. *International Journal of Fisheries and Aquatic Studies*, *5*(3), 256–260.

Adesuyi, A. A., Njoku, K. L., Akinola, M. O., & Jolaoso, A. O. (2018). Biomonitoring of heavy metals level in wetland plants of Lagos Lagoon, Nigeria. *Journal of Applied Sciences and Environmental Management*, *22*(9), 1489–1498.

Adewumi, A. J., & Laniyan, T. A. (2020). Contamination, sources, and risk assessments of metals in media from Anka artisanal gold mining area, Northwest Nigeria. *Science of the Total Environment*, *718*, 137235.

Adewumi, A. J., & Lawal, A. E. (2023). Heavy metals in paddy soils and their uptake in rice plants collected along Ogbese River, Southwest Nigeria: Implications for contamination and health risk. *Ife Journal of Science*, *24*(3), 569–582.

Adewumi, A. J., Laniyan, T. A., Xiao, T., Yizhang, L., & Ning, Z. (2020). Exposure of children to heavy metals from artisanal gold mining in Nigeria: Evidences from bio-monitoring of hairs and nails. *Acta Geochimica*, *39*(4), 451–470.

Adewumi, A. J., Ogundele, O. D., & Adeseko, A. A. (2022). Heavy metals in soils around a major cement factory in Southern Nigeria: Ecological and human health risks. *Nigerian Journal of Environmental Science and Technology*, *6*(2), 283–294.

Adeyemi, M. O., Olusola, J. A., Akpobasah, O., Adidi, N. E., & Shelle, R. O. D. (2019). Assessment of heavy metals pollution in sediments from Ologe Lagoon, Agbara, Lagos, Nigeria. *Journal of Geoscience and Environment Protection*, *7*(7), 61–73.

Agarwal, S., Albeshr, M. F., Mahboobb, S., Atique, U., Pramanick, P., & Mitra, A. (2022). Bioaccumulation Factor (BAF) of heavy metals in green seaweed to assess the phytoremediation potential. *Journal of King Saud University – Science*, *34*(5), 102078.

Ahmed, I., Mostefa, B., Bernard, A., & Olivier, R. (2018). Levels and ecological risk assessment of heavy metals in surface sediments of fishing grounds along Algerian coast. *Marine Pollution Bulletin*, *136*, 322–333.

Akanmu, R. T., & Onyema, I. C. (2020). Phytoplankton composition and dynamics off the coast of Lagos south-west, Nigeria. *Regional Studies in Marine Science*, *37*, 101356. https://doi.org/10.1016/j.rsma.2020.101356.

Ali, A. Y., Idris, A. M., & Eltayeb, M. A. (2017). Uptake of heavy metals by seven green algae species at the Red Sea coast. *Fresenius Environmental Bulletin*, *26*, 7160–7171.

Ali, H., Khan, E., & Ilahi, I. (2019). Environmental chemistry and ecotoxicology of hazardous heavy metals: Environmental persistence, toxicity, and bioaccumulation. *Journal of Chemistry*, *2019*, 6730305. https://doi.org/10.1155/2019/6730305.

Al-Shwafi, N. A., & Rushdi, A. I. (2008). Heavy metal concentrations in marine green, brown, and red seaweeds from coastal waters of Yemen, the Gulf of Aden. *Environmental Geology*, *55*(3), 653–660.

AMA (2020). Accra's per capita waste generation rate Is 0.70kg per day – report. A report of the Accra metropolitan assembly. Available at https://ama.gov.gh/news-details.php?n=OTU3NTU5MjUxNXMwNTcwcDYxczhxczFzb3Izc3M0MHI2cHMyNDFycw.

Ansari, T. M., Marr, I. L., & Tariq, N. (2004). Heavy metals in marine pollution perspective-a mini review. *Journal of Applied Sciences*, *4*(1), 1–20.

Appiah-Opong, R., Ofori, A., Ofosuhene, M., Ofori-Attah, E., Nunoo, F. K., Tuffour, I., … Fosu-Mensah, B. Y. (2021). Heavy metals concentration and pollution index (HPI) in drinking water along the southwest coast of Ghana. *Applied Water Science*, *11*, 1–10.

Ara, A., & Usmani, J. A. (2015). Lead toxicity: A review. *Interdisciplinary Toxicology*, *8*(2), 55–64.

Aransiola, S. A., Ijah, U. J. J., & Abioye, O. P. (2013). Phytoremediation of lead polluted soil by Glycine max L. *Applied and Environmental Soil Science*, *2013*, Article ID 631619. https://doi.org/10.1155/2013/631619. https://www.hindawi.com/journals/aess/2013/631619/abs/.

Aransiola, S. A., Ijah, U. J. J., Abioye, O. P., & Bala, J. D. (2021). Microbial and heavy metal determination of contaminated soil using Melissa officinalis L. *International Journal of Environmental Planning and Management*, *7*(3), 102–107. http://www.aiscience.org/journal/paperinfo/ijepm?paperid=5369.

Bandara, K. R., & Manage, P. M. (2022). Heavy metal contamination in the coastal environment and trace level identification. IntechOpen. https://doi.org/10.5772/intechopen.106653.

Blankson-Arthur, S. (2013). Heavy metal assessment of marine sediment in selected coastal districts of the Western Region, Ghana. *Advances in Environmental Research*, *2*(2), 155–166.

Bocodaho, O. L., Chouti, W. K., Cakpo, R. A., & Mama, D. (2021). South-West Benin Coastal Lagoon: Waters and sediments' toxicity and contents in heavy metals during high water period (Togbin to the Mono Mouth). *International Journal of Biological and Chemical Sciences*, *15*(3), 1249–1263.

Boitsov, S., Newman, B. K., Muiambo, H. F., Chaúque, E. F. C., Serigstad, B., & Malauene, B. S. (2021). Distribution and possible sources of polycyclic aromatic hydrocarbons (PAHs) and metals in marine surface sediments off northern Mozambique. *Marine Pollution Bulletin*, *163*, 111952.

Bonne, K. P. (2014). Reconstruction of the evolution of the Niger River and implications for sediment supply to the Equatorial Atlantic margin of Africa during the Cretaceous and the Cenozoic. *Geological Society, London. Special Publications*, *386*(1), 327–349.

Bosch, A. C., O'Neill, B., Sigge, G. O., Kerwath, S. E., & Hoffman, L. C. (2016). Heavy metals in marine fish meat and consumer health: A review. *Journal of the Science of Food and Agriculture*, *96*(1), 32–48.

Boudjenah, M., Mokrane, Z., & Soualili, D. (2019). Diversity of Phytoplanktonic populations along the Algerian coastline. *Biodiversity Journal*, *10*(2), 81–92.

Chevrollier, L. A., Koski, M., Søndergaard, J., Trapp, S., Aheto, D. W., Darpaah, G., & Nielsen, T. G. (2022). Bioaccumulation of metals in the planktonic food web in the Gulf of Guinea. *Marine Pollution Bulletin*, *179*, 113662.

Collin, S., Baskar, A., Geevarghese, D. M., Ali, M. N. V. S., Bahubali, P., Choudhary, R., ... Swamiappan, S. (2022). Bioaccumulation of lead (Pb) and its effects in plants: A review. *Journal of Hazardous Materials Letters, 3*(2022): 100064. https://doi.org/10.1016/j.hazl.2022.100064.

Cordes, E. E., Jones, D. O., Schlacher, T. A., Amon, D. J., Bernardino, A. F., Brooke, S. & Witte, U. (2016). Environmental impacts of the deep-water oil and gas industry: A review to guide management strategies. *Frontiers in Environmental Science, 4*, 58.

de Moura, J. F., Roges, E. M., de Souza, R. L., Siciliano, S., & dos Prazeres Rodrigues, D. (2012). Marine environment and public health. *Biodiversity Conservation and Utilization in a Diverse World, 263*,-284.

Dillon, W. P., & Sougy, J. M. (1974). Geology of West Africa and canary and Cape Verde Islands. In: Naim, A. E. M., andStehli, F.G. (eds.), *The Ocean Basins and Margins.* Springer, Boston, MA, pp. 315–390. https://doi.org/10.1007/978-1-4684-3033-2_10.

Dinis, P., Armando, A., & Pratas, J. (2020). Sources of potentially toxic elements in sediments of the Mussulo Lagoon (Angola) and implications for human health. *International Journal of Environmental Research and Public Health, 17*(7), 2466.

Diop, C., Dewaelé, D., Diop, M., Touré, A., Cabral, M., Cazier, F., ... Ouddane, B. (2014). Assessment of contamination, distribution and chemical speciation of trace metals in water column in the Dakar coast and the Saint Louis estuary from Senegal, West Africa. *Marine Pollution Bulletin, 86*(1–2), 539–546.

Ekoa Bessa, A. Z., Ngueutchoua, G., Kwewouo Janpou, A., El-Amier, Y. A., Njike Njome Mbella Nguetnga, O. A., Kankeu Kayou, U. R., ... Armstrong-Altrin, J. S. (2021). Heavy metal contamination and its ecological risks in the beach sediments along the Atlantic Ocean (Limbe coastal fringes, Cameroon). *Earth Systems and Environment, 5*(2), 433–444.

El Agawany, N., Kaamoush, M., El-Zeiny, A., & Ahmed, M. (2021). Effect of heavy metals on protein content of marine unicellular green alga Dunaliella tertiolecta. *Environmental Monitoring and Assessment, 193*(9), 584.

FAO (2017). African package for climate-resilient ocean economies. 5 pp. Available at https://www.fao.org/3/i6441e/i6441e.pdf.

Ferletta, M., Bråmer, P., Semesi, A., & Björk, M. (1996). Heavy metal contents in macroalgae in the Zanzibar Channel, an initial study. In *Symposium on the Biology of Microalgae, Macroalgae, and Seagrasses in the Western Indian Ocean, held 3-10 December 1995 at the University of Mauritius. Sida Marine Science Program, Stockholm* (pp. 332–346).

Fondriest Environmental, Inc. (2014, October 22). Algae, phytoplankton and chlorophyll. *Fundamentals of Environmental Measurements.* Web. https://www.fondriest.com/environmental-measurements/parameters/water-quality/algae-phytoplankton-and-chlorophyll.

Fonge, A. B., Chuyong, B. G., Tening, A. S., Fobid, A. C., & Numbisi, N. F. (2013). Seasonal occurrence, distribution and diversity of phytoplankton in the Douala Estuary, Cameroon. *African Journal of Aquatic Science, 38*(2), 123–133. https://doi.org/10.2989/16085914.2013.769086.

Fosu-Mensah, B. Y., Ofori, A., Ofosuhene, M., Ofori-Attah, E., Nunoo, F. E., Darko, G., & Appiah-Opong, R. (2018). Assessment of heavy metal contamination and distribution in surface soils and plants along the west coast of Ghana. *West African Journal of Applied Ecology, 26*, 167–178.

Gama, P. T., Adams, J. B., Schael, D. M., & Skinner, T. (2005). Phytoplankton chlorophyll a concentration and community structure of two temporarily open/closed estuaries. *Water Research Commission Report, 1255*(1), 05.

Gargouri, D., Azri, C., Serbaji, M. M., Jedoui, Y., & Montacer, M. (2010). Heavy metal concentrations in the surface marine sediments of Sfax Coast, Tunisia. *Environmental Monitoring and Assessment, 175*(1–4), 519–530. https://doi.org/10.1007/s10661-010-1548-7.

Genchi, G., Sinicropi, M. S., Lauria, G., Carocci, A., & Catalano, A. (2020). The effects of cadmium toxicity. *International Journal of Environmental Research and Public Health, 17*(11), 3782.

Göthberg, A., Greger, M., Holm, K., & Bengtsson, B. (2004). Influence of nutrients on uptake and effects of Hg, Cd and Pb in Ipomoea aquatica. *Journal of Environment Quality, 33*, 1247–1255.

GreenCape. (2020). Waste-2020 market intelligence report. https://green-cape.co.za/assets/WASTE_MIR_20200331.pdf.

Grigalunas, T. A., Opaluch, J. J., French, D., Reed, M., & Knauss, D. (1988). A natural resource damage assessment model for coastal and marine environments. *GeoJournal*, 315–321.

Hain, T. K., Adandedjan, D., & Hountogan, M. E. (2022). Phytoplankton communauties of the Porto-Novo Lagoon (South-East Bénin). *International Journal of Fisheries and Aquatic Studies, 10*(1), 130–141. https://doi.org/10.22271/fish.2022.v10.i1b.2633.

Hakima, Z., Mohamed, M., Aziza, M., Mehdi, M., Meryem, E. B., Bendahhou, Z., & Jean-Francois, B. (2017). Environmental and ecological risk of heavy metals in the marine sediment from Dakhla Bay, Morocco. *Environmental Science and Pollution Research, 24*(9), 7970–7981. https://doi.org/10.1007/s11356-017-8367-0.

Halpern, B. S., Frazier, M., Afflerbach, J., Lowndes, J. S., Micheli, F., O'Hara, C., Selkoe, K. A., & Selkoe, K. A. (2019). Recent pace of change in human impact on the world's ocean. *Scientific Reports, 9*(1), 11609.

Hong, Y. S., Song, K. H., & Chung, J. Y. (2014). Health effects of chronic arsenic exposure. *Journal of Preventive Medicine and Public Health, 47*(5), 245.

Idris, A. M., Eltayeb, M. A. H., Potgieter-Vermaak, S. S., Van Grieken, R., & Potgieter, J. H. (2007). Assessment of heavy metals pollution in Sudanese harbours along the Red Sea Coast. *Microchemical Journal, 87*(2), 104–112.

Issifou, L., Atanle, K., Radji, R., LAWSON, H.L., Adjonou, K., Edorh, M.T. and Kokou, K. (2014). Checklist of tropical algae of Togo in the Guinean Gulf of West-Africa. *Scientific Research and Essays, 9*(22), 932–958. https://doi.org/10.5897/SRE2014.6113.

Jezierska, B., & Witeska, M. (2006). The metal uptake and accumulation in fish living in polluted waters. In Twardowska, I., Allen, H.E., Häggblom, M.M., and Stefaniak, S. (eds.), *Soil and Water Pollution Monitoring, Protection and Remediation. NATO Science Series. 69* (pp. 107–114). Springer, Dordrecht. https://doi.org/10.1007/978-1-4020-4728-2_6

Jin, P., Zhang, J., Wan, J., Overmans, S., Gao, G., Ye, M. and Xia, J. (2021). The combined effects of ocean acidification and heavy metals on marine organisms: A meta-analysis. *Frontiers in Marine Science, 8*, 801889.

Jounaid, H., Chakri, N., Remmal, T., Elamrani, B., & Amraoui, F. (2021). Heavy metal contamination in surface water of Mohammedia wetland, Morocco. In: El Moussaoui, S., El Talibi, H., Cherkaoui Dekkaki, H., and Etebaai, I. (eds.),*E3S Web of Conferences* (Vol. 298, p. 05001). EDP Sciences.

Koukal, B., Guéguen, C., Pardos, M., & Dominik, J. (2003). Influence of humic substances on the toxicity effects of cadmium and zinc to green alga Pseudokirchneriella subcapitata. *Chemosphere, 53*(8), 953–961.

Kröner, A., Mabogunje, A. L., Nicol, D. S.H.W., Clarke, J. I., McMaster, D. N., Steel, R. W., Smedley, A., Middleton, J. F. M., Gardiner, R. K. A. and Dickson, K. B. (2023, February 20). *Africa. Encyclopedia Britannica.* https://www.britannica.com/place/Africa.

Lobus, N. V. (2022). Biogeochemical role of algae in aquatic ecosystems: Basic research and applied biotechnology. *Journal of Marine Science and Engineering, 10*(12), 1846.

MacArtain, P., Gill, C. I. R., Brooks, M., Campbell, R., & Rowland, I. R. (2007). Nutritional value of edible seaweeds. *Nutrition Reviews, 65*(12), 535–543.

Macrotrends (2023a). Accra, Ghana metro area population 2000–2023. Available at https://www.macrotrends.net/cities/21107/accra/population.

Macrotrends (2023b). Dakar, Senegal metro area population 2000–2023. Available at https://www.macrotrends.net/cities/22439/dakar/population.

Macrotrends (2023c). Lagos, Nigeria metro area population 2000–2023. Available at https://www.macrotrends.net/cities/22007/lagos/population.

Macrotrends (2023d). Abidjan, cote d'voire metro area population 2000–2023. Available at https://www.macrotrends.net/cities/21602/abidjan/population.

Mahdi, A. M., Doumenq, P., Awaleh, M. O., Syakti, A. D., Asia, L., & Chiron, S. (2017). Levels and sources of heavy metals and PAHs in sediment of Djibouti-city (Republic of Djibouti). *Marine Pollution Bulletin, 120*(1–2), 340–346. 10.1016/j.marpolbul.2017.05.055.

Muzuka, A. (2008). Distribution of heavy metals in the coastal marine surficial sediments in the Msasani Bay-Dar es Salaam harbour area. *Western Indian Ocean Journal of Marine Science, 6*(1), 73–83.

Mwatsahu, S. H., Wanjau, R., Tole, M., & Munga, D. (2020). Heavy metal contamination in water, sediments, and fauna of selected areas along the Kenyan coastline. *Indo Pacific Journal of Ocean Life, 4*(1), 37–47.

Naifar, I., Pereira, F., Zmemla, R., Bouaziz, M., Elleuch, B., & Garcia, D. (2018). *Spatial Distribution and Contamination Assessment of Heavy Metals in Marine Sediments of the Southern Coast of Sfax, Gabes* Gulf, Tunisia. Marine pollution bulletin, 131, 53–62.

NASA (2023). Water cycle. Available at https://science.nasa.gov/earth-science/oceanography/ocean-earth-system/ocean-water-cycle.

Natij, L., Damsiri, Z., Khalil, K., Loudiki, M., Ettahiri, O., & Elkalay, K. (2014). Phytoplankton abundance and diversity in the coastal waters of Oualidia lagoon, south Moroccan Atlantic in relation to environmental variables. *International Journal of Advanced Research, 2*, 1022–1032.

Ndiaye, B., Ndiaye, M., Perez Cid, B., et al. (2021). Metals Content in Green Algae *Ulva lactuca* from Dakar coast (Senegal). *Analytical Chemistry An Indian Journal, 21*(3), 157.

Ndiokwere, C. L. (1984). An investigation of the heavy metal content of sediments and algae from the River Niger and Nigerian Atlantic coastal waters. *Environmental Pollution Series B:, Chemical and Physical, 7*(4), 247–254.

Nihal, G., El Khair, E. M. A., & Mohamed, M. (2014). Phytoplankton community in the Egyptian Mediterranean coastal waters. *Indian Journal of Geo-Marine Sciences, 43*(10), 1981–1988.

Nour, H. E., & El-Sorogy, A. S. (2017). Distribution and enrichment of heavy metals in Sabratha coastal sediments, Mediterranean Sea, Libya. *Journal of African Earth Sciences, 134*, 222–229.

Nour, H. E., & El-Sorogy, A. S. (2020). Heavy metals contamination in seawater, sediments and seashells of the Gulf of Suez, Egypt. *Environmental Earth Sciences, 79*, 1–12.

Ogundele, O. D., Adewumi, A. J., & Oyegoke, D. A. (2023). Phycoremediation: Algae as an effective agent for sustainable remediation and waste water treatment. *Environmental and Earth Sciences Research Journal, 10*(1), 7–17. https://doi.org/10.18280/eesrj.100102.

Ohr, L. M. (2019). Nutrients from the sea. Available at https://www.ift.org/news-and-publications/food-technology-magazine/issues/2019/november/columns/nutrients-from-the-sea.

Okoro, H. K., Fatoki, O. S., Adekola, F. A., Ximba, B. J., & Snyman, R. G. (2014). Geochemical assessment of sediment in Cape Town harbour, South Africa. *Bulletin of the Chemical Society of Ethiopia, 28*(1), 17–28.

Peterson, H. G., Healey, F. P., & Wagemann, R. (1984). Metal toxicity to algae: A highly pH dependant phenomenon. *Canadian Journal of Fisheries and Aquatic Sciences, 41*(6), 974–979.

Plum, L. M., Rink, L., & Haase, H. (2010). The essential toxin: Impact of zinc on human health. *International Journal of Environmental Research and Public Health, 7*(4), 1342–1365.

Potter, K., Douglas, J., Bricj, E., DeFries, R. S., Asner, G. P., & Houghton, R. A. (2004). Impacts of agriculture on aquatic ecosystems in the humid United States. In: Defries, R. S., Asner, G. P., and Houghton, R. A. (eds.),*Ecosystems and Land Use Change: American Geophysical Union* (pp. 31–40). https://doi.org/10.1029/153GM04

Redwan, M., & Elhaddad, E. (2022). Heavy metal pollution in Manzala Lake sediments, Egypt: Sources, variability, and assessment. *Environmental Monitoring and Assessment, 194*(6), 436. 10.1007/s10661-022-10081-0.

Rekik, A., Ben Salem, Z., Ayadi, H., & Elloumi, J. (2016). Spring phytoplankton variability along a south coast of Sfax at the water-sediment interface (Tunisia, eastern Mediterranean Sea). *Journal of Coastal Life Medicine, 4*(2), 121–127.

Saleh, H. N., Amin, H. A., Omar, M. Y., Mostafa, A. R., & Ebraham, Y. E. (2020). Environmental assessment of water quality and heavy metals pollution of seawater in Tobruk Bay-Libya. In Ezziyyani, M. (Eds.), *Advanced Intelligent Systems for Sustainable Development (AI2SD'2019) Volume 3-Advanced Intelligent Systems for Sustainable Development Applied to Environment, Industry and Economy* (pp. 306–318). Springer International Publishing.

Salgado, L. T., Andrade, L. R., & Filho Amado, G. M. (2005). Localization of specific monosaccharides in cells of the brown alga *Padina gymnospora* and the relation to heavy-metal accumulation. *Protoplasma, 225*(1–2), 132–128.

Sekadende, B. C., Michael, A., Painter, S. C., Shayo, S., Noyon, M., & Kyewalyanga, M. S. (2021). Spatial variation in the phytoplankton community of the Pemba Channel, Tanzania, during the south-east monsoon. *Ocean & Coastal Management, 212*, 105799.

Serfor-Armah, Y., Nyarko, B., Osae, E., Carboo, D., & Seku, F. (1999). Elemental analysis of some green and brown seaweeds from the coastal belt of Ghana. *Journal of Radioanalytical and Nuclear Chemistry, 242*(1), 193–197.

Seu-Anoï, N. M., Ouattara, A., Koné, Y. J., & Gourène, G. (2011). Seasonal distribution of phytoplankton in the Aby lagoon system, Ivory Coast, West Africa. *African Journal of Aquatic Science, 36*(3), 321–330.

Sharifuzzaman, S. M., Rahman, H., Ashekuzzaman, S. M., Islam, M. M., Chowdhury, S. R., & Hossain, M. S. (2016). Heavy metals accumulation in coastal sediments. *Environmental Remediation Technologies for Metal-Contaminated Soils*, 21–42.

Singh, A., Sharma, A., Verma, R. K., Chopade, R. L., Pandit, P. P., Nagar, V., & Sankhla, M. S. (2022). Heavy metal contamination of water and their toxic effect on living organisms. In: Junqueira Dorta, D., and Palma de Oliveira, D. (eds.), *The Toxictiy of Environmental Pollutant.* IntechOpen. https://doi.org/10.5772/intechopen.105075.

Sparks, C., Odendaal, J., & Snyman, R. (2017). Metal concentrations in intertidal water and surface sediment along the west coast of the Cape Peninsula, Cape Town, South Africa. *Water S. Part A, 43*(1), 17–24.

Staista (2023). Forecast of the total population of Africa from 2020 to 2050 (*in millions*). Available at https://www.statista.com/statistics/1224205/forecast-of-the-total-population-of-africa/.

Stern, R. J., & Johnson, P. R. (2019). Constraining the opening of the Red Sea: Evidence from the Neoproterozoic margins and Cenozoic magmatism for a volcanic rifted margin. In Rasul, N., and Stewart, I. (eds.) *Geological Setting, Palaeoenvironment and Archaeology of the Red Sea* (pp. 53–79). Springer, Cham. https://doi.org/10.1007/978-3-319-99408-6_4.

Suami, R. B., Sivalingam, P., Al Salah, D. M., Grandjean, D., Mulaji, C.K., Mpiana, P.T. & Poté, J. (2020). Heavy metals and persistent organic pollutants contamination in river, estuary, and marine sediments from Atlantic Coast of Democratic Republic of the Congo. *Environmental Science and Pollution Research*, *27*(16), 20000–20013.

Tchounwou, P. B., Yedjou, C. G., Patlolla, A. K., & Sutton, D. J. (2012). Heavy metal toxicity and the environment. *Molecular, Clinical and Environmental Toxicology: Volume 3: Environmental Toxicology*, *3*, 133–164.

Theconversion (2021). Informal waste management in Lagos is big business: Policies need to support the trade. Available at https://theconversation.com/informal-waste-management-in-lagos-is-big-business-policies-need-to-support-the-trade-151583.

Tnoumi, A., Angelone, M., Armiento, G., Caprioli, R., Crovato, C., De Cassan, M. & Zourarah, B. (2020). Assessment of trace metals in sediments from Khnifiss Lagoon (Tarfaya, Morocco). *Earth*, *2*(1), 16–31.

Tshithukhe, G., Motitsoe, S. N., & Hill, M. P. (2021). Heavy metals assimilation by native and non-native aquatic macrophyte species: A case study of a river in the Eastern Cape Province of South Africa. *Plants*, *10*(12), 2676.

Tuo, A. D., Soro, M. B., Trokourey, A., & Bokra, Y. (2012). Assessment of waters contamination by nutrients and heavy metals in the Ebrié Lagoon (Abidjan, Ivory Coast). *Research Journal of Environmental Toxicology*, *6*(5), 198.

Tuo, A. D., Yeo, K. M., Soro, M. B., Trokourey, A., & Bokra, Y. (2013). Contamination of surface sediments by heavy metals in Ebrié lagoon (Abidjan, Ivory Coast). *International Journal of Chemical Technology*, *5*(1), 10–21.

USGS (2019). How much water is there on Earth? Available at https://www.usgs.gov/special-topics/water-science-school/science/how-much-water-there-earth.

van Vleck Anderson, R. (1936). *Geology in the Coastal Atlas of Western Algeria* (Vol. 4). Geological Society of America.

Vlachos, V., Critchley, A. T., Bannatyne, T. E., & Von Holy, A. (1998). Metal concentrations in seaweeds from KwaZulu-Natal, South Africa-a first report. *South African Journal of Botany*, *64*(4), 233–237.

Wang, H., Fan, Z., Kuang, Z., Yuan, Y., Liu, H., & Huang, H. (2021). Heavy metals in marine surface sediments of Daya Bay, Southern China: Spatial distribution, sources apportionment, and ecological risk assessment. *Frontiers in Environmental Science*, *559*,-570.

Wang, W. X., & Dei, R. C. H. (2001). Effects of major nutrient additions on metal uptake in phytoplankton. *Environmental Pollution*, *111*(2), 233–240.

WWAP (United Nations World Water Assessment Programme) (2016). *The United Nations World Water Development Report 2016: Water and Jobs*. UNESCO.

Yahaya, T. O., Illo, Z. Z., Abdulgafar, I. B., Salihu, M. G., Gomo, C. B., Abdulrahim, A., & Abdulkadir, A. Concentrations and health risk parameters of heavy metals in water samples from Epe lagoon in Lagos State, Nigeria. *Dutse Journal of Pure and Applied Sciences (DUJOPAS)*, *8*(2b), 149–157.

Yehdhih, M., El Hadi, H., Baghdad, B., Chakiri, S., Hamoud, A., Zerdeb, A., & Moussa, K. (2022). Assessment of heavy metals pollution in the marine sediments of the Lévrier Bay (Nouadhibou, Mauritania). *Ecological Engineering & Environmental Technology*, *23*(5), 84–90.

Yehouenou, E. A. P., Adamou, R., Azehoun, P. J., Edorh, P. A., & Ahoyo, T. (2013). Monitoring of heavy metals in the complex Nokoué Lake-Cotonou and Porto-Novo Lagoon ecosystem during three years in the Republic of Benin. *Research Journal of Chemical Sciences*, *3*(5), 1–4.

Zongo, S. B., Ngohang, F. E., Mabert, B. D. C. K., Nzaba, E. N., Djounga, F. A., Ondo, J. P., & Stieglitz, T. (2022). The marine benthic algae diversity of Gabon: Case of the rocky foreshore of cap Estérias. *Open Journal of Marine Science*, *12*(4), 127–140. https://doi.org/10.4236/ojms.2022.124008.

20 Bioactive Molecules of Medicinal Value in Marine Green Algae

Adekunle Babajide Rowaiye, Ugonna Morikwe, Solomon Oni, Grace Onyukwo Abakpa, Itoroabasi Joseph Okon, Olubukunola Osisami, and Gordon C. Ibeanu

20.1 INTRODUCTION

Marine algae are abundant sources of unique secondary metabolites that are produced as a defense against predators and hostile environmental conditions. These metabolites are extremely bioactive and could be used as leads for the development of novel pharmaceuticals. The use of seaweeds by humans is extensive; uses include food, livestock feed, food additives, agar, pigments, gels, and cosmetics (El Gamal 2010; Dharmananda 2002; Yao et al. 2022).

The consumption of marine algae is global and is especially a culture among Oriental people. For example, it is estimated that in Japan, seaweed is prepared in many cuisines and frequently consumed in quantities that are significantly higher than 1 gram per day, either as food or in supplements (Yao et al. 2022; Dharmananda 2002). This is due to the fact that it presents a rich blend of nutrients, such as protein, vitamins (carotenes, ascorbic acid), macronutrients (calcium, chlorine, sulfur, sodium, magnesium, potassium, and phosphorus), and micronutrients (iodine, copper, iron, selenium, zinc, fluoride, manganese, molybdenum, and cobalt). Seaweed also has a low fat and high fiber content (Dharmananda 2002).

The marine algae can be classified into two major groups: first, the macroalgae or seaweeds, which can be further botanically classified into brown, green, and red, and second, the microalgae, which include the bacillariophyta (diatoms), the blue-green algae, and the dinoflagellates (El Gamal 2010). The marine green algae, which are the focus of this review, consist of over 4,000 species, and they are a group of diverse photosynthetic organisms that contain chlorophyll which is responsible for their color. Their main classes include Chlorodendrophyceae, Chlorodendrophyceae, Ulvophyceae, Prasinophyceae, and Trebouxiophyceae (Tragin and Vaulot 2018). Green algae can grow in freshwater or marine environments, moist earth, trees, or rocks. These algae can be unicellular, existing as single cells; colonial, existing in colonies; or multicellular (Kandale et al. 2011). The siphonalean or coenocytic organization is a specific body structure unique to the green algae (Jayasankar and Reeta 2022).

Like other marine organisms, the green algae are hosts to a plethora of bioactive molecules, such as peptides, proteins, lipids, polysaccharides, polyphenolic compounds, etc. These molecules could be used as, or serve as templates for the production of, novel drugs for various diseases and disorders. While clinically approved drugs have been developed from other marine organisms, natural products for the marine green algae are at different stages of development.

This study is aimed at investigating the various bioactive molecules that exist specifically in the marine green algae and their associated medicinal effects on humans or animals with empirical data.

 DOI: 10.1201/9781003369738-24

20.2 BIOACTIVE MOLECULES IN MARINE GREEN ALGAE

Algae are significant marine plants that are abundant in bioactive compounds. Although the terrestrial environment is a huge source of biologically active molecules, the finding of new metabolites is decreasing (Jensen and Fenical 2000). Thus, greater study on fresh sources of biologically active molecules is required. In this regard, the vast biodiversity seen in marine habitats provides a fertile ground for the discovery of new compounds with a variety of significant bioactivities. The marine green alga is a source of structurally diverse compounds with a variety of bioactivities, such as anti-inflammatory and anti-cancer (Table 20.1) (Mahmoud et al. 2021). Marine green algae, consisting of seaweeds and microalgae, are excellent sources of primary metabolites, which include polysaccharides, dietary fiber, amino acids, proteins, essential fatty acids, and secondary metabolites, such as alkaloids, polyketides, polyphenols, terpenes, and anthraquinones (Bharathai et al. 2022; Ghareeb et al. 2020; Mahmoud et al. 2021).

20.2.1 Primary Metabolites

20.2.1.1 Lipids

Marine green algae contain sterols and long-chain polyunsaturated fatty acids (PUFAs) (Vigor et al. 2020). Their mechanisms of sterol synthesis have been characterized. Specifically, *Chlorella sorokiniana* and symbiotic *Chlorella variabilis* NC64A produce ergosterol, as compared with plants that produce phytosterols such as β -sitosterol, campesterol, and stigmasterol (Voshall et al. 2021). PUFAs are essential fatty acids that have several positive effects on human health (Jain et al. 2022). Very long-chain polyunsaturated fatty acids (VLC-PUFAs), recognized for their positive impacts on human health, can be produced in particular by marine species. In fact, VLC-PUFAs are the building blocks for several signaling molecules in humans that are implicated in the intricate processes of inflammation (Ramos-Romero et al. 2021; Vigor et al. 2020). Marine microalgae strains obtained from the coast of the Arabian Sea were discovered to have high concentrations of PUFAs. *Planophila* sp., *Chlorella* sp., and *Nannochloropsis oculata* produced a large amount of PUFA, with the proportion of alpha-Linolenic acid (ALA) to total fatty acids reaching its highest levels of 36%, 50%, and 50%, respectively (Jain et al. 2022). Another strain, *Monoraphidium* (Selenastraceae, Chlorophyta), which is cold-adapted and found in an Antarctic ice-covered lake, was also discovered to have a high content of PUFAs (Řezanka et al. 2017). Other PUFAS, such as the ω-3 PUFAs docosahexaenoic acid and eicosapentaenoic acid, can be found in marine algae (Ramos-Romero et al. 2021).

20.2.1.2 Proteins and Peptides

Certain proteins found in the marine green algae have demonstrated pharmacological importance. For example, the hydrophilic glycoproteins obtained from the edible green alga *Capsosiphon fulvescens* have exhibited anti-aging properties (Oh et al. 2019). A new sulfated glycoprotein derived from the Moroccan green seaweed *Codium decorticatum* was able to confer immunity in tomatoes (Aitouguinane et al. 2022). Also, Arabinogalactan-like glycoproteins were derived from *Ulva lactuca* (Chlorophyta), and their structure suggests a specialized adaptation to the marine environment (Přerovská et al. 2021). Marine green algae also contain bioactive lectins (Vigor et al. 2020). For example, two new lectins, namely, CiL-1 and CiL-2, have been extracted from the green alga *Codium isthmocladum*. Both proteins, weighing 12,027 ± 5 Da and 12, 264 ± 5 Da respectively, could agglutinate human and rabbit red blood cells (Carneiro et al. 2020). Another novel lectin, HRL40, was extracted from the green alga *Halimeda renschii*. HRL40. This protein, with a molecular weight of 46.56 kDa, has a high binding affinity for mannose and inhibits the H3N2 influenza virus (Mu et al. 2017).

Marine green algae also contain bioactive peptides with diverse properties such as antimicrobial, anti-inflammatory, antioxidant, and anti-tumor activities. N-methylsansalvamide, a novel cyclic

TABLE 20.1
Bioactivities of the Molecules Present in Several Species of Marine Green Algae

Species	Name of molecule/extract	Activity (reference)
Ulva lactuca	UPO[a]	Anti-aging (Liu et al. 2019)
Enteromorpha prolifera	EPO[a]	Anti-aging (Liu et al. 2019)
Ulva prolifera	Fraction P1-1[a]	Anti-aging (Cai et al. 2016)
Dunaliella tertiolecta	Protein hydrolysate and microalgae residual biomass[b]	Anti-aging (Norzagaray-Valenzuela et al. 2017)
Nannochloropsis sp.	Protein hydrolysate and microalgae residual biomass[b]	Anti-aging (Norzagaray-Valenzuela et al. 2017)
Chaetomorpha sp.	Dichloracetic acid, oximes, and L-terpinol[a]	Anti-cancer (Haq et al. 2019)
Dunaliella tertiolecta	Violaxanthin[a]	Anti-cancer (Pasquet et al. 2011)
Bryopsis plumosa	BPL1, BPL2, and BPL3[b]	Anti-cancer (Lee et al. 2022)
Derbesia marina	Methanolic extract	Antidiabetic (Lee et al. 2007)
Codium adhaerens	Methanolic extract	Antidiabetic (Lee et al. 2007)
Caulerpa racemosa	Ethyl acetate extract	Antidiabetic (Aroyehun et al. 2020)
Ulva lactuca	ULP[a]	Antidiabetic (Chen et al. 2022)
Enteromorpha linza	Methanolic extract	Antihypertensive (Cha et al. 2006)
Codium contractum	Aqueous extract	Antihypertensive (Cha et al. 2006)
Enteromorpha compressa	Aqueous extract	Antihypertensive (Cha et al. 2006)
Codium fragile	Methanolic extract	Antihypertensive (Kolsi et al. 2017)
Ulva intestinalis	FGMPLDR[b]	Antihypertensive (Sun et al. 2019)
Ulva intestinalis	MELVLR[b]	Antihypertensive (Sun et al. 2019)
Chlorella sorokiniana	PN-1[b]	Antihypertensive (Lin et al. 2018)
Chlorella ellipsoidea	Val-Glu-Gly-Tyr[b]	Antihypertensive (Ko et al. 2012)
Cladophora koeiei	Ethanolic extract	Anti-hyperlipidemic (Abdel-Raouf et al. 2018)
Enteromorpha prolifera	EPP[a]	Anti-hyperlipidemic (Wassie et al. 2021)
Ulva lactuca	ULP[a]	Anti-hyperlipidemic (Sathivel et al. 2008)
Capsosiphon fulvescens	Hydrophilic glycoproteins[b]	Neuroprotective (Oh et al. 2019)
Monostroma angicava	MSP[a]	Anticoagulative (Liu et al. 2018)
Monostroma nitidum	MS-1[a]	Anticoagulative (Cao et al. 2019)
Chaetomorpha aerea	PCA[a]	Anticoagulative (Qin et al. 2023)
Ulva rigida	Ulvan[a]	Anticoagulative (Adrien et al. 2019)
Ulothrix flacca	UP2-1[a]	Anticoagulative (Li et al. 2018)
Cladophora oligoclada	Novel sulfated polysaccharide[a]	Anticoagulative (He et al. 2021)
Codium decorticatum	Dichloromethane and acetone/methanol extracts	Anti-inflammatory (Zbakh et al. 2020)
Caulerpa racemosa	Squalene[a]	Anti-inflammatory (Fernando et al. 2018).
Dunaliella bardawil	Beta-carotene[a]	Anti-inflammatory (Lee et al. 2013)
Chlorella marina	Lycopene[a]	Anti-inflammatory (Lee et al. 2013)
Dunaliella tertiolecta	Phytosterols[a]	Anti-inflammatory (Lee et al. 2013)
Caulerpa cupressoides	Sulfated polysaccharide[a]	Anti-inflammatory (Lee et al. 2013)
Ulva spp.	Hydrolysates[b]	Anti-inflammatory (Cian et al. 2018)
Pyropia yezoensis	PPY1[b]	Anti-inflammatory (Lee et al. 2015)
Caulerpa cupressoides	Lectins[b]	Anti-inflammatory (Vanderlei et al. 2010)
Porphyra yezoensis	PGP[b]	Anti-inflammatory (Shin et al. 2011)
Cladophora rupestris	Crude hydroalcoholic extract	Antioxidative (Surget et al. 2017)
Codium fragile	Crude hydroalcoholic extract	Antioxidative (Surget et al. 2017)
Ulva lactuca	Phenolics[a]	Antioxidative (Sirbu et al .2019)

(Continued)

TABLE 20.1 (CONTINUED)
Bioactivities of the Molecules Present in Several Species of Marine Green Algae

Species	Name of molecule/extract	Activity (reference)
Enteromorpha intestinalis	Phenolics[a]	Antioxidative (Sirbu et al .2019)
Cladophora vagabunda	Phenolics[a]	Antioxidative (Sirbu et al .2019)
Chaetomorpha spp.	Phenolics[a]	Antioxidative (Haq et al. 2019)
Ulva rigida	Ethanolic precipitate	Antioxidative (Bourguiba et al. 2017)
Ulva lactuca	UPO[a]	Antioxidative (Liu et al. 2019)
Enteromorpha prolifera	EPO[a]	Antioxidative (Liu et al. 2019)
Tetradesmus obliquus	WPRGYFL[b]	Antioxidative (Montone et al. 2018)
Tetradesmus obliquus	GPDRPKFLGPF[b]	Antioxidative (Montone et al. 2018)
Tetradesmus obliquus	WYGPDRPKFL[b]	Antioxidative (Montone et al. 2018)
Tetradesmus obliquus	SDWDRF[b]	Antioxidative (Montone et al. 2018)
Ulva pertusa	Ulvan[a]	Antiviral (Chi et al. 2020)
Caulerpa racemosa	Caulerpin[a]	Antiviral (Esteves et al. 2019)
Caulerpa taxifolia	Ethyl acetate extract	Antibacterial (Rajivgandhi et al. 2018)
Caulerpa racemosa	Ethyl acetate extract	Antibacterial (Ramachandran et al. 2019)
Caulerpa racemosa	Caulerprenylols B[a]	Antifungal (Liu et al. 2013)
Tetraselmis suecica	Antimicrobial peptide[b]	Antimicrobial (Guzman et al. 2019)

[a]Natural compound; [b]peptides and proteins.

depsipeptide with poor *in vitro* cytotoxicity on the NCI human tumor cell line, was produced by the *Fusarium* sp. CNL-619 that was extracted from the green alga *Avrainvillea* sp. obtained from Bovoni Cay of the United States Virgin Islands (Zhang et al. 2016). Pharma Mar SA introduced the cyclic depsipeptide Kahalalide F, which was derived from the green alga *Bryopsis* sp., into phase I trials as a lead drug against prostate cancer (Blunt et al. 2005). Elisidepsin is a synthetic peptide derived from the Kahalalide F family, which attained a phase II clinical trial for its anti-cancer activity, but the trial was discontinued (Serova et al. 2013; Liu et al. 2021).

JM47, a cyclic tetrapeptide, was obtained from a marine Fusarium strain (MOBCOF-1), which was isolated from *Codium fragile* subsp. *Atlanticum*, a green alga collected off the east coast of Scotland. The structure of JM47 was found to be Cyclo(Ala-Ala-Aoh-Pro), where Aoh is (2*S*,9*S*)-2-amino-8-oxo-9-hydroxydecanoic acid (Jiang et al. 2002).

20.2.1.3 Polysaccharides

Sulfated polysaccharides are essential building blocks of the cell walls of seaweeds (Ciancia et al. 2020). Galactans (such as agarans and carrageenans), ulvans, and fucans are some of the main sulfated polysaccharides produced by seaweeds (Ji et al. 2011). Sulfated polysaccharides like ulvans, found in the Ulvophyceae family of marine green algae (Vigor et al. 2020), are metabolites with numerous biological properties, including hypoglycemic, antiviral, immunostimulative, anti-cancer, and anti-inflammatory activities (Ghareeb et al. 2020). Also, WF1 and WF3, which are two sulfated polysaccharides, were extracted from the marine green alga *Monostroma nitidum*, and their structural properties were identified. Both compounds exhibited significant anticoagulant properties (Mao et al. 2008). Also present in marine green algae are cellulose, mannan, sulfated rhamnan, and a sulfated glucuronorhamnoxylan polysaccharide (Yao et al. 2022; Choi et al. 2019).

20.2.2 Secondary Metabolites

20.2.2.1 Alkaloids

Alkaloids are a diverse group of widely dispersed cyclic molecules that include basic nitrogen in a state of negative oxidation. Alkaloids are divided into seven sub-classes based on their structure and bioactivity, including the amino imidazole, indole, pyridoacridine, isoquinoline, pyrrolizidine, guanidine, and steroidal alkaloids (Ghareeb et al. 2020). A sizable and continuously expanding class of secondary metabolites, marine bis-indole alkaloids, continues to provide a wide range of structural templates for various biological targets. In particular, Caulerpin, the bis-indole alkaloid that has been extracted from the marine green alga *Caulerpa* in a number of locations around the world, has been studied for its potential as a medicinal agent with anti-tumor, anti-herpes, antitubercular, anti-herpes virus, antimicrobial, and immuno-stimulating properties (Lunagariya et al. 2019; Bharathai et al. 2022).

Several other indole derivatives have been isolated from the fungi found on various marine green algae. Seven novel cytochalasan derivatives, known as Cytoglobosins A–G, were discovered in the culture of the endophytic fungus *Chaetomium globosum* QEN-14, which is produced from the sea green alga *Ulva pertusa* obtained from the Qingdao coastline of China. The A-549 tumor cell line was susceptible to cytotoxic activity from Cytoglobosins C and D (Ahmed et al. 2021). Isolated from *Aspergillus flavus* C-F-3, associated with the marine green alga *Enteromorpha tubulosa*, obtained from Putian Pinghai in China, iso-αcyclopiazonic acid has proved marginally cytotoxic to a number of human tumor cell lines (Zhang et al. 2016). Aspeverin, a new alkaloid comprising carbamate and cyano moieties, was extracted from an algicolous *Aspergillus versicolor* dl-29 and moderately inhibited the growth of the phytoplankton *Heterosima akashiwo* associated with the green alga *Codium fragile* obtained from Dalian in China (Ji et al. 2013).

20.2.2.2 Polyketides

Malonyl-CoA and methylmalonyl-CoA are two examples of acyl-thioester units that condense to generate polyketides, a wide family of medicinally significant natural compounds. These metabolites have a variety of structural and biological properties (Ridley et al. 2007). They comprise a wide range of chemical scaffolds such as polyols, macrolides, and polyethers (Ghareeb et al. 2020). Anti-tumor, antibacterial, anti-inflammatory, and antioxidant activities are only a few of the diverse properties that polyketides exhibit (Ghareeb et al. 2020). The fermentation broth of *Streptomyces* sp. OUCMDZ-3434 associated with the marine green alga *Enteromorpha prolifera* contains novel phenolic polyketides such as Wailupemycin J, R- and S-Wailupemycin K, Wailupemycin L, and 3-O-Methylwailupemycin G, as well as the known 5-Deoxyenterocin and Enterocin. These compounds have demonstrated antidiabetic, anti-tumor, and antiviral activities (Liu et al. 2017).

Coniothyrium cereale is a fungus isolated from the alga *Enteromorpha* sp. obtained from the Baltic Sea. When cultured in a saline medium, *C. cereale* produced Cereoanhydride and Conioimide, which represent novel structural types of polyketides. Conioimide has significant and selective inhibitory activity against the human leukocyte elastase enzyme (Elsebai et al. 2012). The endophytic fungus *Talaromyces* sp. CX11 associated with marine green algae yielded three novel spirocyclic polyketides, Talaromyacins A, B, and C. Extensive spectroscopic examination, Snatzke's methods, and quantum chemical calculations were used to identify their structures, including their absolute configurations. The recognized Sequoiamonascin A, for which a structural change is necessary, and Talaromyacin A are identical (Cao et al. 2023).

Anomalins A and B are two xanthone derivatives produced by the cultivation of the fungus *Wardomyces anomalus* OS4T3-2-1, which was isolated from the green alga *Enteromorpha* sp. that was obtained on Fehmarn Island in the Baltic Sea. Significant inhibition of p56(lck) tyrosine kinase and antioxidant activity was reported in Anomalin A (Zhang et al. 2016). Also, Arugosins G and H were produced by growing the endophytic fungus *Emericella nidulans var. acristata* 14 15E, which was isolated from a Mediterranean green alga. Arugosin H proved effective against the green alga *Chlorella fusca* and the fungus *Mycotypha microspora* (Kralj et al. 2016).

Also found in seaweeds are anthraquinones, which are an interesting chemical group of polyketides that are common in fungi and plants, represent the largest class of naturally occurring quinones, and have the basic structure of C6-C2-C6. The biological properties of anthraquinones, which are benzoquinone derivatives, include antioxidant, anti-inflammatory, anticoagulant, antibacterial, cytotoxic, and antitubercular activities (Ghareeb et al. 2020).

20.2.2.3 Terpenes

Terpenes are the byproducts of the biosynthesis process that also produces open-chain isoprene units. Terpenes are classified into six different groups according to the number of isoprene units they contain: monoterpenes (acyclic and cyclic), diterpenes, triterpenes, tetraterpenes (carotenoids), sesterterpenes, and sesquiterpenes. A variety of marine creatures produce terpene metabolites, which exhibit biological properties such as antiviral and anti-inflammatory activity (Ghareeb et al. 2020). Four novel hydroxylated sclerosporin derivatives, 8-hydroxysclerosporin, 11-hydroxysclerosporin, 12-hydroxysclerosporin, and 15-hydroxysclerosporin, were produced by the fungus *Cadophora malorum* SY3-1-1MIT, which was extracted from the green alga *Enteromorpha* sp. When tested on 3T3-L1 mouse adipocytes, the fat accumulation–inhibitory effect of 8-hydroxysclerosporin was weak (Zhang et al. 2016). It has been reported that the bioactive sesquiterpene (±)-dihydrorhipocephalin, which was extracted from Caribbean marine green algae of the species *Udotea* and *Penicillus*, has undergone its first complete synthesis (Blunt et al. 2005). The sesquiterpene albican-11,14-diol, produced by the endophytic *A. versicolor* d129 from the green alga *Codium fragile*, which was collected off the Dalian coast in China, had strong efficacy against brine shrimp, *Staphylococcus aureus*, and *Escherichia coli* (Zhang et al. 2016).

Twelve novel terpene esters were extracted from marine green algae obtained from the Saronicos Gulf of Greece. The fouling microalga *Phaeodactylum tricornutum* was significantly inhibited in its growth by the solvent extract of *Caulerpa prolifera*, which also showed moderate to significant activity against three uncharacterized marine bacterial species (Blunt et al. 2005). Also, *Aspergillus ustus* cf-42, a fungus isolated from the fresh tissue of the marine green alga *Codium fragile* obtained from Zhoushan Island in China, produced five novel sesterterpenes: (6)-16,17-dihydro-21-deoxyophiobolin G, (6)-21-deoxyophiobolin G, and ophiobolins U–W. The growth of *E. coli* was only mildly inhibited by ophiobolin U (Zhang et al. 2016).

20.2.2.4 Polyphenols

Phenolic compounds are widely dispersed phytochemicals that can be found in most plant tissues, including those of fruits and vegetables. They are produced through the phenylpropanoid and shikimic acid pathways as secondary metabolites (de la Rosa et al. 2019). Marine green algae are rich in polyphenolic compounds and these could have several bioactivities, such as anti-cancer, antioxidative, and antibacterial (against both Gram-negative and Gram-positive bacteria). These compounds provide a template for the development of new drugs (Besednova et al. 2020). The endophytic fungus *Chaetomium globosum*, associated with the marine green alga *Chaetomorpha media* obtained from the Konkan coastline of India, is a novel source of Chrysin. Chrysin is a dihydroxyflavone with anti-cancer activity (Kamat et al. 2020). Similarly, the bioactivity potential of three algal species from Romanian coastal waters near the Black Sea—*Enteromorpha intestinalis*, *Ulva lactuca*, and *Cladophora vagabunda*—was assessed. *C. vagabunda* (325.8 mg/100 g fresh weight [f.w.]) had the greatest total phenolic content, followed by *U. lactuca* (285.5 mg/100 g f.w.), and *E. intestinalis* (268.8 mg/100 g f.w.) had the lowest rate. All these green seaweeds demonstrated antibacterial activity (Sirbu et al. 2019).

20.3 SOME MEDICINAL PROPERTIES OF MARINE GREEN ALGAE

20.3.1 Antimicrobial Activity

Although numerous antiviral medications have been licensed for clinical usage during the past ten years, significant problems such as inadequate efficacy, toxicity, and excessive drug costs persist (Hao et al. 2019). Glycans have made exceptional contributions to the production of new drugs in very recent decades (Grice and Mariottini 2018). Compared with human glycans, marine glycans exhibit a wider range of structural variations, and most importantly, with minimal toxicity and low cost, they have shown powerful antiviral activity via numerous molecular pathways. As a result, preclinical and clinical trials for a number of marine glycan-derived drugs are being conducted (Hao et al. 2019).

Glycans serve crucial functions in the stages of viral infection and are important parts of the surfaces of both host cells and the majority of viruses (Hao et al. 2019). Experimental data suggest that marine-sourced sulfated glycans demonstrate activity against a broad scope of enveloped and non-enveloped viruses (Hans et al. 2021) that cause human, murine, swine, avian, and plant infections (Grice and Mariottini 2018). Some of these sulfated polysaccharides, such as carrageenan, agar, ulvan, fucoidan, and alginates, prevent the various stages of viral infection inside the host cell. By modifying the host cell's immunological response, these molecules are able to inhibit initial viral entry into the cell or prevent its transcription and translation (Hans et al. 2021). Ulvan, a major water-soluble cell wall polysaccharide obtained from the marine green macroalga *Ulva pertusa*, has an intricate molecular structure with demonstrated antiviral activity. At 100 μg/mL, ulvan-F1 (38.5 kDa) and ulvan (1068.2 kDa) both show the ability to suppress the infection and replication of the vesicular stomatitis virus, with respective inhibition rates of 40.13% and 40.75%, respectively (Chi et al. 2020). The alkaloid Caulerpin, extracted from the marine green alga *Caulerpa racemosa*, has also been shown to demonstrate antiviral activity against Chikungunya virus (CHIKV) infections. In the study, Vero cells infected with CHIKV were given various doses of Caulerpin. It was observed that the isolated compound had a very strong and promising EC_{50} inhibitory effect of 0.8 μM in the antiviral activity. By testing the virucidal activity, the results showed that Caulerpin was particularly effective against CHIKV and able to inhibit almost 90% of the viral invasion when administered in doses of 5 μM. It was also observed that Caulerpin continued to have a 100% inhibitory capacity for CHIKV viral replication when added at times 0, 1, 2, and 3 hours post infection (Esteves et al. 2019). Apart from alkaloids, marine macroalgae are an abundant source of polyphenols, and this collection of compounds includes a variety of biopolymers with distinctive biological properties and structures that have strong anti-infective activity. An example is the phlorotannins, which have shown great promise in the development of novel antiviral medications (Besednova et al. 2021).

It has been established that extracts using various solvents (especially ethyl acetate) of 5 out of the 20 strains of endophytic actinomycetes isolated from the internal tissue of the marine green alga *Caulerpa taxifolia* demonstrated strong antibacterial activity against multi-drug-resistant (MDR) uropathogens such as *Escherichia coli*, *Klebsiella pneumoniae*, *Pseudomonas aeruginosa*, and *Enterobacter* sp. (Rajivgandhi et al. 2018). In another study, potential antibacterial compounds were identified in endophytic actinomycetes from the marine alga *Caulerpa racemosa*. The compounds showed inhibitory activity against MDR Gram-negative bacteria including *E. coli*, *P. aeruginosa*, *Proteus mirabilis*, *K. pneumoniae*, and *Enterobacter* sp., which was confirmed by urinary tract infection (UTI) panel antibiotic discs. The ethyl acetate extract of 5 strains out of the 100 endophytic actinomycetes displayed greater inhibitory activity against all of the chosen uropathogens (Ramachandran et al. 2019). These findings could be very important in addressing the constantly emerging multi-drug-resistant bacteria (Rajivgandhi et al. 2018).

Antifungal properties have also been exhibited by marine green algae extracts. Caulerprenylols A and B, two new prenylated para-xylenes, were isolated from the marine green alga *Caulerpa racemosa* that was found along the Chinese shoreline in Zhanjiang. Caulerprenylol B exhibits a unique indane ring structure. When compared with amphotericin B used as a positive control, Caulerprenylols A and B showed broad-spectrum antifungal activity in *in vitro* bioassays against *Trichophyton rubrum*, *Candida glabrata*, and *Cryptococcus neoformans* with minimum inhibitory concentration (MIC_{80}) values ranging between 4 and 64 g/mL (Liu et al. 2013).

Marine organisms such as microalgae have evolved defense mechanisms to withstand exposure to harmful bacteria, viruses, and fungi in the aquatic environment (Guzman et al. 2019). Antimicrobial peptides (AMPs), which are also known as host defensive peptides, are essential for innate immunity (Cheung et al. 2015). It has been discovered that the acid extract of the green microalga *Tetraselmis suecica* has demonstrated antibacterial activity; de novo peptide sequences were discovered in this microalga (Guzman et al. 2019). These AMPs can offer various distinctive sources of possible candidates for anti-infective drugs (Cheung et al. 2015).

20.3.2 Antioxidative Activity

Free radicals must be controlled at extremely low levels because they are important in physiological processes such as immunomodulation, apoptosis, ion transport, cell signaling, and neuromodulation, and can be damaging. Oxidative stress results from excessive free radical production in cells, which outweighs the usual role of endogenous and exogenous antioxidants. Free radicals accumulate in the body and destroy vital biological components like lipids, proteins, and DNA molecules because of their highly reactive nature (Rowaiye et al. 2020). Globally, marine green algae are known to be sources of a wide range of bioactive molecules, and the pharmaceutical industry has extracted a number of these compounds for the potential development of novel drugs (Sirbu et al. 2019).

Growing in challenging abiotic conditions and receiving a lot of UV radiation, macroalgae have evolved defense mechanisms and are now known to be a significant source of macromolecules and secondary metabolites with antioxidant activity (Tziveleka et al. 2021). The antioxidant potential of phenolic compounds in two marine green macroalgae, *Codium fragile* and *Cladophora rupestris*, was investigated. The phenolic compounds were obtained from the ethyl acetate fraction (EAF) of the crude hydroalcoholic extract (CE) of the two species. The results showed that the EAF of both species had high antioxidant activity, which was measured using radical scavenging activity and reducing power assay, and this activity was strongly linked with the phenolic content of each fraction (Surget et al. 2017). Also, three isolated species from Romanian coastal waters near the Black Sea, *Ulva lactuca*, *Enteromorpha intestinalis*, and *Cladophora vagabunda*, were evaluated for their free radical scavenging potential. Using high-performance liquid chromatography with diode array detection (HPLC-DAD) methods, six distinct phenolic compounds were found in all of the seaweeds examined, and these were responsible for their antioxidant properties. As follows, their total phenolic content correlated with their antioxidant activity: *C. vagabunda* (325.8 mg/100 g f.w. and 64.15 nmol/g sample), *U. lactuca* (285.5 mg/100 g f.w. and 22.8 nmol/g sample), and *E. intestinalis* (268.8 mg/100 g f.w. and 10.4 nmol/g sample) (Sirbu et al. 2019). According to antioxidant assays, the ethanol extracts of another marine green alga, *Chaetomorpha* sp., contained a total of 21.92 ± 0.43 mg GAE (Gallic acid equivalent)/g phenolic content, 189.14 ± 0.99 mg QE (quercetin acid equivalent)/g flavonoid content, and 21.81 ± 0.04 mg GAE/g tannins content (Haq et al. 2019).

In a study performed by Bourguiba et al, the antioxidant potential of the ethanolic precipitate of *Ulva rigida* (EP) was evaluated by determining the DPPH-scavenging capacity and cytoprotective activity in zebrafish embryos and yeast cells with H_2O_2-induced damage. The EP demonstrated significantly higher DPPH-scavenging activity (92.5%) at 1 µg/mL than gallic acid, the positive control (84%), at 18 µg/mL. When pretreated at various concentrations (0.005, 0.3, 0.5, or 1 g/mL), EP effectively prevented H_2O_2-induced death in zebrafish embryos and yeast cells. Specifically, when administered at 0.3 µg/mL, EP significantly inhibited H_2O_2-induced modification of superoxide dismutase and catalase activities. At 50 µM, the antioxidant N-acetylcysteine (NAC) produced a similar effect to that of EP, suggesting that EP of *U. rigida* had a protective effect against oxidative stress caused by H_2O_2 (Bourguiba et al. 2017).

Polysaccharides obtained from marine algae have also demonstrated significant antioxidant properties (Zhong et al. 2019). In another study, the oligosaccharides of *Enteromorpha prolifera* (EPO) and *Ulva lactuca* (ULO) were employed to study anti-aging effects and their associated underlying mechanisms in SAMP8 mice. The results revealed that with the administration of EPO and ULO, the concentration of advanced glycation end products and malondialdehyde decreased, while the concentrations of catalase, glutathione, dismutase, superoxide, and telomerase, as well as the total antioxidant capacity, were increased. The levels of the neurotransmitter modulators choline acetyltransferase (ChAT) and brain-derived neurotrophic factor (BDNF) increased, and the hippocampus neurons were protected. Furthermore, there was downregulation of the FOXO1 and p53 genes and overexpression of the Sirt1 gene, suggesting that EPO and ULO may have benefits in the treatment of aging in SAMP8 mice (Liu et al. 2019).

Found in their protein hydrolysates, microalgae contain a variety of bioactive peptides with free radical scavenging potential, as determined by antioxidant assays (Montone et al. 2018). Twenty-five sequenced peptides with putative antioxidant and angiotensin-converting enzyme (ACE)-inhibitory properties were discovered in *Tetradesmus obliquus* microalgae. Four of these peptides—GPDRPKFLGPF, WPRGYFL, SDWDRF, and WYGPDRPKFL—were chosen for production and *in vitro* testing for particular bioactivity. These peptides showed good levels of ACE-inhibitory and antioxidant activities (Montone et al. 2018). Three classes of water-soluble phycobilin proteins—phycocyanin, phycoerythrin, and allophycocyanin—which make up to 60% of the total soluble cellular proteins in microalgae have earned widespread recognition for their potent antioxidant properties. Their structure and the order of the amino acid sequences affect the antioxidant activity, which is typically linked to metal ion chelation, lipid peroxidation inhibition, and free radical scavenging (Vladkova et al. 2022).

20.3.3 Anti-Inflammatory Activity

Inflammation is a pathological condition that is initiated, modulated, and resolved by a complex interaction of immune cells, inflammasomes, multiple signaling pathways, cytokines, and inflammatory metabolites (Rowaiye et al. 2022). Many natural compounds found in marine green algae have been identified to be anti-inflammatory agents, as they are known to inhibit the process of inflammation (Souza et al. 2020). Interestingly, these unique and varied bioactive molecules could serve as a template for the development of new drugs, cosmetics, and nutraceutical products (Li et al. 2021; Kapoor et al. 2019).

A study was performed to determine the anti-inflammatory effects of extracts of the green alga *Codium decorticatum* on several cancer cell lines. The extracts used include the n-hexane extract (Extr 1), dichloromethane extract (Extr 2), and acetone/methanol extract (Extr 3). The results revealed that the pro-inflammatory cytokine interleukin-8 (IL-8) was significantly suppressed by Extr 2 and Extr 3 in lipopolysaccharide (LPS)- and tumor necrosis factor (TNF)α-stimulated endothelial cells. Specifically, Extr 2 inhibited E-selectin and IL-8 mRNA expression, which was stimulated by LPS. Also, Extr 2 demonstrated cytotoxic activity against the HeLa cell line, triggering apoptosis at high doses (IC_{50} = 37.22 g/mL). These findings suggest that the seaweed *C. decorticatum* possesses anti-inflammatory effects in endothelial cells as well as an anti-cancer effect on cervix cancer cells, suggesting that it may be used as a possible therapy for inflammatory conditions (Zbakh et al. 2020).

The bioactive components of the 70% ethanol extract (CRE) of the macroalga *Caulerpa racemosa* which contains ethyl acetate (CREE) and hexane (CREH) fractions showed strong anti-inflammatory properties. Squalene, which was isolated from CREH using bioassay-guided purification, exhibited strong anti-inflammatory properties. In lipopolysaccharide-induced RAW macrophages, squalene decreased the levels of prostaglandin E2 (PGE2), nitric oxide (NO), cyclo-oxygenase (COX-2), inducible nitric oxide synthase (iNOS), and a few important pro-inflammatory cytokines (Fernando et al. 2018). Several other metabolites that have been isolated from marine green algae have anti-inflammatory properties, including beta-carotene from *Dunaliella bardawil*, lycopene from *Chlorella marina*, phytosterols from *Dunaliella tertiolecta*, and a sulfated polysaccharide from *Caulerpa cupressoides* (Lee et al. 2013).

In vitro experiments have shown that biologically active peptides (BAP) derived from marine green algae can reduce inflammatory responses (Cian et al. 2018). The ability of a seaweed protein hydrolysate produced by simultaneously hydrolyzing several *Ulva* species to reduce inflammation has been reported. In this study, the immunomodulatory effect of *Ulva* spp. hydrolysates and their peptide fractions on rat macrophages, splenocytes, and T lymphocytes with or without *in vitro* gastrointestinal digestion and/or ultrafiltration were investigated. Through the TLR4-NFκB/MAPK pathways, the hydrolysates and their peptide fractions inhibited interferon (IFN)-γ production in LPS- or concanavalin A-induced cells. Also, the production of IL-10 was increased, and this was enhanced by *in vitro* gastrointestinal digestion and ultrafiltration (Cian et al. 2018). PPY1 is another anti-inflammatory peptide, which can be obtained from the marine green alga *Pyropia yezoensis.* A study has shown that PPY1 has strong inhibitory effects on the generation of inflammatory mediators in LPS-induced RAW 264.7 macrophages. PPY1 significantly reduced NO, intracellular reactive oxygen species (ROS), cyclooxygenase (COX)-2, IL-1, iNOS, and NF-κB in a dose-dependent manner. Furthermore, PPY1 demonstrated significant reduction in the expression of JNK and p38/MAPK, suggesting that PPY1 has the potential to play a key role as an anti-inflammatory agent (Lee et al. 2015).

Most research points to lectins as good examples of seaweed proteins having anti-inflammatory properties. In fact, some *in vivo* studies demonstrate that these proteins exert their anti-inflammatory effects via diverse pathways (Echave et al. 2022). For instance, lectins obtained from the green alga *Caulerpa cupressoides* have been evaluated for their *in vivo* anti-inflammatory efficacy in numerous studies. *C. cupressoides* lectins were administered intravenously to mice with inflammation induced by high doses of carrageenan. Similarly to dexamethasone, the lectin therapy greatly decreased inflammatory signs, including neutrophil counts and paw edema. In addition, the lectins seemed to lessen nociception without significantly altering the weight of the organs or the transaminase levels (Vanderlei et al. 2010). Similarly, lecithin obtained from *C. cupressoides* demonstrated its anti-inflammatory effect in rats with temporomandibular arthritis (da Conceicao Rivanor et al. 2014) and paw edema induced using a number of pro-inflammatory agents (de Queiroz et al. 2015). As revealed by enzymatic and immunohistochemistry staining methods, the anti-inflammatory activity of lectins is largely due to the significant suppression of the primary cytokines COX-2, IL-1β, IL-6, and TNF-α, and reducing leukocyte concentration, as shown by reduced myeloperoxidase (MPO) levels (da Conceicao Rivanor et al. 2014). Furthermore, it was discovered that lectin therapy did not block hemeoxygenase-2 (HO-2) but was capable of reversing tissue damage and exerting an analgesic effect through a non-opioid mechanism (de Queiroz et al. 2015).

Another seaweed protein with anti-inflammatory properties can be found in *Porphyra yezoensis.* Sodium docecyl sulfate polyacrylamide gel electrophoresis (SDS-PAGE) revealed a 3.5-kDa glycoprotein called *P. yezoensis* glycoprotein (PGP). Through the Toll-like receptor 4 (TLR4) signaling pathway, by reducing the formation of TLR4-IRAK4 and TLR4-TRIF complexes, PGP modulated Toll-like receptor 4 (TLR4) signaling and inhibited NF-κB (by blocking IκB phosphorylation) in oxidative events induced in LPS-stimulated RAW 264.7 mouse macrophages. This consequently blocked the production of ROS and NO and the expression of the pro-inflammatory mediators COX-2, iNOS, IL-1β, and TNF-α. In a dose-dependent manner, PGP also inhibited the phosphorylation of ERK1/2 and JNK, which are also significant mediators of the MAPK inflammatory signaling pathway (Shin et al. 2011).

20.3.4 Anticoagulative Activity

Blood changes from a liquid to a gel during the process of coagulation, causing a clot to develop, resulting in hemostasis, which is followed by repair of the damaged vessel. Blood clot hydraulic permeability largely depends on fibrin and platelet density (Wufus et al. 2013). The acquired condition known as disseminated intravascular coagulation (DIC) is characterized by widespread intravascular activation of coagulation, which can be brought on by both infection and trauma. The primary pathophysiological processes of DIC are tissue factor-dependent coagulation activation by pro-inflammatory cytokines, inadequate modulation of anticoagulant pathways, and suppression of fibrinolysis by plasminogen activator inhibitor 1 (Gando et al. 2016).

Sulfated polysaccharides from marine algae have received a lot of attention for the development of drugs and are excellent prospective sources of naturally occurring anticoagulant agents (Qin et al. 2023). In a study by Liu et al., *Monostroma* sulfated polysaccharide (MSP) was isolated from *Monostroma angicava.* Also, low-molecular-weight fragments of MSP, namely MSP-F1, MSP-F2, MSP-F3, MSP-F4, MSP-F5, and MSP-F6, were obtained through carefully regulated acid degradation. The *in vitro* anticoagulant activity of MSP and MSP-F1 to MSP-F6 was evaluated by measuring the activated partial thromboplastin time (APTT), thrombin time (TT), and prothrombin time (PT). MSP and MSP-F4 were further examined for their *in vitro* thrombolytic activity and *in vivo* fibrin(ogen)olytic activity using the D-dimer, plasminogen activator inhibitior-1, fibrin degradation products, and clot lytic rate tests. The results demonstrated that MSP and MSP-F1-MSP-F4 have potent anticoagulant activity at molecular weights of 24–240 kDa. The anticoagulant activity of the MSP-Fs decreased as their molecular weight dropped, and stronger anticoagulant activity necessitated a molecular weight of over 12 kDa. Significant *in vivo* anticoagulative, thrombolytic, and fibrin(ogen)olytic activities were all exhibited by MSP and MSP-F4, suggesting that these biomolecules may be used as drugs or as beneficial dietary supplements to improve human health (Liu et al. 2018).

In another marine green alga, *Monostroma nitidum*, a homogeneous sulfated polysaccharide named MS-1 was isolated using aqueous extraction and anion-exchange and size-exclusion chromatography. MS-1 demonstrated potent anticoagulant activity *in vitro* and *in vivo*, as measured by the APTT and TT assays, and significantly reduced the aggregation of platelets. The strong potentiation of thrombin is primarily accountable for the anticoagulant activity of MS1. Also, MS1 expedited the inhibition of coagulation factor Xa and thrombin by potentiating antithrombin III. MS-1 exhibited strong antithrombotic activities *in vitro* and *in vivo* (electrically induced carotid artery thrombosis in rats), suggesting that it might be a potential drug for the prevention and treatment of thromboembolic disease (Cao et al. 2019).

Qin et al. obtained another homogeneous polysaccharide from *Chaetomorpha aerea* (PCA) using alkali extraction, anion-exchange chromatography, and size-exclusion chromatography. According to tests for APTT, TT, and fibrinogen level, PCA demonstrated a potent *in vitro* anticoagulant effect. PCA also exhibited a significant *in vivo* anticoagulant activity, as seen in the weak inhibition of the common coagulation factors II and X and strong inhibition of the intrinsic coagulation factors XII, XI, IX, and VIII. Strong thrombin inhibition, potentiated by antithrombin III or heparin cofactor II, was thought to be the mechanism of PCA's anticoagulation. It also appeared to have an inhibitory impact on coagulation factor Xa, which was mediated by antithrombin III. These findings indicate that PCA could be a viable anticoagulant drug for thrombotic disease therapy (Qin et al. 2023).

In another study, the anticoagulant properties of the green alga *Ulva linza,* which was gathered from Abo Qir Bay on the northern shore of Alexandria in Egypt, was evaluated by measuring the APTT and PT. The extracts of this alga were deemed to be natural products that may be helpful in the treatment of various cardiovascular diseases (Mohy El-Din and Alagawany 2019). Similarly, *Ulva rigida* yielded a compound, ulvan, which is a sulphurated polysaccharide found in its cell wall. A chemically sulfurated ulvan fraction had better anticoagulant properties than Lovenox®, a widely used commercial anticoagulant, against both extrinsic and intrinsic pathways of the coagulation cascade (Adrien et al. 2019). Using ion-exchange and size-exclusion chromatography, a water-soluble low-molecular-weight polysaccharide known as UP2-1 was identified and purified from the marine green alga *Ulothrix flacca*, which is consumed in China. UP2-1 showed significant anticoagulant activity (Li et al. 2018).

A new sulfated polysaccharide from the green alga *Cladophora oligoclada* also possesses *in vitro* and *in vivo* anticoagulative activity, as revealed by the modulation of TT, APTT, and fibrinogen levels. The sulfated polysaccharide also significantly demonstrated *in vitro* thrombolytic activity. The investigation of the mechanism of the anticoagulant activity showed that the sulfated polysaccharide selectively inhibited common coagulation factors but significantly inhibited all intrinsic coagulation factors, which were less than 1.0% at 50 g/mL. Furthermore, the sulfated polysaccharide significantly increased the suppression of factor Xa mediated by antithrombin III (AT-III), and it also significantly triggered the inhibition of thrombin by potentiating heparin cofactor II or AT-III (He et al. 2021).

20.3.5 Neuroprotective Activity

Neuroinflammation is the hallmark of neurodegenerative diseases, and it plays a role in the initiation and progression of these age-related diseases (Pereira and Valado 2021). The onset and development of some neurodegenerative disorders, including Parkinson's disease (PD), Alzheimer's disease (AD), Huntington's disease (HD), and amyotrophic lateral sclerosis (ALS), are closely correlated with neuroinflammation (Parameswari and Lakshmi 2022). Astrocytes and microglia are activated as part of the brain's defense against injured tissues and harmful infections, and their persistent activation leads to neuroinflammation, which can worsen or cause dementia (Bauer et al. 2021).

Algal metabolites have protective effects against oxidative stress, mitochondrial dysfunction, neuroinflammation, and impaired proteostasis, which are known to play a role in the pathophysiology of neurodegenerative disorders and the complications that follow cerebral ischemia and brain injuries. These properties include anti-inflammatory, antioxidative, cholesterol homeostasis, protein clearance, and anti-amyloidogenic potential (Hannan et al. 2020). Algal compounds confer neuroprotection against a variety of neurotoxic stressors such as oxygen/glucose deprivation, hydrogen peroxide, amyloid, glutamate, or 1-methyl-4-phenylpyridinium (MPP+), as was evident in several preclinical studies, and as a result, hold promise in the treatment of brain disorders (Hannan et al. 2020).

Several anti-inflammatory compounds with potential protective effects against neuroinflammation have recently been identified from marine algae (Barbalace et al. 2019). Due to their anti-inflammatory activities, marine compounds such as xanthines, bryostatin, and 11-dehydrosinulariolide have neuroprotective benefits against neurodegenerative diseases (Catanesi et al. 2021). To combat infections and heal damaged tissues, these compounds can activate microglia and astrocytes in the brain (Barbalace et al. 2019). In the therapy and prevention of neurodegenerative disorders and multiple sclerosis as well as other chronic diseases, it has been demonstrated that green algae have beneficial therapeutic effects (Pereira and Valado 2021; Parameswari and Lakshmi 2022).

Extracts of seaweed polysaccharides exhibit notable neuroprotective and even therapeutic properties. Examples of these polysaccharides are fucoidan, ulvan, and their derivatives, which have been of research interest in the treatment of Alzheimer's disease (Bauer et al. 2021; Dhahri et al. 2021). These natural polysaccharides have attracted a lot of interest for their diverse bioactivity, which includes antioxidant, anti-neuroinflammatory, anticholinesterase, and anti-amyloidogenic activities (Dhahri et al. 2021).

In a study conducted by Oh et al., the neuroprotective effect of the edible green alga *Capsosiphon fulvescens* was established. Hydrophilic glycoproteins obtained from chronic oral administration of *C. fulvescens* prevented aging-induced spatial memory impairment by inhibiting glycogen synthase kinase (GSK)-3 beta–mediated endoplasmic reticulum (ER) stress in dorsal hippocampus (Oh et al. 2019).

20.3.6 Anti-Hyperlipidemic Activity

Hyperlipidemia is a chronic condition that raises the risk of cardiovascular disease and stroke when lipids or fat levels are abnormally high (Wassie et al. 2021). In a study by Abdel-Raouf et al., the anti-hyperlipidemic activity of the ethanolic extract of the marine green alga *Cladophora koeiei* was tested. The results revealed that the extract reduced triglyceride (TG), low-density lipoprotein (LDL-C), blood urea nitrogen (BUN), total cholesterol (TC), and creatinine levels. Nevertheless, total protein, globulin, albumin, and high-density lipoprotein (HDL-C) levels remained. Also, the extract showed hepatoprotective properties against toxic alcohol by restoring the alanine transaminase (ALT) and aspartate transaminase (AST) enzyme concentrations (Abdel-Raouf et al. 2018).

Recent research in animal models demonstrated that polysaccharides produced from the seaweed green alga *Enteromorpha prolifera* (EPP) demonstrated hypolipidemic properties in a dose-dependent manner. In mice with high-fat diet-induced hyperlipidemia, the anti-hyperlipidemic activity of EPP was achieved by the oral administration of 300 mg per kg body weight of hot water extract of EPP, and this significantly lowered plasma and liver levels of TG, TC, and LDL, and inhibited lipid metabolism (Wassie et al. 2021). It was also found that feeding hyperlipidemic rats EPP increased hepatic acetyl-CoA carboxylase mRNA expression, and improved insulin sensitivity and glucose tolerance (Wassie et al. 2021). It has been postulated that the anti-hyperlipidemic properties of EPP might be linked to a reduction of sterol regulatory element-binding protein-1c (SREBP-1c), which led to the inhibition of cholesterol biosynthesis (Ren et al. 2018).

Similar anti-hyperlipidemic properties were exhibited in rats pretreated for 21 days with *Ulva lactuca* polysaccharide (ULP) at a dose of 30 mg per kg body weight per day (Sathivel et al. 2008).

20.3.7 Antihypertensive Activity

Oxidative stress, stimulation of the sympathetic and renin-angiotensin systems, secretion of inflammatory mediators, and increased adipogenesis are factors that enhance systemic dysfunction, resulting in clinical symptoms of cardiovascular complications (Seca and Pinto 2018). Seaweeds are now widely recognized as a priceless source of new natural compounds that may contain significant prospects for future drug discovery and development, notably in the prophylaxis and/or therapy of cardiovascular diseases. Polysaccharides, peptides, carotenoids, phlorotannins, and sterols, which have been extracted from green macroalgae, significantly reduce blood pressure and prevent obesity (Seca and Pinto 2018).

The application of ACE inhibitors is one of the therapeutic strategies for combating hypertension. The renin-angiotensin system (RAS) performs a crucial role in the modulation of blood volume, electrolytes, and blood pressure, and the pathophysiology of cardiovascular diseases (Kolsi et al. 2017). The *in vitro* ACE-inhibiting properties of aqueous and methanol extracts of ten marine green algae collected along the Jeju coast of Korea were investigated. The findings revealed that that the 70 °C methanolic extract of *Enteromorpha linza* and 20 °C aqueous extracts of *Codium contractum* and *Enteromorpha compressa* demonstrated moderate ACE-inhibitory activities (Cha et al. 2006). In another *in vitro* study, the antihypertensive effect of *Codium fragile* extract (CFE) was evaluated. According to the findings of the study, CFE demonstrated ACE-inhibitory activity in a dose-dependent manner. At a concentration of 0.8 mg/mL, the methanolic extract of *Codium fragile* strongly inhibited ACE *in vitro* by 66.84% ± 2.74%, with a low IC_{50} value of 0.59 mg/mL (Kolsi et al. 2017).

Seaweed-derived peptides that suppress ACE activity offer a possible source of novel antihypertensives (Sun et al. 2019). Sun et al. also purified ACE-inhibitory peptides from five different protein hydrolysates of the marine green macroalga *Ulva intestinalis*. The trypsin hydrolysates had the strongest ACE-blocking activity when compared with the other hydrolysates. Two novel pure ACE-inhibitory peptides (FGMPLDR, with a molecular weight of 0.83 kDa, and MELVLR, with a molecular weight of 0.76 kDa) were produced after fractionation and purification using gel exclusion chromatography, ultrafiltration, and reverse-phase HPLC. Their IC_{50} values were 236.85 μM (0.179 mg/mL) and 219.35 μM (0.183 mg/mL), respectively. The findings of this study offer a theoretical foundation for the highly beneficial use of *U. intestinalis* for the production of ACE-inhibiting peptides obtained from foods (Sun et al. 2019). In another study, the ACE-suppressing effects of protein hydrolysates (PN-1) obtained from the hot aqueous extract of *Chlorella sorokiniana* were evaluated. After 6 hours of *per os* administration of a dose of 171.4 mg of PN-1 powder per kg body weight, the diastolic and systolic blood pressures of spontaneously hypertensive rats (SHRs) were decreased by 21 and 20 mm, Hg respectively (Lin et al. 2018).

The protein from the marine weed *Chlorella ellipsoidea* was hydrolyzed by the enzyme alcalase and separated into three fractions with molecular weight below 5 kDa, between 5 and 10 kDa, and above 10 kDa. The highest inhibitory activity against ACE was found in the fraction below 5 kDa, which was then used for further purification procedures. After purification, a strong ACE inhibitor peptide was found to be Val-Glu-Gly-Tyr with a molecular weight of 467.2 Da and an IC_{50} value of 128.4 μM. According to Lineweaver–Burk plots, the purified peptide competitively inhibits ACE and is stable against the digestive enzymes trypsin, pepsin, and α-chymotrypsin. Additionally, the oral administration of a purified peptide can substantially lower systolic blood pressure, as demonstrated by the antihypertensive effect in SHRs. Based on the findings, the marine alga *C. ellipsoidea* appears to be a promising material for the manufacture of antihypertensive nutraceutical ingredients (Ko et al. 2012).

20.3.8 Antidiabetic Activity

Diabetes mellitus (DM) is a metabolic disease distinguished by chronic hyperglycemia, which is linked to a number of potential disorders such as nephropathy, atherosclerosis, and retinopathy. Presently, oral antidiabetic medications and insulin injections are the main treatments available for the management of DM. Unfortunately, prolonged use of synthetic medications results in numerous negative side effects; hence the need for efficient naturally based antidiabetic therapy (Unnikrishnan and Jayasri 2018; Gunathilaka et al. 2021).

Many bioactive molecules of nutritional and medicinal value obtained from marine seaweeds have demonstrated antidiabetic activity through a variety of mechanisms, most notably by inhibiting a number of enzymes, including α -amylase, aldose reductase, α -glucosidase, protein tyrosine phosphatase 1B (PTP1B), and dipeptidyl peptidase-4 (Gunathilaka et al. 2021). For example, the *in vitro* PTP1B inhibitory property of methanolic extracts from various green seaweeds, such as *Derbesia marina* and *Codium adhaerens*, revealed that they have 80.6% and 71.2% inhibitory activity, respectively (Lee et al. 2007).

In a recent study, the antidiabetic properties of the edible green alga *Caulerpa racemosa* (Forsskal) obtained from fish ponds located in Johor, Malaysia were examined (Aroyehun et al. 2020). In diabetic-induced rats, the ethyl acetate extract of *C. racemosa* showed significant antidiabetic activity. In rats induced with diabetes, the high (200 mg/kg) and low (100 mg/kg) doses of cultivated *C. racemosa* extract significantly reduced blood glucose concentration, prevented weight loss, and decreased plasma AST and ALT concentrations as an indication of hepatoprotective properties, and the recorded albumin concentration was comparable to positive controls (Aroyehun et al. 2020).

In another study, gut microbiota variation data was used for the structural characterization of *Ulva lactuca* polysaccharide (ULP) and to determine its hypoglycemic impact in mice with diabetes associated with aging. The results show that ULP regulated the expression levels of FoxO1, MMP2, p16Ink4a, STAT3, GLP-1/GLP-1R, and GLUT4 genes to retard aging and diabetes, and this was correlated with the increased abundance of *Aerococcus*, *Bifidobacterium*, *Enterococcus*, *Clostridium sensustricto* 1, *Dubosiella*, *Romboutsia*, *Faecalibaculum*, *Corynebacterium*, *Kurthia*, and *Vibrio* species. Particularly, *Clostridium sensustricto* 1, *Dubosiella, Romboutsia, Bifidobacterium*, and *Turicibacter* spp. may operate as crucial intermediaries in the prevention of diabetes and aging. Hence, ULP has the ability to act as a bioceutical to postpone or stop the onset of type 2 diabetes linked to aging (Chen et al. 2022).

Several bioactive peptides from protein hydrolysates obtained from the solvent extracts of seaweeds exhibit their antidiabetic activity through renin, ACE, and food oxidation inhibition mechanisms (Admassu et al. 2018).

20.3.9 Anti-Cancer Activity

Marine algal polysaccharides (MAPs) have demonstrated significant anti-cancer properties (Yao et al. 2022). The anti-proliferative potential of the *Chaetomorpha* sp. of the Cladophoraceae family of green algae was examined in a recent study. By using the 3-(4,5-Dimethylthiazol-2-cyl)-2,5-D iphenyltetrazolium bromide (MTT) test, anti-tumor efficacy against MDA-MB-231 and MCF-7 breast cancer cell lines was evaluated. The energy-dispersive X-ray spectroscopy (EDX) examination revealed that oxygen, silicon, and calcium were the three main elements present. The ethanol extracts of *Chaetomorpha* demonstrated strong anti-cancer potential by inhibiting the development of the MDA-MB-231 cell line and having a low IC_{50} value (225.18 ± 0.61 g/mL). The bioactive anti-tumor component of *Chaetomorpha* sp., as shown by gas chromatography-mass spectrometry (GC/MS) analysis, is dichloracetic acid (DCA). The other anti-cancer molecules discovered include L-terpinol and oximes (Haq et al. 2019).

Dunaliella tertiolecta is a photosynthetic green microalga belonging to the class *Chlorophyceae*. It is known for its high production of pigments, such as violaxanthin, a β-carotene (Pasquet et al. 2011), and also zeaxanthin, a xanthophyll carotenoid, known for its antioxidant properties (Martínez et al. 2022). Mass spectrometric and spectrophotometric analyses from the dichloromethane extract of *D. tertiolecta* identified violaxanthin as having anti-tumor activity on MCF-7 breast cancer cells by inducing dose-dependent growth inhibition at levels as low as 0.1 μg/mL (0.17 μM) (Pasquet et al. 2011).

Lectin is a protein that binds to carbohydrates and recognizes particular cells by attaching to the polysaccharides on their surface. Typically, different glycosylation patterns differentiate tumor cells from non-cancerous cells, allowing their identification. As a result, lectin has been recommended as a potent anti-tumor substance (Lee et al. 2022). Mannose-specific lectins isolated and identified from marine green algae have different structural scaffolds related to their anti-cancer properties (Barre et al. 2019). For example, lectins derived from *Bryopsis plumosa* (BPL1, BPL2, and BPL3) were investigated for activity against lung cancer cell lines H460, A549, and H1299. Comparing BPL2 with BPL1 and BPL3, BPL2 demonstrated higher anti-cancer activity in a concentration-dependent way. BPL2 level and incubation period had an impact on cell survival. Additionally, BPL2 reduced the invasion and motility of malignant tumor cells and downregulated genes concerned with the epithelial-to-mesenchymal transition (Twist, Vimentin, and Zeb 1). Overall, BPL2 is a promising option for the creation of an anti-cancer drug (Lee et al. 2022).

Polysaccharides from marine green algae have also been reported to demonstrate anti-cancer activities. In a recent study, the anti-cancer activity of SPS-CF, a sulfated glucuronorhamnoxylan polysaccharide isolated from the green alga *Capsosiphon fulvescens*, was evaluated both *in vitro* and *in vivo* against colorectal cancer. The findings of this study showed that SPS-CF effectively slows the growth of colorectal tumors both *in vitro* and *in vivo* by inducing the death of tumor cells through apoptosis, indicating that it may be a potent component of foods or medications used to treat or prevent colon cancer in humans (Choi et al. 2019). Also, the polysaccharide ulvan, obtained from the green alga *Ulva lactuca*, demonstrated significant cytotoxicity against three cancer cell lines comprising human breast cancer (IC_{50} 25.09 ± 1.36 μg/mL), hepatocellular carcinoma (IC_{50} 29.67 ± 2.87 µg/mL), and cervical cancer (IC_{50} 36.33 ± 3.84 µg/mL) (Thanh et al. 2016).

20.3.10 Anti-Aging Activity

Several potential anti-aging compounds have been obtained from marine macroalgae. These natural compounds could serve as substitutes for synthetic anti-aging compounds in nutraceuticals, functional meals, and cosmetics (Wang et al. 2021). A number of pathways, including Sirtuin, insulin-like growth factor (IGF), AMP-activated protein kinase (ampk), nuclear factor erythroid 2-related factor 2 (nrf2), and autophagy, play important roles in controlling aging. The modulation of these pathways can delay the aging process and lengthen lifespan (Cao et al. 2020). A growing body of research demonstrates that specific metabolites obtained from seaweed extract can alter these aging-regulating mechanisms at the molecular and cellular levels or even lengthen the lives of different laboratory animal models. Nevertheless, few of these investigations have been done on higher animals or humans (Cao et al. 2020).

Enteromorpha prolifera (EPO) and *Ulva lactuca* (ULO) oligosaccharides were employed to study the anti-aging and causal mechanisms in SAMP8 mice. These oligosaccharides increased plasma concentrations of glutathione, superoxide dismutase, catalase, and telomerase as well as the capacity of all antioxidants, while lowering the concentration of advanced glycation end products and malondialdehyde. TNF-α, IFN-γ, and IL-6 levels dropped after receiving ULO and EPO treatment; however, choline acetyltransferase (ChAT) and brain-derived neurotrophic factor (BDNF) levels rose, and hippocampus neurons were protected (Liu et al. 2019). EPO and ULO may have therapeutic benefits in the treatment of aging in SAMP8 mice, due to the downregulation of the FOXO1 and p53 genes and overexpression of the Sirt1 gene. These findings showed that oligosaccharides from *E. prolifera* and *U. lactuca* are excellent candidates for usage as anti-aging ingredients in functional foods and drugs (Liu et al. 2019). Also, the anti-aging effect of *Ulva prolifera* on human skin fibroblasts has also been established. Fraction P1-1 of a purified polysaccharide obtained from *Ulva prolifera* demonstrated protection in human skin fibroblasts induced by H_2O_2 by scavenging its free radicals (Cai et al. 2016).

In one study, Norzagaray-Valenzuela et al. examined the anti-aging properties of protein hydrolysates and microalgae residual biomass (MRB) from three distinct green microalgae species, *Tetraselmis suecica*, *Dunaliella tertiolecta*, and *Nannochloropsis* sp., by evaluating their inhibiting activity against two enzymes that degrade the extracellular matrix, hyaluronidase and elastase. The results show that *D. tertiolecta* and *Nannochloropsis* biomasses were found to have higher anti-elastase activity than other microalgae species, but anti-hyaluronidase activity did not differ significantly between them. *Nannochloropsis* exhibited the greatest anti-hyaluronidase and anti-elastase activity for the protein hydrolysates. These findings suggest that a few MRBs and their protein hydrolysates showed promising anti-aging properties, highlighting their commercial significance in the nutraceutical and nutricosmetic sectors (Norzagaray-Valenzuela et al. 2017).

20.4 CONCLUSION

By evolution, biotic and abiotic factors have made the marine green algae a well-established natural resource of bioactive molecules responsible for various medicinal properties such as antimicrobial, antioxidant, anti-inflammatory, anti-cancer, anticoagulant, antidiabetic, antihypertensive, and more. This rich source of chemical and biological diversity has the ability to translate into health-promoting applications such as drugs, nutritional supplements, and cosmetics. Indeed, there remains a largely untapped reservoir of natural products in the marine green algae. Therefore, further studies should be directed toward the clinical development of more novel algal-derived products and to further enhance the efficiency and safety of existing products through applying greater technology in the extraction and delivery processes.

REFERENCES

Abdel-Raouf N, Al-Enazi NM, Ibraheem IB, Alharbi RM, Alkhulaifi MM (2018) Antibacterial and anti-hyperlipidemic activities of the green alga Cladophora koeiei. *Beni Suef Univ J Basic Appl Sci.* 7(1):158–64. https://doi.org/10.1016/j.bjbas.2017.10.002

Admassu H, Gasmalla MA, Yang R, Zhao W (2018) Bioactive peptides derived from seaweed protein and their health benefits: Antihypertensive, antioxidant, and antidiabetic properties. *J Food Sci.* 83(1):6–16. https://doi.org/10.1111/1750-3841.14011

Adrien A, Bonnet A, Dufour D, Baudouin S, Maugard T, Bridiau N (2019) Anticoagulant activity of sulfated ulvan isolated from the green macroalga Ulva rigida. *Mar Drugs.* 17(5):291. https://doi.org/10.3390/md17050291

Ahmed NT, Noor S, Rahman MM, Mazid MA (2021) Cytotoxic compounds derived from marine algicolous and spongicolous endophytic fungi: A review. *Dhaka Univ J Pharm Sci.* 20(2):247–65. https://doi.org/10.3329/dujps.v20i2.57175

Aitouguinane M, El Alaoui-Talibi Z, Rchid H, Fendri I, Abdelkafi S, Ould El-Hadj MD, Boual Z, Dubessay P, Michaud P, Le Cerf D, Rihouey C (2022) A novel sulfated glycoprotein elicitor extracted from the Moroccan green seaweed Codium decorticatum induces natural defenses in tomato. *Appl Sci.* 12(7):3643. https://doi.org/10.3390/app12073643

Aroyehun AQ, Razak SA, Palaniveloo K, Nagappan T, Rahmah NS, Jin GW, Chellappan DK, Chellian J, Kunnath AP (2020) Bioprospecting cultivated tropical green algae, Caulerpa racemosa (Forsskal) J. Agardh: A perspective on nutritional properties, antioxidative capacity and anti-diabetic potential. *Foods.* 9(9):1313. https://doi.org/10.3390/foods9091313

Barbalace MC, Malaguti M, Giusti L, Lucacchini A, Hrelia S, Angeloni C (2019) Anti-inflammatory activities of marine algae in neurodegenerative diseases. *Intl J Mol Sci.* 20(12):3061. https://doi.org/10.3390/ijms20123061

Barre A, Simplicien M, Benoist H, Van Damme EJ, Rougé P (2019) Mannose-specific lectins from marine algae: Diverse structural scaffolds associated to common virucidal and anti-cancer properties. *Mar Drugs.* 17(8):440. https://doi.org/10.3390/md17080440

Bauer S, Jin W, Zhang F, Linhardt RJ (2021) The application of seaweed polysaccharides and their derived products with potential for the treatment of Alzheimer's disease. *Mar Drugs.* 19(2):89. https://doi.org/10.3390/md19020089

Besednova NN, Andryukov BG, Zaporozhets TS, Kryzhanovsky SP, Fedyanina LN, Kuznetsova TA, Zvyagintseva TN, Shchelkanov MY (2021) Antiviral effects of polyphenols from marine algae. *Biomedicines.* 9(2):200. https://doi.org/10.3390/biomedicines9020200

Besednova NN, Andryukov BG, Zaporozhets TS, Kryzhanovsky SP, Kuznetsova TA, Fedyanina LN, Makarenkova ID, Zvyagintseva TN (2020) Algae polyphenolic compounds and modern antibacterial strategies: Current achievements and immediate prospects. *Biomedicines.* 8(9):342. https://doi.org/10.3390/biomedicines8090342

Bharathai K, Kaviya MS, Sangeetha T, Vijayarani K (2022) Pharmacological importance of seaweeds. In Balasubramanian B, Liu WC, Sattanathan G (eds.), *Aquaculture Science and Engineering* (pp. 347–71). Singapore: Springer. https://doi.org/10.1007/978-981-19-0817-0_13

Blunt JW, Copp BR, Munro MH, Northcote PT, Prinsep MR (2005) Marine natural products. *Nat Prod Rep.* 22(1):15–61. https://doi.org/10.1039/B415080P

Bourguiba I, Zahlila A, Bouaïcha N, Amri M, Mezghani S (2017) Antioxidant effect of the marine green alga Ulva rigida ethanolic precipitate in yeast cells and zebrafish embryos. *S Afr J Bot.* 113:253–60. https://doi.org/10.1016/j.sajb.2017.09.001

Cai C, Guo Z, Yang Y, Geng Z, Tang L, Zhao M, Qiu Y, Chen Y, He P (2016) Inhibition of hydrogen peroxide induced injuring on human skin fibroblast by Ulva prolifera polysaccharide. *Int J Biol Macromol.* 91:241. https: //doi.org/10.1016/j.ijbiomac.2016.05.071

Cao L, Lee SG, Lim KT, Kim HR (2020) Potential anti-aging substances derived from seaweeds. *Mar Drugs.* 18(11):564. https://doi.org/10.3390/md18110564

Cao S, He X, Qin L, He M, Yang Y, Liu Z, Mao W (2019) Anticoagulant and antithrombotic properties in vitro and in vivo of a novel sulfated polysaccharide from marine green alga Monostromanitidum. *Mar Drugs.* 17(4):247. https://doi.org/10.3390/md17040247

Cao X, Ge Y, Lan D, Wu X, Wu B (2023) Spirocyclic polyketides from the marine fungus *Talaromyces* sp. CX11. *Fitoterapia.* 164:105359.https://doi.org/10.1016/j.fitote.2022.105359

Carneiro RF, Duarte PL, Chaves RP, da Silva SR, Feitosa RR, de Sousa BL, da Silva Alves AW, de Vasconcelos MA, da Rocha BA, Teixeira EH, Sampaio AH (2020) New lectins from Codium isthmocladum Vickers show unique amino acid sequence and antibiofilm effect on pathogenic bacteria. *J Appl Phycol.* 32(6):4263–76. https://doi.org/10.1007/s10811-020-02198-x

Catanesi M, Caioni G, Castelli V, Benedetti E, d'Angelo M, Cimini A (2021) Benefits under the sea: The role of marine compounds in neurodegenerative disorders. *Mar Drugs.* 19(1):24. https://doi.org/10.3390/md19010024

Cha SH, Ahn GN, Heo SJ, Kim KN, Lee KW, Song CB, Kim SM, Jeon YJ (2006) Screening of extracts from marine green and brown algae in Jeju for potential marine angiotensin-I converting enzyme (ACE) inhibitory activity. *J Korean Soc Food Sci Nutr.* 35(3):307–14. https://doi.org/10.3746/jkfn.2006.35.3.307

Chen Y, Ouyang Y, Chen X, Chen R, Ruan Q, Farag MA, Chen X, Zhao C (2022) Hypoglycaemic and anti-ageing activities of green alga Ulva Lactuca polysaccharide via gut microbiota in ageing-associated diabetic mice. *Intl J Biol Macromol.* 212:97–110. https://doi.org/10.1016/j.ijbiomac.2022.05.109

Cheung RC, Ng TB, Wong JH (2015) Marine peptides: Bioactivities and applications. *Mar Drugs.* 13(7):4006–43. https://doi.org/10.3390/md13074006

Chi Y, Zhang M, Wang X, Fu X, Guan H, Wang P (2020) Ulvan lyase assisted structural characterization of ulvan from Ulva pertusa and its antiviral activity against vesicular stomatitis virus. *Intl J Biol Macromol.* 157:75–82. https://doi.org/10.1016/j.ijbiomac.2020.04.187

Choi JW, Lee J, Kim SC, You S, Lee CW, Shin J, Park YI (2019) Glucuronorhamnoxylan from Capsosiphon fulvescens inhibits the growth of HT-29 human colon cancer cells in vitro and in vivo via induction of apoptotic cell death. *Int J Biol Macromol.* 124:1060–8. https://doi.org/10.1016/j.ijbiomac.2018.12.001

Cian RE, Hernández-Chirlaque C, Gámez-Belmonte R, Drago SR, Sánchez de Medina F,Martínez-Augustin O (2018) Green alga Ulva spp. hydrolysates and their peptide fractions regulate cytokine production in splenic macrophages and lymphocytes involving the TLR4-NFκB/MAPK pathways. *Marine Drugs.* 16(7):235. https://doi.org/10.3390/md16070235

Ciancia M, Fernández PV, Leliaert F (2020) Diversity of sulfated polysaccharides from cell walls of coenocytic green algae and their structural relationships in view of green algal evolution. *Front Plant Sci.* 11:554–85. https://doi.org/10.3389/fpls.2020.554585

da Conceição Rivanor RL, Chaves HV, do Val DR, de Freitas AR, Lemos JC, Rodrigues JA, Pereira KM, de Araújo IW, Bezerra MM, Benevides NM (2014) A lectin from the green seaweed Caulerpa cupressoides reduces mechanical hyper-nociception and inflammation in the rat temporomandibular joint during zymosan-induced arthritis. *Intl Immunopharmacol.* 21(1):34–43. https://doi.org/10.1016/j.intimp.2014.04.009

de la Rosa LA, Moreno-Escamilla JO, Rodrigo-García J, Alvarez-Parrilla E (2019) Phenolic compounds. In Yahia, E. M. (ed), *Postharvest Physiology and Biochemistry of Fruits and Vegetables* (pp. 253–71). Woodhead Publishing.

de Queiroz INL, Quinderé ALG, Rodrigues JAG, de Sousa Oliveira Vanderlei E, Ribeiro NA, da Conceição Rivanor RL, Ribeiro KA, Coura CO, Pereira KMA, Chaves HV, et al. (2015) Dual effects of a lectin from the green seaweed *Caulerpa cupressoides* var. *lycopodium* on inflammatory mediators in classical models of inflammation. *Inflamm Res.* 64(12):971–82. https://doi.org/10.1007/s00011-015-0880-3

Dhahri M, Alghrably M, Mohammed HA, Badshah SL, Noreen N, Mouffouk F, Rayyan S, Qureshi KA, Mahmood D, Lachowicz JI, Jaremko M (2021) Natural polysaccharides as preventive and therapeutic horizon for neurodegenerative diseases. *Pharmaceutics.* 14(1):1. https://doi.org/10.3390/pharmaceutics14010001

Dharmananda S (2002) *The Nutritional and Medicinal Value of Seaweeds Used in Chinese Medicine.* ITM.

Echave J, Otero P, Garcia-Oliveira P, Munekata PE, Pateiro M, Lorenzo JM, Simal-Gandara J, Prieto MA (2022) Seaweed-derived proteins and peptides: Promising marine bioactives. *Antioxidants.* 11(1):176. https://doi.org/10.3390/antiox11010176

El Gamal AA (2010) Biological importance of marine algae. *Saudi Pharm J.* 18(1):1–25. https://doi.org/10.1016/j.jsps.2009.12.001

Elsebai MF, Nazir M, Kehraus S, Egereva E, Ioset KN, Marcourt L, Jeannerat D, Gütschow M, Wolfender JL, König GM (2012) Polyketide skeletons from the marine alga-derived fungus Coniothyrium cereale. *Eur J Org Chem.* 31(31):6197–203. https://doi.org/10.1002/ejoc.201200700

Esteves PO, de Oliveira MC, de Souza Barros C, Cirne-Santos CC, Laneuvlille VT, Palmer Paixao IC (2019) Antiviral effect of caulerpin against chikungunya. *Nat Prod Commun.* 14(10):1934578X19878295. https://doi.org/10.1177/1934578X19878295

Fernando IP, Sanjeewa KK, Samarakoon KW, Lee WW, Kim HS, Jeon YJ (2018) Squalene isolated from marine macroalgae Caulerpa racemosa and its potent antioxidant and anti-inflammatory activities. *J Food Biochem.* 42(5):e12628. https://doi.org/10.1111/jfbc.12628

Gando S, Levi M, Toh CH (2016) Disseminated intravascular coagulation. *Nat Rev Dis Primers.* 2(1):1. https://doi.org/10.1038/nrdp.2016.37

Ghareeb MA, Tammam MA, El-Demerdash A, Atanasov AG (2020) Insights about clinically approved and Preclinically investigated marine natural products. *Curr Res Biotechnol.* 2:88–102. https://doi.org/10.1016/j.crbiot.2020.09.001

Grice ID, Mariottini GL (2018) Glycans with antiviral activity from marine organisms. In Kloc M, Kubiak J (eds.), *Marine Organisms as Model Systems in Biology and Medicine. Results and Problems in Cell Differentiation* (Vol. 65). Springer. https://doi.org/10.1007/978-3-319-92486-1_20

Gunathilaka T, Keertihirathna LR, Peiris D (2021) Advanced pharmacological uses of marine algae as an anti-diabetic therapy. *Nat Med Plants.* https://doi.org/10.5772/intechopen.96807

Guzmán F, Wong G, Román T, Cárdenas C, Alvárez C, Schmitt P, Albericio F, Rojas V (2019) Identification of antimicrobial peptides from the microalgae Tetraselmissuecica (Kylin) Butcher and bactericidal activity improvement. *Mar Drugs.* 17(8):453. https://doi.org/10.3390/md17080453

Hannan MA, Dash R, Haque MN, Mohibbullah M, Sohag AA, Rahman MA, Uddin MJ, Alam M, Moon IS (2020) Neuroprotective potentials of marine algae and their bioactive metabolites: Pharmacological insights and therapeutic advances. *Mar Drugs.* 18(7):347. https://doi.org/10.3390/md18070347

Hans N, Malik A, Naik S (2021) Antiviral activity of sulfated polysaccharides from marine algae and its application in combating COVID-19: Mini review. *Biores Technol Rep.* 13:100623. https://doi.org/10.1016/j.biteb.2020.100623

Hao C, Yu G, He Y, Xu C, Zhang L, Wang W (2019) Marine glycan–based antiviral agents in clinical or pre-clinical trials. *Rev Med Virol.* 29(3):e2043. https://doi.org/10.1002/rmv.2043

Haq SH, Al-Ruwaished G, Al-Mutlaq MA, Naji SA, Al-Mogren M, Al-Rashed S, Ain QT, Al-Amro AA, Al-Mussallam A (2019) Antioxidant, anticancer activity and phytochemical analysis of green algae, Chaetomorpha collected from the Arabian Gulf. *Sci Rep.* 9(1):18906. https://doi.org/10.1038/s41598-019-55309-1

He M, Yang Y, Shao Z, Zhang J, Feng C, Wang L, Mao W (2021) Chemical structure and anticoagulant property of a novel sulfated polysaccharide from the green alga Cladophora oligoclada. *Mar Drugs.* 19(10):554. https://doi.org/10.3390/md19100554

Jain P, Minhas AK, Shukla S, Puri M, Barrow CJ, Mandal S (2022) Bioprospecting indigenous marine microalgae for polyunsaturated fatty acids under different media conditions. *Front Bioengr Biotechnol.* 10:842797. https://doi.org/10.3389%2Ffbioe.2022.842797

Jayasankar, R. (2022) Classification of algae with special reference to seaweed. In *ICAR-CMFRI -Winter School on Recent Development in Taxonomic Techniques of Marine Fishes for Conservation and Sustainable Fisheries Management.* ICAR-Central Marine Fisheries Research Institute, Kochi (pp. 521–6).

Jensen PR, Fenical W (2000) Marine microorganisms and drug discovery: Current status and future potential. *Drugs Sea.* 2000:6–29.

Ji NY, Liu XH, Miao FP, Qiao MF (2013) Aspeverin, a new alkaloid from an algicolous strain of Aspergillus versicolor. *Org Lett.* 15(10):2327–9. https://doi.org/10.1021/ol4009624

Jiang Z, Barret MO, Boyd KG, Adams DR, Boyd ASF, Burgess JG (2002) JM47, a cyclic tetrapeptide HC-toxin analogue from a marine Fusarium species. *Phytochemistry.* 60(1):33–8. https://doi.org/10.1016/S0031-9422(02)00061-4

Jiao G, Yu G, Zhang J, Ewart HS (2011) Chemical structures and bioactivities of sulfated polysaccharides from marine algae. *Mar Drugs.* 9(2):196–223. https://doi.org/10.3390/md9020196

Jimenez-Lopez C, Pereira AG, Lourenço-Lopes C, García-Oliveira P, Cassani L, Fraga-Corral M, Prieto MA, Simal-Gandara J (2021) Main bioactive phenolic compounds in marine algae and their mechanisms of action supporting potential health benefits. *Food Chem.* 341:128262. https://doi.org/10.1016/j.foodchem.2020.128262

Jubair N, Rajagopal M, Chinnappan S, Abdullah NB, Fatima A (2021) Review on the antibacterial mechanism of plant-derived compounds against multidrug-resistant bacteria (MDR). *Evid Based Complement Altern Med.* 2021:3663315. https://doi.org/10.1155/2021/3663315

Kamat S, Kumari M, Sajna KV, Jayabaskaran C (2020) Endophytic fungus, Chaetomium globosum, associated with marine green alga, a new source of chrysin. *Sci Rep.* 10(1):18726. https://doi.org/10.1038/s41598-020-72497-3

Kandale A, Meena AK, Rao MM, Panda P, Mangal AK, Reddy G, Babu R (2011) Marine algae: An introduction, food value and medicinal uses. *J Pharm Res.* 4(1):219–21.

Kapoor S, Nailwal N, Kumar M, Barve K (2019) Recent patents and discovery of anti-inflammatory agents from marine source. *Recent Pat Inflam Allergy Drug Discov.* 13(2):105–14. https://doi.org/10.2174/1872213X13666190426164717

Ko SC, Kang N, Kim EA, Kang MC, Lee SH, Kang SM, Lee JB, Jeon BT, Kim SK, Park SJ, Park PJ (2012) A novel angiotensin I-converting enzyme (ACE) inhibitory peptide from a marine Chlorella ellipsoidea and its antihypertensive effect in spontaneously hypertensive rats. *Process Biochem.* 47(12):2005–11. https://doi.org/10.1016/j.procbio.2012.07.015

Kolsi RB, Salah HB, Hamza A, Belguith K (2017) Characterization and evaluating of antioxidant and antihypertensive properties of green alga (Codium fragile) from the coast of Sfax. *J Pharmacogn Phytochem.* 6(2):186–91.

Kralj A, Kehraus S, Krick A, Eguereva E, Kelter G, Maurer M, Wortmann A, Fiebig HH, König GM (2006) Arugosins G and H: Prenylated polyketides from the Marine-Derived Fungus Emericella n idulans var. a cristata. *J Nat Prod.* 69(7):995–1000. https://doi.org/10.1021/np050454f

Lee HA, Kim IH, Nam TJ (2015) Bioactive peptide from Porphyra yezoensis and its anti-inflammatory activities. *Intl J Mol Med.* 36(6):1701–6. https://doi.org/10.3892/ijmm.2015.2386

Lee HJ, Kim YA, Lee JI, Lee BJ, Seo YW (2007) Screening of Korean marine plants extracts for inhibitory activity on protein tyrosine phosphatase 1B. *J Appl Biol Chem.* 50(2):74–7.

Lee JC, Hou MF, Huang HW, Chang FR, Yeh CC, Tang JY, Chang HW (2013) Marine algal natural products with anti-oxidative, anti-inflammatory, and anti-cancer properties. *Cancer Cell Int.* 13(1):1–7. https://doi.org/10.1186/1475-2867-13-55

Lee JH, Lee SB, Kim H, Shin JM, Yoon M, An HS, Han JW (2022) Anticancer activity of mannose-specific lectin, BPL2, from marine green alga Bryopsis plumosa. *Mar Drugs.* 20(12):776. https://doi.org/10.3390/md20120776

Li CQ, Ma QY, Gao XZ, Wang X, Zhang BL (2021) Research progress in anti-inflammatory bioactive substances derived from marine microorganisms, sponges, algae, and corals. *Mar Drugs.* 19(10):572. https://doi.org/10.3390/md19100572

Li P, Wen S, Sun K, Zhao Y, Chen Y (2018) Structure and bioactivity screening of a low molecular weight Ulvan from the green alga *Ulothrix flacca. Mar Drugs.* 16(8):281. https://doi.org/10.3390/md16080281

Lin YH, Chen GW, Yeh CH, Song H, Tsai JS (2018) Purification and identification of angiotensin I-converting enzyme inhibitory peptides and the antihypertensive effect of Chlorella sorokiniana protein hydrolysates. *Nutrients.* 10(10):1397. https://doi.org/10.3390/nu10101397

Liu AH, Liu DQ, Liang TJ, Yu XQ, Feng MT, Yao LG, Fang Y, Wang B, Feng LH, Zhang MX, Mao SC (2013) Caulerprenylols A and B, two rare antifungal prenylated para-xylenes from the green alga Caulerpa racemosa. *Bioorg Med Chem Lett.* 23(9):2491–4. https://doi.org/10.1016/j.bmcl.2013.03.038

Liu H, Chen Z, Zhu G, Wang L, Du Y, Wang Y, Zhu W (2017) Phenolic polyketides from the marine alga-derived *Streptomyces* sp. OUCMDZ-3434. Tetrahedron. 73(36):5451–5. https://doi.org/10.1016/j.tet.2017.07.052

Liu X, Du P, Liu X, Cao S, Qin L, He M, He X, Mao W (2018) Anticoagulant properties of a green algal rhamnan-type sulfated polysaccharide and its low-molecular-weight fragments prepared by mild acid degradation. *Mar Drugs.* 16(11):445. https://doi.org/10.3390/md16110445

Liu XY, Liu D, Lin GP, Wu YJ, Gao LY, Ai C, Huang YF, Wang MF, El-Seedi HR, Chen XH, Zhao C (2019) Anti-ageing and antioxidant effects of sulfate oligosaccharides from green algae Ulva Lactuca and Enteromorpha prolifera in SAMP8 mice. *Intl J Biol Macromol.* 139:342–51. https://doi.org/10.1016/j.ijbiomac.2019.07.195

Lu WY, Li HJ, Li QY, Wu YC (2021) Application of marine natural products in drug research. *Bioorg Med Chem.* 35:116058. https://doi.org/10.1016/j.bmc.2021.116058

Lunagariya J, Bhadja P, Zhong S, Vekariya R, Xu S (2019) Marine natural product bis-indole alkaloid caulerpin: Chemistry and biology. *Mini Rev Med Chem.* 19(9):751–61. https://doi.org/10.2174/1389557517666170927154231

Mahmoud AM, Bin-Jumah M, Abukhalil MH (2021) Antiinflammatory natural products from marine algae. *Inflamm Natural Prod.* 175–203. https://doi.org/10.1016/B978-0-12-819218-4.00012-2

Mao WJ, Fang F, Li HY, Qi XH, Sun HH, Chen Y, Guo SD (2008) Heparinoid-active two sulfated polysaccharides isolated from marine green algae Monostroma nitidum. *Carbohydr Polym.* 74(4):834–9. https://doi.org/10.1016/j.carbpol.2008.04.041

Martínez KA, Saide A, Crespo G, Martín J, Romano G, Reyes F, Lauritano C, Ianora A (2022) Promising antiproliferative compound from the green microalga Dunaliellatertiolecta against human cancer cells. *Front Mar Sci.* 9:778108. https://doi.org/10.3389/fmars.2022.778108

Mohy El-Din SM, Alagawany NI (2019) Phytochemical constituents and anticoagulation property of marine algae Gelidiumcrinale, Sargassum hornschuchii and Ulva linza. *Thalassas An Intl J Mar Sci.* 35(2):381–97. https://doi.org/10.1007/s41208-019-00142-6

Montone CM, Capriotti AL, Cavaliere C, La Barbera G, Piovesana S, ZeneziniChiozzi R, Laganà A (2018) Peptidomic strategy for purification and identification of potential ACE-inhibitory and antioxidant peptides in Tetradesmus obliquus microalgae. *Anal Bioanal Chem.* 410(15):3573–86. https://doi.org/10.1007/s00216-018-0925-x

Mu J, Hirayama M, Sato Y, Morimoto K, Hori K (2017) A novel high-mannose specific lectin from the green alga Halimeda renschii exhibits a potent anti-influenza virus activity through high-affinity binding to the viral hemagglutinin. *Mar Drugs.* 15(8):255. https://doi.org/10.3390/md15080255

Norzagaray-Valenzuela CD, Valdez-Ortiz A, Shelton LM, Jiménez-Edeza M, Rivera-López J, Valdez-Flores MA, Germán-Báez LJ (2017) Residual biomasses and protein hydrolysates of three green microalgae species exhibit antioxidant and anti-aging activity. *J Appl Phycol.* 29(1):189–98. https://doi.org/10.1007/s10811-016-0938-9

Oh JH, Nam TJ (2019) Hydrophilic glycoproteins of an edible green alga Capsosiphon fulvescens prevent aging-induced spatial memory impairment by suppressing GSK-3β-mediated ER stress in dorsal hippocampus. *Mar Drugs.* 17(3):168. https://doi.org/10.3390/md17030168

Parameswari RP, Lakshmi T (2022) Microalgae as a potential therapeutic drug candidate for neurodegenerative diseases. *J Biotechnol.* 358:128–39. https://doi.org/10.1016/j.jbiotec.2022.09.009

Pasquet V, Morisset P, Ihammouine S, Chepied A, Aumailley L, Berard JB, Serive B, Kaas R, Lanneluc I, Thiery V, Lafferriere M, Piot JM, Patrice T, Cadoret JP, Picot L (2011) Antiproliferative activity of violaxanthin isolated from bioguided fractionation of Dunaliellatertiolecta extracts. *Mar Drugs.* 9(5):819–31. https://doi.org/10.3390/md9050819

Pereira L, Valado A (2021) The seaweed diet in prevention and treatment of the neurodegenerative diseases. *Mar Drugs.* 19(3):128. https://doi.org/10.3390/md9050819

Pradhan B, Ki JS (2022) Phytoplankton toxins and their potential therapeutic applications: A journey toward the quest for potent pharmaceuticals. *Mar Drugs.* 20(4):271. https://doi.org/10.3390/md20040271

Přerovská T, Henke S, Bleha R, Spiwok V, Gillarová S, Yvin JC, Ferrières V, Nguema-Ona E, Lipovová P (2021) Arabinogalactan-like glycoproteins from Ulva Lactuca (Chlorophyta) Show unique features compared to land plants AGPs. *J Phycol.* 57(2):619–35. https://doi.org/10.1111/jpy.13121

Qin L, Yang Y, Mao W (2023) Anticoagulant property of a sulfated polysaccharide with unique structural characteristics from the green alga Chaetomorphaaerea. *Mar Drugs.* 21(2):88. https://doi.org/10.3390/md21020088

Rajivgandhi G, Ramachandran G, Maruthupandy M, Saravanakumar S, Manoharan N, Viji R (2018) Antibacterial effect of endophytic actinomycetes from marine algae against multi drug resistant gram-negative bacteria. *Exam Mar Biol Oceanogr.* 1(4):1–8. http://doi.org/10.31031/EIMBO.2018.01.000522

Ramachandran G, Rajivgandhi G, Maruthupandy M, Manoharan N (2019) Extraction and partial purification of secondary metabolites from endophytic actinomycetes of marine green algae Caulerpa racemosa against multi drug resistant uropathogens. *Biocatal Agric Biotechnol.* 17:750–7. https://doi.org/10.1016/j.bcab.2019.01.016

Ramos-Romero S, Torrella JR, Pagès T, Viscor G, Torres JL (2021) Edible microalgae and their bioactive compounds in the prevention and treatment of metabolic alterations. *Nutrients.* 13(2):563. https://doi.org/10.3390/nu13020563

Ren R, Gong J, Zhao Y, Zhuang X, Ye Y, Huang F, Lin W (2018) Sulfated polysaccharide from Enteromorpha prolifera suppresses SREBP-1c and ACC expression to lower serum triglycerides in high-fat diet-induced hyperlipidaemic rats. *J Funct Foods.* 40:722–8. https://doi.org/10.1016/j.jff.2017.12.010

Řezanka T, Nedbalová L, Lukavský J, Střížek A, Sigler K (2017) Pilot cultivation of the green alga Monoraphidium sp. producing a high content of polyunsaturated fatty acids in a low-temperature environment. *Algal Res.* 22:160–5. https://doi.org/10.1016/j.algal.2016.12.017

Ridley CP, Lee HY, Khosla C (2008) Evolution of polyketide synthases in bacteria. *Ecology.* 105(12):4595–600. https://doi.org/10.1073/pnas.0710107105

Rowaiye AB, Onuh OA, Oli AN, Okpalefe OA, Oni S, Nwankwo EJ (2020) The pandemic COVID-19: A tale of viremia, cellular oxidation and immune dysfunction. *Pan Afr Med J.* 36(1). https://doi.org/10.11604/pamj.2020.36.188.23476

Sathivel A, Raghavendran HR, Srinivasan P, Devaki T (2008) Anti-peroxidative and anti-hyperlipidemic nature of Ulva Lactuca crude polysaccharide on d-galactosamine induced hepatitis in rats. *Food Chem Toxicol.* 46(10):3262. https://doi.org/10.1016/j.fct.2008.07.016

Seca AM, Pinto DC (2018) Overview on the antihypertensive and anti-obesity effects of secondary metabolites from seaweeds. *Mar Drugs.* 16(7):237. https://doi.org/10.3390/md16070237

Serova M, De Gramont A, Bieche I, Riveiro ME, Galmarini CM, Aracil M, Jimeno J, Faivre S, Raymond E (2013) Predictive factors of sensitivity to elisidepsin, a novel kahalalide F-derived marine compound. *Mar Drugs.* 11(3):944–59. https://doi.org/10.3390/md11030944

Shin ES, Hwang HJ, Kim IH, Nam TJ (2011) A glycoprotein from Porphyra yezoensis produces anti-inflammatory effects in liposaccharide-stimulated macrophages via the TLR4 signaling pathway. *Intl J Mol Med.* 28(5):809–15. https://doi.org/10.3892/ijmm.2011.729

Sirbu R, Stanciu G, Tomescu A, Ionescu AM, Cadar E (2019) Evaluation of antioxidant and antimicrobial activity in relation to total phenolic content of green algae from Black Sea. *Rev Chim.* 70(4):1197–203.

Souza CR, Bezerra WP, Souto JT (2020) Marine alkaloids with anti-inflammatory activity: Current knowledge and future perspectives. *Mar Drugs.* 18(3):147. https://doi.org/10.3390/md18030147

Sun S, Xu X, Sun X, Zhang X, Chen X, Xu N (2019) Preparation and identification of ACE inhibitory peptides from the marine macroalga Ulva intestinalis. *Mar Drugs.* 17(3):179. https://doi.org/10.3390/md17030179

Surget G, Roberto VP, Le Lann K, Mira S, Guérard F, Laizé V, Poupart N, Cancela ML, Stiger-Pouvreau V (2017) Marine green macroalgae: A source of natural compounds with mineralogenic and antioxidant activities. *J Appl Phycol.* 29(1):575–84. https://doi.org/10.1007/s10811-016-0968-3

Thanh TT, Quach TM, Nguyen TN, Vu Luong D, Bui ML, Tran TT (2016) Structure and cytotoxic activity of ulvan extracted from green seaweed Ulva Lactuca. *Int J Biol Macromol.* 93(Pt A):695–702. https://doi.org/10.1016/j.ijbiomac.2016.09.040

Tragin M, Vaulot D (2018) Green microalgae in marine coastal waters: The Ocean Sampling Day (OSD) dataset. *Sci Rep.* 8(1):1–2. https:/doi.org/10.1038/s41598-018-32338-w

Tziveleka LA, Tammam MA, Tzakou O, Roussis V, Ioannou E (2021) Metabolites with antioxidant activity from marine macroalgae. *Antioxidants.* 10(9):1431. https://doi.org/10.3390/antiox10091431

Unnikrishnan PS, Jayasri MA (2018) Marine algae as a prospective source for antidiabetic compounds–a brief review. *Curr Diabetes Rev.* 14(3):237–45. https://doi.org/10.2174/1573399812666161229151407

Vanderlei ES, Patoilo KK, Lima NA, Lima AP, Rodrigues JA, Silva LM, Lima ME, Lima V, Benevides NM (2010) Antinociceptive and anti-inflammatory activities of lectin from the marine green alga Caulerpa cupressoides. *Intl Immunopharmacol.* 20(9):1113–8. https://doi.org/10.1016/j.intimp.2010.06.014

Vigor C, Oger C, Reversat G, Rocher A, Zhou B, Linares-Maurizi A, Guy A, Bultel-Poncé V, Galano JM, Vercauteren J, Durand T (2020). Isoprostanoid profiling of marine microalgae. *Biomolecules.* 10(7):1073. https://doi.org/10.3390/biom10071073

Vladkova T, Georgieva N, Staneva A, Gospodinova D (2022) Recent progress in antioxidant active substances from marine biota. *Antioxidants.* 11(3):439. https://doi.org/10.3390/antiox11030439

Voshall A, Christie NT, Rose SL, Khasin M, Van Etten JL, Markham JE, Riekhof WR, Nickerson KW (2021) Sterol Biosynthesis in four green algae: A bioinformatic analysis of the ergosterol versus phytosterol decision point. *J Phycol.* 57(4):1199–211. https://doi.org/10.1111/jpy.13164

Wang X, Zhang Z, Zhang S, Yang F, Yang M, Zhou J, Hu Z, Xu X, Mao G, Chen G, Xiang W (2021) Antiaging compounds from marine organisms. *Food Res Intl.* 143:110313. https://doi.org/10.1016/j.foodres.2021.110313

Wassie T, Niu K, Xie C, Wang H, Xin W (2021) Extraction techniques, biological activities and health benefits of marine algae Enteromorpha prolifera polysaccharide. *Front Nutr.* 8:747928. https://doi.org/10.3389/fnut.2021.747928

Wufsus AR, Macera NE, Neeves KB (2013) The hydraulic permeability of blood clots as a function of fibrin and platelet density. *Biophys J.* 104(8):1812–23. https://doi.org/10.1016/j.bpj.2013.02.055

Yadav D, Srivastava S, Tripathi YB (2019) *Acorus calamus*: A review. *Intl J Sci Res Biol Sci.* 6(4):62–7. https://doi.org/10.26438/ijsrbs/v6i4.6267

Yao W, Qiu HM, Cheong KL, Zhong S (2022) Advances in anti-cancer effects and underlying mechanisms of marine algae polysaccharides. *Intl J Biol Macromol.* 221:472–85. https://doi.org/10.1016/j.ijbiomac.2022.09.055

Yao L, Liang Y, Sun M, Song S, Wang H, Dong Z, Feng T, Yue H (2022) Characteristic volatile fingerprints of three edible marine green algae (Ulva spp.) in China by HS-GC-IMS and evaluation of the antioxidant bioactivities. *Food Res Intl.* 162:112109. https://doi.org/10.1016/j.foodres.2022.112109

Zbakh H, Salhi G, Bochkov V, Ciudad CJ, Noe V, Hassoun M, Riadi H (2020) Insights on the anti-inflammatory and antitumor activities of extracts from the marine green alga Codium decorticatum. *Eur J Integr Med.* 37:101170. https://doi.org/10.1016/j.eujim.2020.101170

Zhang P, Li X, Wang BG (2016) Secondary metabolites from the marine algal-derived endophytic fungi: Chemical diversity and biological activity. *Planta Med.* 82(9/10):832–42. https://doi.org/10.1055/s-0042-103496

Zhong Q, Wei B, Wang S, Ke S, Chen J, Zhang H, Wang H (2019) The antioxidant activity of polysaccharides derived from marine organisms: An overview. *Mar Drugs.* 17(12):674. https://doi.org/10.3390/md17120674

Index

A

B

C

N

W

X

Z